José Ignacio Calvo Díez, Antonio Hernández Giménez
Membrane Characterization

Also of interest

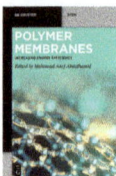

Polymer Membranes.
Increasing Energy Efficiency
Abdulhamid (Ed.), 2024
ISBN 978-3-11-079599-8, e-ISBN (PDF) 978-3-11-079603-2

Mixed-Matrix Membranes.
Preparation Methods, Applications, Challenges and Performance
Pirouzfar, Tehrani, Su, Hasanzad, Hosseini, 2024
ISBN 978-3-11-128245-9, e-ISBN (PDF) 978-3-11-128434-7

Engineering Materials Characterization
Kumar, Zindani, 2023
ISBN 978-3-11-099760-6, e-ISBN (PDF) 978-3-11-099759-0

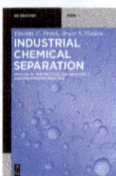

Industrial Chemical Separation.
Historical Perspective, Fundamentals, and Engineering Practice
Frank, Holden, 2023
ISBN 978-3-11-069502-1, e-ISBN (PDF) 978-3-11-069505-2

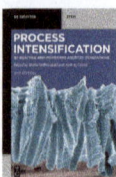

Process Intensification.
by Reactive and Membrane-Assisted Separations
Skiborowski, Górak, 2022
ISBN 978-3-11-072045-7, e-ISBN (PDF) 978-3-11-072046-4

José Ignacio Calvo Díez,
Antonio Hernández Giménez

Membrane Characterization

Porometries and Porosimetries

DE GRUYTER

Authors
Prof. José Ignacio Calvo Díez
Department of Applied Physics ETSIIAA
University of Valladolid
Avda. Madrid 57
34004 Palencia
Spain
joseignacio.calvo@uva.es

Prof. Antonio Hernández Giménez
Department of Applied Physics Faculty of Sciences
University of Valladolid
P.º de Belén, 7
47011 Valladolid
Spain
antonio.hernandez@uva.es

ISBN 978-3-11-079218-8
e-ISBN (PDF) 978-3-11-079219-5
e-ISBN (EPUB) 978-3-11-079231-7

Library of Congress Control Number: 2024940913

Bibliographic information published by the Deutsche Nationalbibliothek
The Deutsche Nationalbibliothek lists this publication in the Deutsche Nationalbibliografie;
detailed bibliographic data are available on the Internet at http://dnb.dnb.de.

© 2024 Walter de Gruyter GmbH, Berlin/Boston
Cover image: Andrey Prokhorov/iStock/Getty Images Plus.
Typesetting: Integra Software Services Pvt. Ltd.

www.degruyter.com

Preface

For most people involved in membrane processes, it should be more or less clear that the size of the pores present in a membrane is a parameter of great interest in determining its possible applications. Although, it seems clear as well that other aspects of membranes and their interactions with the fluids to separate with adequate selectivity are quite relevant for a given industrial application. Therefore, what actually defines the usefulness of a membrane is its selectivity towards the different components to be separated. In many cases, this selectivity will depend on issues such as the hydrophilicity/hydrophobicity of the membrane, the volatility of the species to be permeated, the mutual electrical interaction or the diffusivity of the molecules through the barrier material. However, there are also a multitude of processes in which this selectivity is clearly related to the size of the channels or holes through which some of the species or some of the molecules present in a fluid must pass through the membrane barrier. More importantly, some other species or molecules will not be allowed this passage, thus creating the desired selectivity.

In such a way, that even when sieving is not the main mechanism of transport through the membrane, the size of the pores present in a certain membrane will allow researchers and industry a useful classification of the membranes and their possible applications.

It is therefore natural that both researchers in membrane technologies and membrane manufacturers are deeply interested in determining, as precisely as possible, the size of the pores present in their membranes. To do this, they generally make use of various characterization techniques which, together, can be termed in a general manner as porometries, and which allow them to determine the average size of these pores as well as their statistical distribution in size.

But the decision on the characterization technique that will bring us such useful information could be somewhat complex, as there are many porometries available in the market, many companies marketing such porometries, and, moreover, not every characterization technique is able to give us information. Nevertheless, without a clear knowledge of the principles, features and possible problems that one can face when using them, it would be difficult to interpret and securely understand their results.

This book is aimed at those who, when choosing a membrane characterization technique to determine information on the size of the pores present, have doubts about which is the most appropriate technique or simply seek to understand the principles of the technique they use (or their lab wants to acquire) in order to be able to wisely read the results obtained more precisely and get useful and reliable conclusions.

https://doi.org/10.1515/9783110792195-202

Acknowledgments

Creating a text like this, whose usefulness will be judged by potential readers, always requires a significant effort in research, compilation, digestion, and writing. Although the subjects are within the authors' scientific knowledge and practice, they still need to be updated and verified against the latest published contributions on the topic.

In the text, we have tried to include explanatory figures and typical examples of the devices, data and treatments associated with all the techniques analysed. Many of these figures are drawn from our own experience, compiled in various publications over more than 40 years of research.

In other cases, we have used diagrams, data or figures published by other authors, specialists in the various sections. All these figures have been conveniently referenced, and we thank their authors for the information published.

However, there are several people whom we would also like to thank explicitly for their contributions to the success of this book.

- Pedro Prádanos and Laura Palacio, colleagues, and friends, of our research group at the University of Valladolid, have been a mainstay in the growth of the group over the years. We also thank Pedro Prádanos for Figure 2.10, while Figures 2.5, 2.6, 2.8 and 9.2 come from the works of Laura Palacio. Finally, we would like to acknowledge the collaboration of all the members, doctoral students, master students, visitors and collaborators who have been shaping the work of the Surfaces and Porous Materials Group of the University of Valladolid over the years and who have made this group one of the most recognized experts in the characterization of all types of membranes and porous materials.
- Aldo Bottino, from the Universitá di Genova (Italy), introduced us to the LLDP technique, and his collaboration was a fundamental boost to our capacity and experience in porosimetric techniques.
- René Peinador, from IFTS (France), previously a collaborator of our group and a PhD student of one of us, has continued to collaborate in his current position in the improvement of porosimetric techniques. The data in Figures 7.11 and 7.12, together with the schematics in Figures 7.8–10, come from his laboratory.
- Melike Begun, formerly at Nanyang Technological University in Singapore and currently at Technische Universiteit Delft (Netherlands), collaborated with the authors on a review of characterization techniques, published in the Journal of Membrane Science, which has inspired the organization of this text. Figures 5.10–5.12 are adapted from several of her publications with Jia Wei Chew, at present at the Chalmers University of Technology in Sweden.
- Omar Sada, from the Universitá di Genova, spent some time with our group. Fruitful discussions with him allowed us to improve the chapter on SEM microscopy.

https://doi.org/10.1515/9783110792195-203

- Our thanks to Walter De Gruyter GmbH, who proposed this project to us, accompanying and guiding our efforts through the whole process. Special thanks to Karin Sora (Editor-in-Chief of the publisher's STEM collection) and Ria Sengbusch, our content editor.
- Last but not least, we would like to thank our families, who have patiently endured that during the two long years of writing this book, we could not dedicate as much time to them as we would have wished and as they deserve.

Thank you to all of them and those, not named here, who have contributed to this outstanding experience.

José Ignacio Calvo
Antonio Hernández

Contents

Chapter 1
Introduction

Much of the scientific literature on membranes begins with an introduction highlighting the great impetus that these materials have experienced in "recent years" and accounting for the large number of emerging applications that are being incorporated day by day into the already broad spectrum of industrial processes in which membranes have found use.

The truth is that these statements, perhaps somewhat repetitive, are certainly true, but merely somehow exaggerated, mostly concerning the assumed pace of development. The real fact is that the growth of membrane technology has been steadily constant for more than 100 years, from the first synthetic membrane commercialized. Thus, it should also be noted that, in differently configured modules and applying different types of gradients, membranes have found an outstanding use in more and more separation processes. More importantly, as more fields of application are found, the initial idea that the membrane is simply a filter that allows particles of a certain size to pass through it has been pointed out and complemented to avoid falling into a gross oversimplification.

But, clearly, we need to start presenting a reasonably flexible definition of what a membrane is, a definition able to comprise all possibilities now developed after so many years of research and development. Then, a membrane can be defined as: "any region that acts as a barrier between two fluids, restricting or promoting the movement of one or more components of one or both fluids across it". This definition is sufficiently generic to adapt reasonably to all types of synthetic membranes that have been developed over the last century and those that have appeared in the last decades or in the near future.

Historically, it is difficult to date the beginning of membrane technology. Some authors cite the pioneering work of the French clergyman and physicist Abbe Nollet, who carried out several experiments with semipermeable membranes. Certainly, these were natural materials (pig bladders), but their application in filtration or, more precisely, diffusion-mediated separations, was obvious. This early, somewhat occasional and mostly descriptive starting point in membrane science and technology was followed in the nineteenth century by more systematic studies on the laws of diffusion (Fick, Traube and Pfeffer) or osmotic pressure (van't Hoff and Raoult).

Thus, step by step, various membrane separation experiments were incorporated into the scientific knowledge. Among those, we must mention the seminal work of Richard Zsigmondy (Nobel Prize in Chemistry in 1925 for his work on colloids), who was the first person (along with his fellow scientist Wilhelm Bachmann) capable of developing a reproducible method to obtain a synthetic filter (cellulose nitrate-based membrane) suitable for sterilizing small quantities of fluid. After the patent arising from the work of Zsigmondy, Sartorius produced the first commercially available mem-

https://doi.org/10.1515/9783110792195-001

brane cartridge for laboratory use. Certainly, other pioneers such as Fick and Bechhold (a name that will appear frequently in some chapters of this book and the man who named ultrafilters to these original attempts) had started in this way through the elaboration and study of collodion membranes.

But the major milestone in membrane technology can be considered the discovery by Loeb, in collaboration with Sourirajan of UCLA (at the beginning of 1960s), of a synthetic asymmetric cellulose acetate membrane [1]. The key advantage of this membrane was that it had two clearly differentiated parts that made it asymmetrical: on the one hand, a very thin active layer on which membrane selectivity was based and, on the other hand, a porous support with significantly larger pores suitable for providing the membrane with mechanical stability.

This Loeb-Sourirajan membrane was a benchmark as it allowed for the first time the desalination of water by a membrane process (reverse osmosis) in significant volumes (we are not talking more about laboratory-scale applications as in the case of Zsigmondy's development) and, more importantly, at a competitive cost with conventional separation techniques.

From that moment on, membrane technology progressed and grew clearly and steadily over the years, finding new developments, new types of membranes, assembled in new modules containing more useful membrane area, which gave rise to new applications in industrial separation processes.

However, in this process of development and evolution, it became increasingly necessary to know the characteristics of the new membranes developed or the new materials suitable for developing them. In this way, after a proper and complete effort of characterization of a given membrane, it would be possible to foresee its best use in a certain separation process.

This process of evaluating and quantifying the main characteristics of a candidate material to be commercialized as a membrane is called characterization and comprises a systematic study of all the properties of this material that may be of interest in terms of its subsequent use. Thus, we can define characterization, without loss of generality, as: "the acquisition of the most complete knowledge of its constitution, structure and functional behaviour, obtained through the combined and critical use of adequate methods and techniques" [2].

Generally, the complete characterization process of a membrane or any other type of porous material involves determining a set of parameters, which are usually classified into two groups: structural and functional, depending on whether they are more related to the proper knowledge of its structure or to evaluate its performance in a given separation process.

Firstly, functional parameters include:

– Permeability: basically, we refer to water permeability in all membrane processes, including liquid solutions as feed, but it should also be necessary to determine the gas (or gases) permeability for membranes intended for gas separation purposes.

- Selectivity (generally analysed for one or several species of interest), including the determination of retention coefficients and separation factors for such species.
- The existence and distribution of charges on the surface or inside the membrane (sometimes accompanied by the determination of membrane zeta potentials or ion exchange capacity).
- Effective diffusion coefficients.
- Adsorption characteristics, which also influence the ability of the membrane to become fouled in each process.
- Appropriate tests to assure membrane compatibility for a given separation process include chemical compatibility or mechanical and physical resistance tests.

On the other hand, structural characterization mainly involves the experimental determination of the following parameters [3]:
- Statistical distribution of pore sizes since pores of a single size are rarely observed in a membrane.
- Pore morphology and average pore size are generally expressed by a shape factor and a value of radius or equivalent pore diameter, respectively.
- Pore surface (volume) density, that is, the number of pores per unit of surface area (volume) of the membrane.
- Volume porosity is defined as the fraction of the total membrane volume that is occupied by pores or voids. In some cases, it may also be appropriate to characterize a surface porosity, defined in parallel, but in terms of surface, to the above.
- Roughness shows the differences in height that may be present on the surface of a membrane.
- Tortuosity, as pores are generally not cylindrical, means that the area occupied on the surface does not correspond to the volume occupied inside the membrane. In that sense, the existence of pore interconnectivity is also of great interest to understand the actual pathway to be followed by the molecular species to be separated.

To summarize, the main parameters to be determined, in a complete and thorough membrane characterization, are summarized in Tab. 1.1.

Of all the structural parameters, those related to the membrane pores hold a preeminent place. The reason is that selectivity remains the defining characteristic of a membrane. And, still in most applications, this selectivity is based on the relationship between the size of the molecules we wish to retain and the size of the channels through which these molecules could pass or be retained. If we consider these channels as pores, then the size of the pores in a membrane affects the size of the molecules that the membrane can separate.

So, to go further in the scope of this text, we can start by distinguishing what we mean by pore: it can be defined as an "Interstice between particles or molecules that constitute a solid body". Although this definition does not fit exactly with liquid membranes, it is sufficient for most cases of interest. We do not really speak of pores when these interstices are too small and must consider them simply as binding zones between the different molecules that make up the membrane material.

Tab. 1.1: Parameters to be determined for structural and functional characterization of a membrane filter.

Structural characterization	Functional characterization
Pore sizes: – PSD – Average pore size	Permeability (liquid/gas)
– Pore morphology – Shape factor	– Selectivity (for selected species of interest) – Retention coefficients – Separation factors
Pore density – Volume – Surface	– Charges – Zeta potentials – Ion exchange capacity
Porosity – Volume – Surface	Diffusion coefficients
Roughness	Adsorption characteristics
Tortuosity	Fouling behaviour
Thickness: – Active layer – Support	General tests: – Chemical compatibility – Mechanical resistance – Physical testing

In that sense, membranes are usually divided into porous and dense membranes. Dense membranes are those with no discernible pores, while those presenting measurable pores are considered porous membranes. This distinction was usually made on the basis of the resolving power of electron microscopes (prior to the advent of higher resolution microscopes such as atomic force microscopy (AFM)). Thus, any interstice larger than 2 nm is considered a pore. This limit is also quite convenient as it matches the IUPAC classification for mesopores and macropores [4]. In fact, IUPAC considers micropores as those with pore sizes below 2 nm (so, dense in our definition), whereas mesopores are those with pores between 2 and 50 nm and, finally, macropores are those with pores larger than 50 nm.

At this point, it is worth making a clarification on a point that can sometimes lead to confusion. Although many pores are similar to cylindrical structures with a circular section area, the pore size is not considered to be the radius of the circle but the diameter (which is to be equated with the largest cross-sectional dimension of a molecule that would pass through the pore). So, frequently, when we speak about pore size, we are referring to the diameter (or maximum dimension) of the pore entrance.

Thus, the techniques that are able to give us information about the size of the pores of a membrane can be called porosimetric techniques, or simply, porometries. Therefore, porometry can be defined as a "porous materials' characterization technique able to measure or evaluate the size of the mean pore of a membrane filter".

Sometimes the term "porosimetry" is also used as equivalent to porometry, also understood as the measurement of the pores of a membrane filter or, in general, of any porous material. However, it is common to reserve the name "porosimetry" for one of the techniques that we will see later, the mercury intrusion porosimetry (HgP), leaving the term "porometry" for the rest of the techniques. Obviously, except for historical reasons of nomenclature, there is no real motive to distinguish porometry from porosimetry, both terms being intrinsically equivalent.

Based on the above definition, it is clear that any porometry we choose must be able to determine the pore size of a membrane or filter. However, as we have already mentioned, these pores will rarely have a single size. Being heteroporous materials in nature, they will present a set of different sizes, each with frequencies (number of pores of the given size) distributed around a central value. The distribution of the frequencies of occurrence for each of the pore sizes present in a membrane is called the pore size distribution (PSD) and will typically follow a statistical law. Relatively homoporous membranes usually present a distribution that fits well with the Gaussian function (known as normal or Gaussian distribution), while for many asymmetric membranes obtained by the phase inversion method, which are clearly heteroporous by nature, usually present PSD that can be considered to be closer to log-normal distributions.

Without going into further detail or proving its mathematical origin, the normal distribution follows a probability density law given by the following expression:

$$\Phi(x) = \frac{1}{\sigma\sqrt{2\pi}} e^{-\frac{(x-\mu)^2}{2\sigma^2}} \tag{1.1}$$

where x is the value of the variable (pore size in our case), μ is the mean value of the variable (in our case, it corresponds to the mean pore size) and σ is the standard deviation (which gives us an idea of the width of the distribution).

Similarly, the log-normal distribution follows a probability distribution that is Gaussian, however, not in the variable x but in its logarithm. Thus, the probability density law will be given by

$$\Phi(x) = \frac{1}{\sigma x\sqrt{2\pi}} e^{-\frac{(\ln x-\mu)^2}{2\sigma^2}} \tag{1.2}$$

with the same meaning for each of the terms.

When we apply a porosimetric characterization technique to a sample, the most complete information would be that which provides us with the complete PSD. From it, we can determine the value of the average pore size, the maximum and minimum pore sizes present in such distribution or even the total porosity (integrating the theoretical expression of the PSD to the whole range of pores).

The porosimetric characterization techniques that we are going to deal with in this book will be able, in general, to provide us, more or less accurately, with the PSD of our membrane, so that we can adjust it to a theoretical function and then obtain, as a final result, the average pore size values or any other necessary information related to such PSD.

Before dealing with each of the existing techniques in the following chapters, it would be useful to classify them according to the different criteria. Thus, a first classification (and probably the most pertinent) could be to distinguish between direct and indirect methods. The former would be those that allow us to directly measure the size of the pores, while in indirect methods we will determine some other magnitude that we must then relate to the size by means of appropriate models.

It seems clear that the only really direct methods, sensitive to measuring the pores, are from visualization. Thus, we can include in this group all microscopies that are capable of visualizing the pores of a membrane and, consequently, of measuring their size. Certainly, these microscopic techniques will be useful or applicable to several membranes based on the resolution they achieve. Accordingly, scanning electron microscopy and transmission electron microscopy, along with scanning tunnelling microscopy or AFM, will be included as microscopic porosimetric techniques.

There are also some spectroscopic techniques that show the defects or voids in a material and, consequently, inform us the size of these voids. Among them, the positron annihilation spectroscopic technique or sometimes referred to as positron annihilation lifetime spectroscopy will be discussed here. Another spectroscopic technique, not so frequently used for the characterization of porous membranes, is the synchrotron radiation. Also, in recent years, another spectroscopic-based technique, the so-called ellipsometric porosimetry (ELLP), has become available, which is particularly useful for the study of thin surfaces of all types of materials, but also can be used to determine the PSD. All these three techniques can be considered as direct ones, but must reasonably be separated from microscopic techniques where pore sizing evaluation is clearly more direct. To be strictly correct, ELLP can only be considered as direct porometry in the study of film thickness while the use of the technique to determine PSD relies on using Kelvin's equation, so it should be considered more appropriately an indirect method.

The rest of the porosimetric techniques that will be reviewed in this book must be considered as indirect techniques and, as such, we will introduce them on the basis of the theoretical model that must be used in each case to interpret the experimental results and correlate them with the information sought.

Three equations are important in order to substantiate some of these indirect techniques (really all of them refer to the original Gibbs equation): Young–Laplace, Kelvin and Gibbs–Thomson equations, and the porosimetric techniques based on each of these equations will be reviewed in the corresponding section.

Thus permporometry, evapoporometry and gas adsorption–desorption will be reviewed jointly based on the Kelvin equation. Thermoporometry and nuclear magnetic resonance cryoporometry will be considered as related to Gibbs–Thomson equation. Finally, liquid displacement porosimetries (GLDP, LLDP and LEIP) along with mercury intrusion porosimetry (HgP) will be reviewed and discussed in relation to the use and limitations of the Young–Laplace equation.

Finally, a last porosimetric technique, which relates the structure and performance of the analysed membrane, could be considered very useful, and quite functional in nature, will be reviewed. This is the solute retention test, which is based on the measurements of the rejection coefficients for certain well-selected solutes and the correlation of such rejection coefficients, with the size of the pores present in the membrane.

Figure 1.1 shows all the porosimetric techniques (marked in blue) that are analysed in this book. They are classified according to the previously explained distinction between direct and indirect methods.

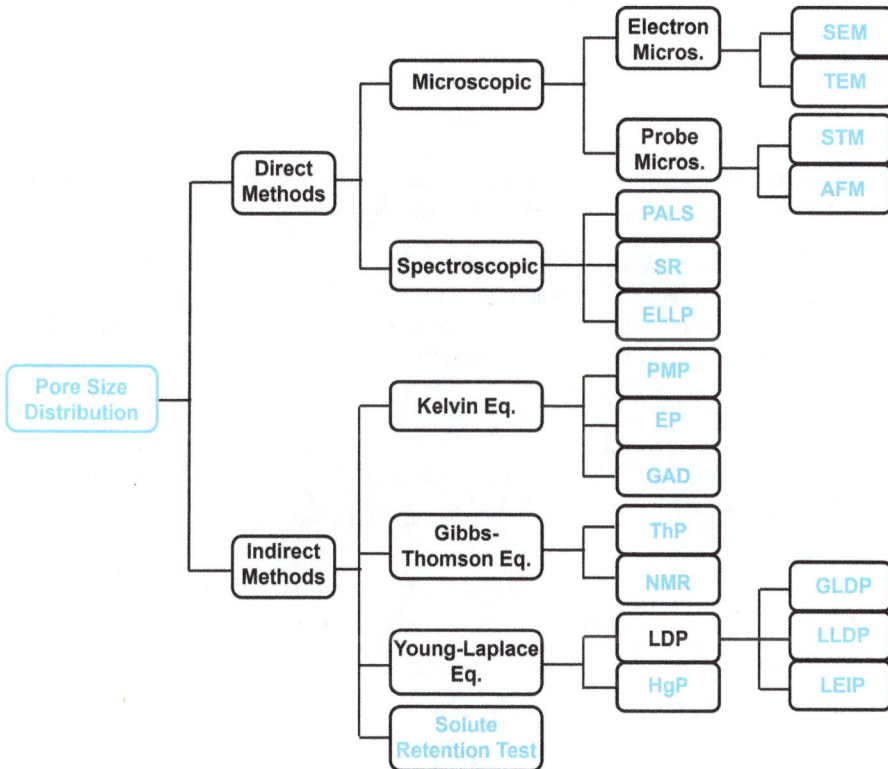

Fig. 1.1: Porosimetric techniques of characterization included in this book, classified as direct or indirect methods. For technique names' abbreviations, refer to the previous pages.

For each of the techniques that we will review in the following chapters, we will indicate the experimental procedure on which they are based, the equations or models that should be used in each case to interpret the experimental data obtained in a typical run, and we will describe briefly about the commercial devices, where they exist, that can be used to carry out the technique. Finally, we will try to make a critical analysis of the results obtained by this technique with its outstanding points and possible drawbacks. In the end, no technique is perfect, and all present advantages and disadvantages must be known and accounted for in the discussion of the results.

Lastly, for all the described techniques, we will mention, depending on their characteristics, the existing possibilities to adapt the technique for the determination of the characteristics of different membrane presentations, that is, different membrane modules for industrial use. The reason is that, where possible, this will allow the analysis of these modules, without destroying them to extract the membrane sample contained in them, so that they can be reused or the porosimetric method considered can analyse filters as part of the quality control prior to their use.

References

[1]   Loeb S., Sourirajan S. Sea Water Demineralization by Means of a Semipermeable Membrane. UCLA Dept. Of Engineering, (July 1960), Sea Water Research Rept, 60–60.

[2]   Calvo J.I., Bottino A., Prádanos P., Palacio L., Hernández A. Membrane characterization: Porosity. In: Encyclopedia of Membrane Science and Technology. Wiley Interscience Pub, New York (USA), (2013), 1–35.

[3]   Palacio L. Caracterización estructural y superficial de membranas microporosas. PhD Thesis, Universidad de Valladolid, (1999), Valladolid (Spain). ISBN: 84-7762-944-7

[4]   D. H. Everett. IUPAC Manual of Symbols and Terminology, Appendix 2, Pt. I. Colloid and Surface Chemistry. Pure Appl Chem, 31 (1972) 578. http://dx.doi.org/10.1351/pac197231040577.

Chapter 2
Microscopic techniques

As mentioned in the brief introduction of the first chapter, few techniques are as straightforward for measuring pore size as direct microscopic observation. It is evident that there is no better way to measure the dimensions of a given object than visualizing it. As it is customarily expressed, "a picture is worth a thousand words".

Microscope was invented at the end of the sixteenth century with the possibility of carving convergent lenses sufficiently precise for the observation of small objects. There is some controversy as to who invented the microscope. Zacharias Janssen (a Dutch lens maker) is said to have built a rudimentary instrument with a power of 9× magnification around 1590 [1]. The English natural philosopher Robert Hooke (a person with such a breadth of knowledge and interests that he is difficult to classify into just one of the known sciences), in 1665, with a compound microscope, managed to observe a kind of network in thin cork sheets, calling the elements of this network cells (*cellula* in Latin). However, it is common to attribute the invention of the optical microscope to the Dutchman Anton van Leeuwenhoek, a cloth manufacturer who managed to build microscopes of up to 200× magnifications, which allowed him to observe the moving protozoa for the first time in 1674.

In any case, the resolving power of an optical microscope is limited by the physical phenomenon of diffraction (linked to the wave nature of light and studied by Abbe in 1873) so that no optical microscope can discriminate objects that are smaller than half a wavelength of the wave with which it is illuminated. In practice, this limits the resolving power of optical microscopes (those that use visible light to illuminate the objects to be viewed) to about 0.2 μm.

Given the size of the usual pores in synthetic membranes (we are generally talking about pores always smaller than a micrometre, and going down to the nanometre), the observation of these pores with an optical microscope, even the most precise and modern one, is beyond the resolving power of these devices. Therefore, the microscopic techniques that can help us to discern the existence and size of the pores of a membrane are the different electron microscopies as well as the more recently developed scanning probe microscopies (SPMs). Fig. 2.1 shows the resolution ranges for the different optical techniques (human eye included) available.

Before dealing with the particularities of each of them, it is useful to make a first disquisition on the real discriminating power of microscopy to determine the size of a pore. It is obvious that we can only measure the size of those pores that we can visualize. Microscopic techniques, by their very nature, not only allow us to visualize the surface of a material (sometimes with sufficient resolution and very sharp images, some of the interior of the pores adjacent to the surface can be glimpsed, but always with a shallow depth of field), but also give us information about the pores on the surface of the membranes.

https://doi.org/10.1515/9783110792195-002

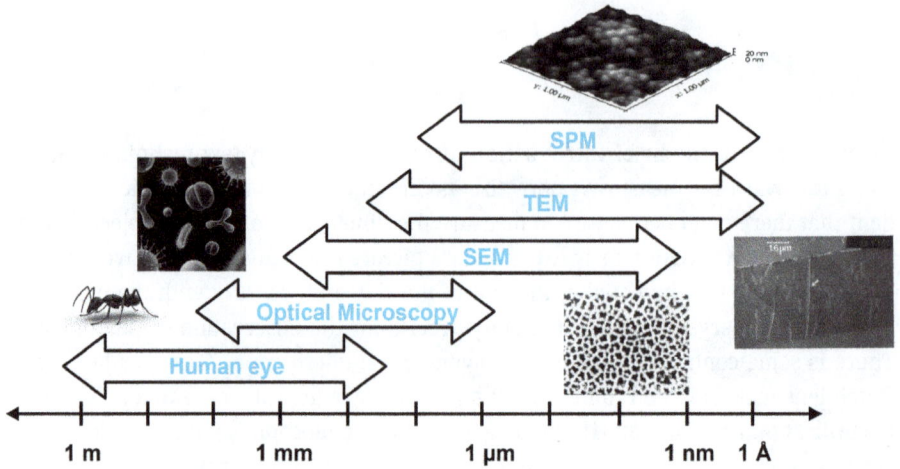

Fig. 2.1: Range of resolution, on a logarithmic scale, for several visual and microscopic techniques.

Another issue is whether these pores are truly representative of the actual pores inside the material. Ultimately the pore (to give rise to effective permeation) must extend from the membrane surface to the other end of the filter. In clearly homoporous membranes (e.g. those obtained by the development of nuclear traces, known as track etching, or by anodic deposition), the pores can reasonably be considered as tubes perpendicular to the surface so that observation of the entrance hole to these tubes will be reasonably representative of their size along their entire length. Similarly, it is also frequent in ultrafiltration (UF) membranes that they show an asymmetric structure, that is, they are formed by a very thin film of material with very small pores (skin or active layer) adhered or cast onto a much more porous and thicker support, whose task is to provide mechanical stability and maintain the permeability of the whole. In this case, microscopic observation of the surface of the active layer will give information on the size of the pores that is really representative in terms of the usefulness of the membrane as a filter, since much larger pores present in the support do not provide selectivity to the whole.

It is obvious that there will be cases in which this observation of the surface is not sufficient to really know the behaviour of the internal pores of the sample. For example, many membranes designed for virus retention [2] have a structure that can be compared to an hourglass. On both surfaces (anterior and posterior) of the membrane, the pores are relatively large, while towards the inner part of the membrane, these pores are slightly narrowed so that their effective size (their equivalent diameter) clearly decreases with respect to the values found on the surface. The reason for this peculiar design lies in the way they are used. Virus retention membranes are designed, as the name suggests, to trap (retain) viruses or other pathogens within them. In this case, the most important thing is that all viruses are trapped, without worrying about the membrane clogging up, as we are generally talking about single-use filters. This is achieved by the hourglass design, which has its selective layer on the inside of the membrane, where the viruses will be trapped by size exclusion.

Even in these more complex cases, a cross section of the membrane can allow us to observe the interior of the pores present in the resulting section. For this purpose, cryogenic fracture techniques are used, which allow us to cleanly cut cross sections of membranes that have been previously frozen using liquid nitrogen. Another possibility to obtain sharp cross sections of the membranes to be viewed comes from the use of a microtome (ultramicrotome, preferably) able to cut very thin slices of the membrane that can then be placed in the microscope. The problem in this approach comes from the nature of the membrane material. In the case of polymeric membranes, even the sharpest cut ultramicrotome can distort the structure of the membrane during the cutting process. To avoid that, the membrane must be hardened by adding an epoxy-type resin, which, once polymerized, allows the sample to be cut without distortion. This procedure is of interest as long as the resin is distinguishable under the microscope from the rest of the material of which the membrane is composed.

In any case, both these types of cuts (if done carefully and assuring minimum distortion of the membrane structure) are very useful for analysing the internal structure of the membrane, discriminating between active and support layers, and analysing the tortuosity of the pores. But they are of little use for obtaining the average pore size of the membrane, let alone a representative pore size distribution (PSD) of the membranes studied.

In the following sections, we will analyse the most common microscopic techniques in the study of membranes and porous materials. Particularly, we will analyse the electron microscopies (both scanning electron microscopy (SEM) and transmission electron microscopy (TEM)) and the newer scanning probe techniques (atomic force microscopy (AFM) and scanning tunnelling microscopy (STM)).

2.1 Electron microscopy: SEM and TEM

As mentioned in the preceding paragraphs, the resolution of the best optical microscopes could be improved (and, in fact, it was done over many years) with more precise lenses, made with fewer imperfections and higher refractive power, and lens combinations that would achieve higher magnifications without losing the image quality. But in the end, any improvement in microscope design or lens materials came up against the barrier of the Abbe diffraction limit, making it impossible to build optical microscopes capable of discriminating details smaller than 0.2 μm.

Overcoming this limit required a further step forward, and it became necessary to wait till the development and understanding of quantum physics. Einstein, through his explanation of the photoelectric effect (published in 1905 and for which he won the Nobel Prize in 1921), showed that a wave can behave as a particle in its interaction with matter. The question that arises naturally is whether symmetrical behaviour is possible.

This question was answered by Louis de Broglie in 1924 with a bold hypothesis, at that time without the support of any experimental basis, the "wave–corpuscle duality". According to de Broglie, "All matter has both wave and corpuscular characteristics and behaves in one way or the other depending on the specific experiment". Mathematically, de Broglie expressed his hypothesis in the following relation:

$$\lambda = h/(m \cdot v) \tag{2.1}$$

with h being Planck's constant, m the mass of the particle and v its velocity. The previous expression tells me what the wavelength of the wave associated with a moving particle is, that is, the matter wave.

The original de Broglie equation has been properly corrected to account for the relativistic velocity [3], so it stands as

$$\lambda = \frac{h}{\sqrt{2m_0 E \left(1 + \frac{E}{2E_0}\right)}} \tag{2.2}$$

where m_0 is the resting mass (at zero velocity) of the electron, and E and E_0 are, respectively, the kinetic energies of the electron and its resting energy.

If we apply this expression to the electron, for example, we can see that for an accelerated electron with an energy of 1 eV (electron volt), the associated wavelength is of the order of 1 nm. This means that the Abbe diffraction of the electron-associated wave will occur between objects less than a nanometre apart, with an improvement of about a factor of 1,000 over the best optical case.

However, despite the elegance and symmetry of de Broglie's hypothesis, it had to be demonstrated experimentally, and this was achieved in only 3 years, when G.P. Thomson succeeded in reproducing Young's double-slit experiment with electrons. In the same year, C.J. Davisson and L.H. Germer performed an outstanding experiment by bombarding a nickel crystal with a beam of electrons and observed that the resulting pattern exhibited the characteristics of diffraction associated with any wave. This experiment can be considered the starting point of electron microscopy.

In any case, once the existence of the matter wave associated with the moving electrons had been demonstrated, it was necessary to obtain suitable lenses that would allow the electron-associated wave to be focused on the object to be viewed. Given the characteristics of the electrons, it is not possible to construct these lenses from any type of solid material that is transparent to the electrons. The solution is provided by the so-called magnetic lenses, elements that generate electric and magnetic fields that allow the electron beam to be directed in the desired direction.

The development of these lenses and of a suitable system for collecting the electrons scattered in a sample (under vacuum conditions) can be considered the birth of electron microscopy, about whose invention there is also some controversy. Thus, Gabor was the first to work with electromagnetic lenses, although he did not go on to design the device. Knoll and Ruska, in 1931, presented an electron microscopic design based on two electromagnetic lenses. At the same time, G. Rühderberg filed a patent for his own design.

After significant technical improvements, departing from these early designs, we can summarize that today's electron microscopes essentially consist of the following elements:

–   Electron gun: the electron source is responsible for generating a beam of high-energy electrons in the SEM. This source, commonly referred to as an electron gun, must be able to emit a bright and stable stream of electrons for consistent focusing on the sample.

    The most common way to achieve the electron beam is through a thermionic emitter, which consists of a resistive material through which an electric current flows. This current heats the material until the electrons on its surface acquire sufficient energy to be ejected in significant quantities (~10^9 e$^-$/s). The material used can be either a tungsten filament or a crystal of cerium hexaboride, CeB$_6$, or lanthanum hexaboride, LaB$_6$ (see Fig. 2.2). The other types of emitters used as electron guns are the so-called field emitters or field emission guns (FEGs), which exploit the tunnel effect in a very thin tungsten tip subjected to a high potential difference. In this way, the electron beam obtained is highly bright and coherent.

    Note: In SEM imaging, we can distinguish two types of coherence: temporal and spatial. The temporal coherence refers to a beam of electrons having all the same energy, therefore, the same wavelength. Spatial coherence refers to electrons emitting from the same point in the gun, which is related to the smaller source size. In practical terms, special coherence is preferred over the temporal one, as it gives electrons hitting the sample at smaller angles and lower diffraction patterns between different emitted electrons.

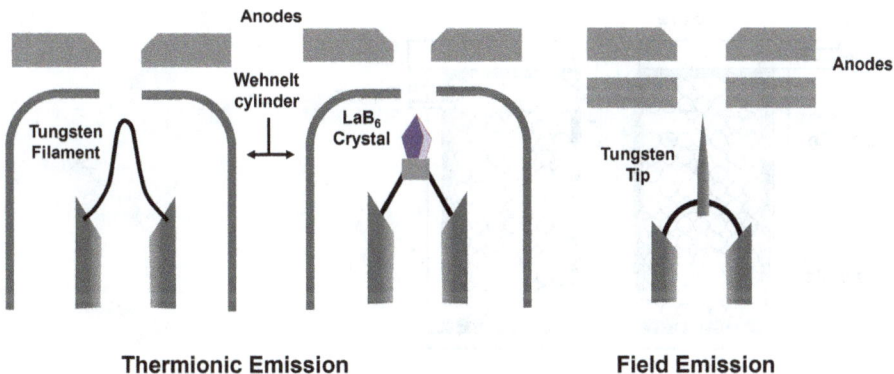

Fig. 2.2: Examples of different electron guns: (A) tungsten wire, (B) LaB$_6$ crystal and (C) field emission gun.

The most frequently used emitters are tungsten filament emitters, which are cheaper and easier to replace in case of breakage but have poorer performance in terms of brightness and coherence. Hexaboride crystal thermionic emitters are more stable and give better results than tungsten filament emitters. Finally, field-effect emitters, which are the most expensive and need to work in higher vacuum conditions, are, on the other hand, the most durable and achieve the most accurate results, especially at high resolutions.

- A set of magnetic lenses designed not only for focusing the electron beam on the sample but also to drive the electrons resulting from the interaction of the beam with the sample, in particular the secondary electrons which, properly focussed and collected, form the main part of the resulting image. An example of a typical magnetic lens to be used for electron microscopy is shown in Fig. 2.3. In fact, the lens cannot collect all the rays from the object, and we often deliberately limit the collection angle with an aperture to control the lens aberration.
- A vacuum system that keeps the entire electron pathway free of interactions with air molecules that could deviate them. The vacuum required depends on the electron gun used, as mentioned before, from the 10^{-2} Pa required for a tungsten filament to the 10^{-9} achieved when a cold FEG is used.
- A detection system (screen) on which the electron beam is projected once it has interacted with the sample, and which allows us to visualize the topography of the sample. The projection on the screen must also be able to be stored in a suitable device for later study (originally photographic plates were used, although nowadays the images are converted into the digitalized information collected in an exportable file).

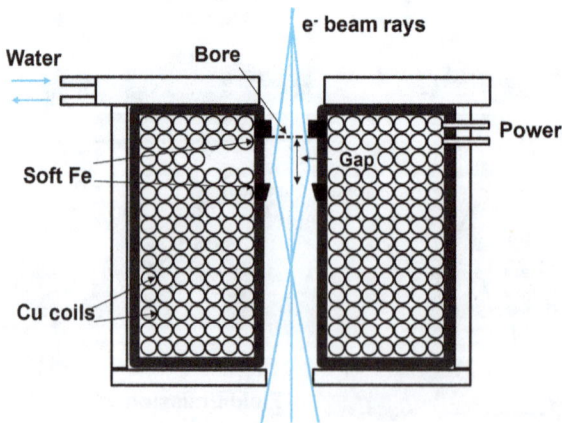

Fig. 2.3: Typical magnetic lens of an electron microscope (adapted from [3]).

The electron beam can be moved (scanned) over the surface of the sample in a slow sweep that allows a relatively large surface area to be imaged.

There are two different ways or modes of working with electron microscopy: SEM and TEM. In SEM, the secondary electrons emitted from the surface of the sample when the electron beam hits the surface are collected. By moving the focal point of the beam in the x and y directions, the sample is scanned to complete the image provided by such secondary electrons.

In TEM, on the other hand, the transmitted electrons pass through the sample. Therefore, it is necessary for the sample to be ultra-thin, if not none of the electrons in the beam will manage to cross the sample to the other side. Clearly, the electrons that manage to pass through the sample give us more information about its internal structure than the surface, which is more properly covered with the electrons collected in an SEM image.

Figure 2.4 shows a comparative diagram of the two modes of operation, which can be easily understood from the different paths followed by the diverse electrons and rays appearing after the initial electron beam interacts with the sample. This figure helps us to understand the different bases of both electron microscopic modes (SEM and TEM).

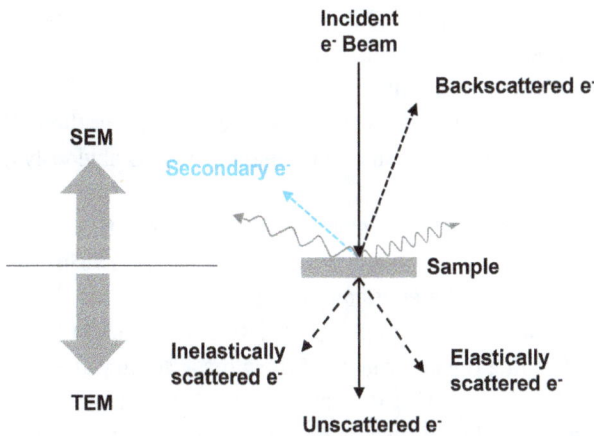

Fig. 2.4: Different particles and rays involved in the operation of electron microscopy.

Effectively, when the electron beam interacts with the sample under study, different effects are produced on the electrons composing the beam. Thus, there is a part of the electrons that are not deflected (unscattered) and these can pass through the sample, provided it is thin enough to allow passage. These transmitted electrons will form the basis of the TEM signal.

But most of the electrons interact with the sample, resulting in electron scattering, which can be elastic or inelastic. An important part of the energy of the electrons will be lost as heat, but there are also other side effects that will allow us to extract important information from the sample.

Secondary electrons, backscattered electrons (BSEs) and X-rays can be produced as a result of this interaction, following different mechanisms:

- The secondary electrons are those produced when an electron from the incident beam passes very close to the nucleus of the atoms of the sample, giving up enough energy to one or more of the inner atom electrons to jump out of the sample. These secondary electrons have very low energy (around 5 eV or lower), so the electrons and the atoms that supply them must be very close to the surface to be produced. This is precisely the reason why secondary electrons provide very valuable topographical information about the sample, and why they are mainly used as source of the image in scanning microscopy. Moreover, since a single incident electron can give rise to several secondary electrons (as it moves more or less deeply through the sample), these electrons will be the most abundant.
- BSEs are produced when an electron in the beam collides head-on with atoms in the sample and is ejected at angles close to 180° to the direction of incidence. These BSEs can also generate more secondary electrons on their way out of the sample. The intensity of the BSE effect is proportional to the atomic number of the sample. The higher the atomic number of atoms means larger the atoms and higher the probability of producing BSE. For this reason, they are used to obtain a mapping with information about the surface composition of the sample.
- Importantly, the BSE production is correlated with the atomic number of the atom in the sample with which the electron collided, so the intensity of the BSE signal will be proportional to the composition of the sample (which is obviously a very useful information).
- Because of the change in the energy of some electrons in the sample, this excess could be emitted in the form of an electromagnetic wave, which, considering the typical energies involved, results in the emission of X-rays. The resulting X-rays are characteristic of each element in the sample (considering the different levels of energy of the electrons forming the outer layer of the atoms), so they are used to obtain spectroscopic information about the composition of the sample. This constitutes the EDS (energy-dispersive spectroscopy) technique complementary to SEM, which will be summarized in a further section.

In this way, SEM, in a single analysis, provides information on the topography, morphology, composition and crystallographic nature of the samples analysed [5].

There are certain differences in the preparation of the samples depending on whether SEM or TEM is used. In the first case, we will seek to detect secondary electrons, so it is advisable to increase their production in order to have a clearer signal. Therefore, the sample must be conductive to guarantee the production of secondary electrons. If this is not the case, as often happens in the case of membranes (usually obtained from polymeric or ceramic materials), it is necessary to coat the sample with a conductive material that increases the production of these electrons. Generally, the coating material is gold or platinum (very thin layers of up to a hundred nanometres, obtained by sputtering), although if X-ray microanalysis for sample composition is required, the coating can be made by carbon wire.

Regarding the preparation of samples for visualization using the TEM technique, the main difference is that, since we will be observing those electrons that pass through the sample, it is necessary to keep the thickness of the sample as small as possible.

In general, specimens with a thickness of less than 100 nm are required for proper TEM visualization. This can be achieved by using an ultramicrotome that can cut sections in the order of several tens of nanometres thick. In the case of polymeric or ceramic samples, cutting should be carried out under cryogenic conditions to minimize damage to the sample structure.

In biological samples, it is useful to perform a carbon replica, consisting of deposition of a thin layer of carbon on the surface to be observed, followed by the removal by chemical washing of the original biological sample, leaving only the carbon replica of adequate thickness for use in TEM.

Although the basic technology of the electron microscope has remained the same since its invention, along the past of the years substantial improvements have been made in the precision, potential and quality of the results of the equipment used. We can highlight two variants of SEM equipment that allow significant improvements in the results or that favour working with biological samples without the need for previous coating.

2.1.1 FESEM

In conventional SEM microscopy, the electron beam production system is usually based on a tungsten filament which, when heated by an electric current, emits thermal electrons. A solid-state crystal, usually lanthanum hexaboride (LaB_6) or cerium hexaboride (CeB_6), can also be used [6]. This type of electron emitter is much more expensive but provides higher emission and longer lifetime.

Another type of SEM microscope that uses neither a tungsten filament nor a solid-state glass as an electron gun is the field-emission scanning electron microscope (FESEM), which allows working with lower accelerating voltage to obtain better spatial resolution. The FESEM works in a similar way to a conventional SEM, scanning the surface of a sample with an electron beam and providing information similar to the SEM, but with different types of detectors. The main difference with respect to an SEM lies in the electron emitting system, which consists of an FEG (formed by a tungsten crystal with a pointed tip or by a Schottky filament) that generates a beam of high and low energy, highly focused electrons. This improves the spatial resolution, allows working at very low accelerating voltage and consequently minimizing the charge on the sample to be observed and the possible impact of the beam on it.

Another important advantage of FESEM is that its improved vacuum system allows several detectors to be placed inside the vacuum column. These detectors, called in-lens detectors, allow the detection of very-low-energy secondary electrons, increasing in that way the resolution of the resulting image.

2.1.2 ESEM

The environmental SEM (ESEM) is an SEM that allows working with wet or uncoated samples, or both. For this purpose, it is necessary that there is gas inside the chamber, as a perfect vacuum would cause the moisture to disappear from the sample resulting in its distortion. Technically, an ESEM microscope is an FESEM whose vacuum chamber is designed to control the vacuum level inside, as well as the humidity and temperature of the sample. Thus, these devices can work in three modes: high vacuum (below 5×10^{-4} Pa), similar to an FESEM; low vacuum (up to 100–150 Pa); and finally in the ESEM mode (up to 3–4,000 Pa). In addition, the temperature control (typically up to 1,500 °C) allows to study the temperature behaviour of the sample as well as to include different types of gases interacting with the sample. The first functional ESEM instrument is considered to have been designed by Gerasimos Danilatos while working at the University of New South Wales.

2.1.3 SEM–EDX

As mentioned above, during the SEM imaging process, X-rays are produced due to the energy loss in some of the electrons in the beam incident or detached from the sample. If we include a suitable X-ray detector in our SEM, we can obtain information known as the EDS, EDX, EDXS or XEDS technique, or simply energy-dispersive X-ray spectroscopy.

The basis of the technique is that the X-rays generated by the incident electron beam come from various electrons present in the sample atoms. These electrons may be in more or less inner shells of the atom, leaving gaps when excited, which are filled by other electrons from another, more external shell. The difference in electron energy levels from both electron shells is what is emitted in the form of X-radiation, and this energy and, consequently, its wavelength, is characteristic of the jump between the two electronic states, which allows us to obtain information on the atomic structure of the sample and, therefore, its composition.

In the case of the EDS technique associated with SEM equipment (SEM–EDS), the source of electrons will be the SEM itself, so it is sufficient to add an X-ray detector, capable of converting the energy of the different X-rays detected into proportional voltage signals and sending this signal to a suitable processor, associating the X-ray spectrum with the defined point of the sample where the electron beam has hit. The number of X-rays observed for a given energy is translated into a series of peaks characteristic of the sample composition. It is clear that the quality of the results obtained with EDS depends on the strength of the signal obtained and the absence of noise in the spectrum. A low level of noise in the signal will be essential for the detection of trace elements (not very abundant in the sample), while a lack of cleanliness (due to contamination of the sample in various elements of the equipment) will result in the presence of spurious peaks.

Finally, compared to conventional EDX detectors, which receive the energy of all the beams coming from one point of the sample, wavelength-dispersive X-ray spectrometry detectors allow focusing on the signal of a single beam. In this way, the analysis is more sensitive and precise, although, obviously, much slower.

2.1.3.1 More information

For a better understanding of the physical phenomena as well as the technology needed to obtain a good result with an SEM, there are several interesting books on the market. Just as an indication, we can mention the book by Goldstein et al. [7], while Williams and Carter [3] focus more on TEM technology.

2.1.4 Electron microscopy suppliers

There are several prestigious manufacturers of electron microscopy equipment, most of which have in their catalogues both SEM and TEM equipment and more advanced models capable of working in FESEM or ESEM modes.

Among them we can mention without claiming to be exhaustive and focusing on those that have been used at some point for membrane characterization:
- JEOL microscopes
- Thermo Fisher Scientific microscopes
- Hitachi microscopes
- Carl Zeiss microscopes
- Nikon Metrology microscopes
- COXEM (Dynamic Korea) microscopes
- Bruker Corp. microscopes
- FEI Co. (Field Electron and Ion Company) microscopes
- TESCAN microscopes

2.2 Probe microscopy: STM and AFM

When we introduced the optical microscopy, we emphasized that the detection limit of every sort of microscope is fixed by its diffraction power, that is, by the size of the diffraction spot that the light wave undergoes when it hits sufficiently small objects. Thus, a microscope cannot discriminate details, objects or pores smaller than this diffraction spot. Likewise, electron microscopy, although it significantly improves the diffraction power of optical microscopy, is subject to fulfil the same diffraction limit as any other type of microscopy based on a wave detection. In the end, diffraction is an intrinsic characteristic of any type of wave, be it light or any type of electromagnetic wave, but also mechanical, vibrational or, certainly, the material waves associated with any particle such as the electron.

To overcome this limitation, the last years of the last century saw the birth of several microscopic techniques based on totally different fundamentals and therefore not constrained to fulfil the above-mentioned diffraction limits.

In this group of novel microscopies, we must include two novel techniques with their distinct variants and working modes: STM and AFM.

Historically, the first to emerge was STM, discovered in 1981 by Gerd Binnig and Heinrich Rohrer (working for IBM Zurich) [8, 9], a discovery that led them to be awarded with the Nobel Prize in Physics in 1986. STM is based on a purely quantum effect, the so-called tunnel effect. When two conducting materials are placed at a very close distance to each other, electrons from one of the materials cannot, in principle, cross the gap between them and pass into the other material. Now, considering that electrons are quantum particles that have associated with them a matter wave, we must interpret this wave (or rather, the square of its amplitude) as the probability of finding the particle in a given place at a given time. Due to the peculiarities of waves, the probability of finding the electron of one material on the other side of the gap (tunnel) separating the two materials, although very small, is not intrinsically zero. Thus, some (very few indeed, but enough for detection purposes) electrons from material A can quantumly pass through this tunnel and appear in material B. If such material B is designed to detect the tunnel jumping electrons, then an image of the electronic shells of the atoms on the surface of material A can be imaged as we move the material B probe over that surface.

In practice, STMs consist of a very thin conductive tip, which is moved very close to the surface to be imaged. By detecting the electric current in the tip, which will be higher when we are close enough to an atom of the material to be analysed, we can visualize the individual atoms contained in the surface (in fact, we are visualizing the outer electron shells of such atoms). In this way, a good STM can achieve a lateral resolution of 0.1 nm, with 0.01 nm depth resolution, sufficient to observe and even manipulate the atoms of the scanned surface.

As soon as the technique appeared, the possibility of using it for the visualization of surfaces on a subnanometric scale became apparent. In this sense, applications of the STM technique to the visualization and characterization of membranes were quickly seen [10]. However, the problem with the technique is that, in order to establish a good tunnel current between the surface and the scanning tip, the membrane material needs to be conductive, a condition that is not fulfilled by most of the synthetic membranes usually employed. Even then, it has been used with some frequency for the characterization of membranes based on graphene and nanoporous carbon [11–14].

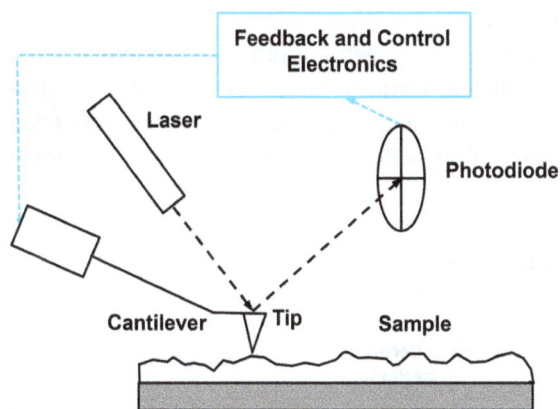

Fig. 2.5: Schematic of the different elements in a typical AFM, [4].

Binning himself, working on the improvement of the STM technique, discovered the so-called AFM. Thus, Binning, Quate, and Gerber, in a 1986 paper, proposed measuring the forces between the tip and the surface as an application concept of STM [9], leading to the appearance of AFM, although the first commercially available AFM did not appear until 1989.

Basically, the AFM microscope consists of a very fine tip at the end of a flexible arm (which is called a *cantilever*). By means of a mechanized and automatized system, this cantilever is moved over the surface to be analysed, parallel to it, and maintaining a very small distance between the tip and the surface to be analysed. At this short distance, the tip interacts with the surface atoms by means of short-range forces. These forces cause the tip, and with it the cantilever to which it is attached, to deflect slightly. The deflection caused by these forces is detected by the optical system on the cantilever, which usually consists of a laser beam focused on the top of the cantilever, and the reflected beam of which is picked up by suitable photodiodes (see Fig. 2.5).

The information collected by these photodiodes is processed electronically and used as a feedback to keep the distance of the tip constant with respect to the sample.

On the other hand, as shown in Fig. 2.6, when the cantilever deflection is simply vertical, the signal collected by the photodetector will consequently be a mere vertical variation of the spot that translates into a variation of the z-direction of the movement. Whereas, in many cases, the interaction with the sample will additionally result in a certain torsion of the cantilever, which translates into a vertical deflection coupled with a lateral deflection. It is the combination of both deflections, collected across the length and width of the sample in a two-dimensional scan, which allows 3D imaging of the sample being analysed.

One of the fundamental elements of a good AFM and basic for a reliable device is the electronic signal collection and feedback control system. This system is essential so that the cantilever (and the tip attached to it) can travel over the entire surface of the sample while always maintaining the chosen parameters.

An anti-vibration system is also essential if high resolutions, even atomic ones, are to be obtained. For this purpose, an anti-vibration table must be used as the base of the microscope, and it is advisable to place all the equipment (and its anti-vibration table support) close to the foundations of the building, where vibrations are expected to be greatly minimized.

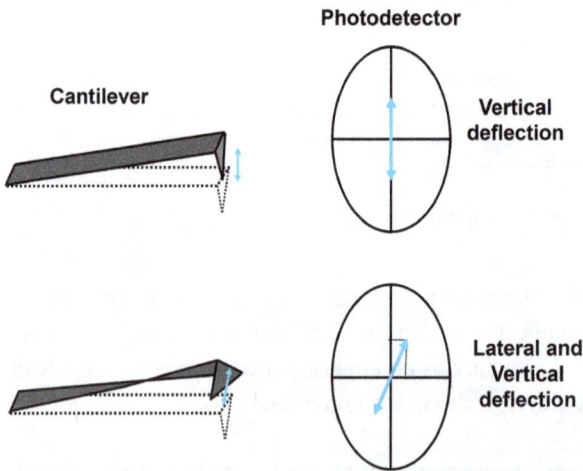

Fig. 2.6: Cantilever deflections and resulting spots in the photodetector of an AFM [4].

A central point in obtaining reliable AFM images with the highest possible resolution lies in the cantilever and the tip attached to it. Figure 2.7 shows an SEM image of a typical cantilever and its tip. The size and thickness of the cantilever influences its resonance frequency, which must be controlled in most AFM modes. Thus, the cantilever must be able to oscillate at high frequencies and relatively large amplitudes without the risk of breaking. Similarly, the tips, which have different shapes for different uses, have in common to the end in the sharpest possible shape (ideally, a tip should end in a single atom, which is technically impossible, but actually very sharp tips can be manufactured) to avoid or minimize convolution of the tip size with imperfections in the sample to be analysed.

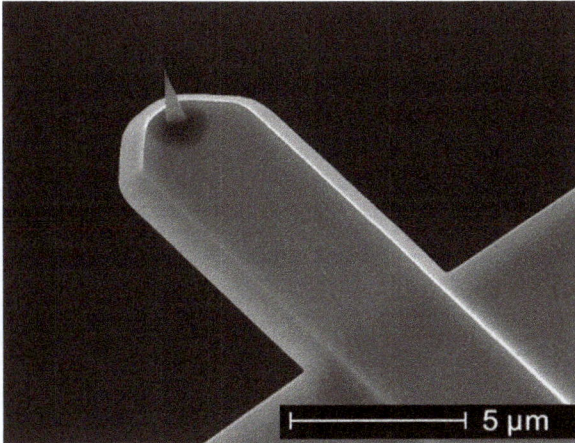

Fig. 2.7: SEM image of a cantilever, showing the sharp tip at the end (USC-F1.5-k0.6 from NanoAndMore GmbH). Available from https://www.nanoandmore.com/AFM-Probe-USC-F1.5-k0.6.

Although both microscopic techniques mentioned in this section have been widely used in the characterization of surfaces and, in particular, membranes, it is clear that the need for a conductive material in the case of STM has meant that the number of applications of this technique in the study of membranes is much lower than in the case of AFM. A simple bibliographic search in Scopus clearly shows that the number of articles dedicated to AFM in connection with membranes is much higher than for STM. Even some of the papers naming the STM technique refer to AFM. This is because both techniques, which have similar origins, are generally known as SPM, since both use a probe (an extremely fine conductive needle in the case of STM or a tip at the end of an oscillating cantilever in the case of AFM).

In the following section, we will analyse in more detail the different working modes of AFM, as well as the other characterization techniques derived from it, just by changing the type of the tip or functionalizing it appropriately.

2.2.1 AFM working modes

Three main modes (contact, non-contact and intermittent contact) are commonly used in AFM. To understand the differences between the various modes, we can look at Fig. 2.8, which shows the interaction potential between the tip and the sample. Typically, the physical potential, which causes the forces acting at short distances (of Van der Waals type), presents a decreasing branch at short distances, reaching a minimum negative value (which corresponds to a certain equilibrium distance) and then starts to grow until it becomes 0 at long distances. Consequently, the interaction forces will be attractive when the distance is greater than the equilibrium distance, while if the distance between the tip and the sample decreases below the equilibrium distance, the forces become repulsive, trying to push the tip away from the surface. The working mode will then depend on the distance between the tip and the sample [15]:

– Contact mode: the tip is kept at a constant vertical distance from the sample, always in the repulsive range of the interaction forces, so that these forces must be balanced by the force applied to the cantilever, as well as the capillary forces due to the thin film of water normally existing on the sample. The feedback control allows this z-distance to be constant, and information on the corrections necessary to maintain this distance is recorded. This information of accumulated vertical z-values as the tip sweeps the surface in the horizontal, x and y, directions allows a complete topography of the analysed sample to be obtained.

– Non-contact mode: in this case, the distance is maintained in the range of the attractive forces, which is evidently clearly larger than in the contact mode. At these distances (typically between 10 and 100 Å), the feedback to maintain a constant distance clearly is not effective, as the height changes caused by the roughness of the sample cannot be detected as easily as with the contact mode; therefore, the feedback is not able to keep the tip–sample distance at a constant value. Thus, in this mode, the cantilever receives a small vertical oscillation of low amplitude and certain frequency. When the tip moves across the sample, the attractive forces it detects are translated into small variations of the oscillation frequency. In this case, the feedback adjusts the height z in order to maintain the frequency at the pre-set value. In this way, we again have topographical information of the z-height in the two horizontal directions, x and y.

– Tapping or intermittent contact mode: in this latter mode, the cantilever is made to oscillate at a certain frequency, close to but below its resonance frequency. The amplitude of the oscillation is set sufficiently high (typically around 100 nm) to ensure that the sharpest end of the tip is in contact with the sample surface during part of the oscillatory motion. In contrast to the non-contact mode, here we do not look at the change in frequency of the oscillation induced by the proximity of the surface, but rather aim to keep constant the amplitude of the oscillation. This is why the intermittent contact mode is also called *amplitude modulation*, while the non-contact mode is also called *frequency modulation*. In this case, the feedback consists of adjusting the distance of the cantilever to the sample (in its equilibrium situation) so that the oscillation amplitude remains fixed as we move across the sample surface. Again, the information gathered from the changes in distance z is plotted as a function of the x and y travel of the cantilever to give a topography of the sample. In the intermittent contact mode, information about the offset between the settled and actual frequency can also be recorded, so that at the end of the sweep, what is called a *phase image* is obtained. This image, although not as accurate as the topographic image, does have better definition of the edges or dips in the surface, so phase images can be very useful in defining pore edges.

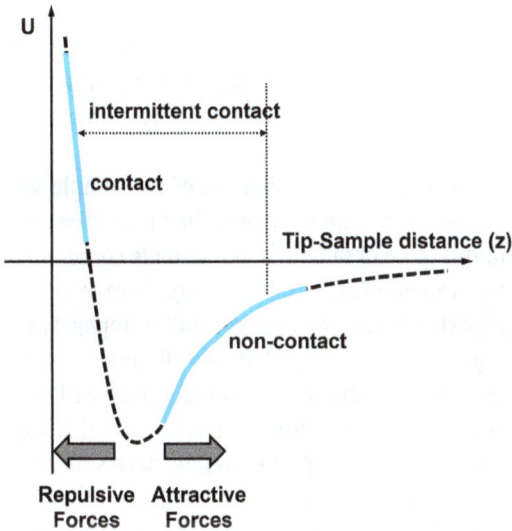

Fig. 2.8: Interaction potential of the short-range forces involved in AFM operation, showing the ranges in which different working modes operate [4].

Another interesting classification of AFM working modes is that which distinguishes between static and dynamic modes. Static modes are those in which the cantilever is not forced to oscillate (basically contact mode), while intermittent mode or non-contact mode would be dynamic working modes.

Although the contact mode is usually preferred for static AFM operation, the non-contact mode can be useful in the determination of long-range magnetic or electrical forces.

Whereas in intermittent (dynamic) mode, contact in the repulsive zone of the forces must be minimized because it prevents the oscillation of the cantilever. Because of this, the cantilevers used in intermittent mode have an elastic constant (typically between 20 and 100 N/m) higher than in static modes (where $k \approx 1$ N/m) [15].

Finally, the non-contact mode minimizes the interaction between the tip and the sample so that the maximum resolution can be obtained, being therefore able to reach atomic resolution under suitable conditions [16]. However, it is more difficult to obtain a good topographic image. Even obtaining sometimes lower resolutions than other working modes, the main advantage is that sample damaging or scratching are avoided.

To have a clear picture of the total applied forces ranges involved we can say that they range from a maximum of 10^{-6} N for the contact mode to the order of 10^{-12} N for the non-contact mode.

Taking all of this in mind, we can summarize the advantages and disadvantages of each AFM working mode as follows:

2.2.1.1 Advantages

– Contact mode: wide range of samples (softer and harder) need to be analysed; elasticity measurements can be performed; in situ measurements can be performed in a liquid cell or in the electrochemical cell; vertical and horizontal resolutions are very high.
– Non-contact mode: there is no modification or contamination of the sample surface. It is also a mode that minimizes the interaction between the tip and the sample, which can allow atomic resolutions to be obtained under suitable conditions.
– Intermittent mode: very stable measurement; very weak pressure force involved; high resolution; provides the best performance for high-resolution topographic measurement; avoids artificial images that occur in AFM. It also eliminates lateral friction forces. It is especially useful to avoid the problems associated with surface contamination when working in ambient conditions [15]. Usually this contamination, resulting in a layer of water deposited on the sample, masks many of the properties of the surface to be analysed, properties that would be hidden in the non-contact mode. Intermittent mode, passing through this layer of contamination, actually interacts with the surface, without the danger of damage to the surface that can be associated with working in the contact mode.

2.2.1.2 Disadvantages

– Contact mode: the tip is in contact with the surface at all times. This can lead to problems such as easy tip destruction (which increases the cost of the analysis), surface modification or entrainment of particles leading to artefacts. In particular, in the contact mode, the intensity of lateral interactions increases, resulting in lower resolution and damage to the sample. Also, the absorbed water layers generate problems of significant capillary forces along with electrostatic surface charge.
– Non-contact mode: high resolutions require the tip to be very close to the surface, and the scanning has to be very slow in order not to lose "contact" with the surface. This makes it more difficult to obtain a good topographic image, since a longer scanning time could be influenced by external vibrations or other undesired influences. In addition, the oscillation of the tip can be slowed down by the existence of water/contamination layers. The water droplets are confused with the topography of the sample.
– Intermittent mode: cannot work in liquid media; no atomic resolution can be obtained by working in this mode; slower sweeps need much longer times to finish the scan (as mentioned in the previous paragraph, this could result in higher probability of sample contamination or environmental changes, disturbances or vibrations affecting the final results).

For all these reasons, we can consider that the contact mode works poorly with polymeric surfaces, while being the most comfortable method for soft and hard surfaces, where it gives good resolution and few problems [6]. On the other hand, different force gradients (magnetic, electrostatic, etc.) can be measured when working in the non-contact mode, which makes it preferable for complementary information, other than topographic, from the surface.

Finally, for the purposes of this book (membrane pore sizing characterization), the tapping mode has become the most widely used mode in the visualization of membrane surfaces, especially polymeric ones, as it avoids the problems of the contact mode. The problems of lower spatial resolution that go along with the use of intermittent mode are not particularly relevant for the analysis of membrane pores. Recall that only pores larger than 2 nm are considered pores, far enough away from the molecular environment to not require excellent resolution when working with the AFM in tapping mode.

2.2.2 Other information available from AFM

The AFM is much more than a tool for visualizing the topography of a surface, which in itself is really important, given the high resolution, even atomic, that can be achieved. As a direct consequence of the AFM's principle of operation, this very special microscope is able to detect any kind of interaction force between various surfaces or between the tip and the scanned surface.

Thus, there are several AFM techniques based on different types of interactions and forces, such as:

- Magnetic force microscopy (MFM): measures the gradient of magnetic force distribution above the sample surface. It is carried out with the LiftMode to follow the topography of the sample at a fixed distance.
- LiftMode: this is a combined two-step technique. Firstly, it measures separately, using tapping mode, the topography of the sample. But simultaneously, it measures another selected property (magnetic, electrical forces, etc.) using the topographical information to keep the tip above the sample surface at a constant height.
- Electric force microscopy: working in the same mode as that of MFM, it focuses on the electrical force domains in the scanned surface.
- Lateral force microscopy: measures frictional forces between the tip and the sample surface. So, it necessarily works in the contact mode.
- Electrochemical STM and AFM: measures changes in the surface and properties of conductive materials immersed in electrolyte solutions, by establishing gradients or electrical intensity–voltage cycles.
- Force modulation microscopy: working in this mode, the AFM can give information about mechanical properties (mostly elasticity or adhesion) of the sample.

- Scanning polarization force microscopy: working in a non-contact mode, it can provide surface topography, surface potential images and dielectric (frequency-dependent) mapping simultaneously.
- Scanning thermal microscopy: in this case, the tip is sensitive to local temperature in the sample (through a nano-scale thermometer); therefore, this mode allows a mapping of the local temperature and thermal conductivity of the sample surface.
- Force–distance measurements: measure repulsive, attractive and adhesive forces between the tip and the sample during the approach, contact and separation of the two parts.
- Force–volume: performs a series of point force–distance measurements over a given area of the sample. With these individual force curves at each point, it displays images of force variations and sample topography.
- Phase imaging: as mentioned previously, recording the information about the phase of the cantilever oscillation gives better definition of pore borders, which could be very useful for membrane pore sizing.

Many of these novel and very useful ways of working in probe microscopy are based on using tips that are designed or functionalized to be especially sensitive to certain surface properties that we want to highlight.

This same functionalization of tips will also be very useful in studies of membrane interaction with various types of substances. Thus, we can analyse the membrane fouling tendency by using tips suitable for the type of fouling substance expected. This way of operation has been used quite often to study material properties, adhesive forces, long-range interaction forces or biomolecular bonding forces [16].

2.2.3 Probe microscopy suppliers

There are several manufacturers of scanning probe microscopy equipment, including STM and AFM.

Among them, we can mention without claiming to be exhaustive and focusing on those that have been used for the characterization of membranes:
- Park Systems Corp.
- Bruker Corp., formerly Digital Inst. (NanoScope series)
- Nanosurf AG
- Horiba Ltd.
- Angstrom Advanced Inc.
- Chengyu Testing Equipment
- OME (Optical Mechanical Electrical) Technology
- Oxford Inst. (Jupiter)

2.3 Computerized image analysis

Obtaining an accurate and detailed image of the surface of our membrane using a suitable microscopic method is a major step towards the structural characterization of our membrane. But obviously, this is not the end of the work. The next step will consist of using such an image to measure accurately (at least as accurately as possible) the pores, grains, imperfections or other relevant features of the structure shown in the picture.

In fact, as soon as microscopes appeared, the visualization provided by microscopes began to be used to determine the morphological properties (including size) of the objects observed. Obviously, this determination was, at first, based on a purely qualitative description of the observed elements [1]. Thus, Van Leeuwenhoek himself approached this estimation of the size of the observed objects by comparing them with others of known sizes and using the Dutch inch as a unit of reference [17]. It is obvious that any estimation made in these early years had to be necessarily subjective (dependent on the operator himself) and, above all, enormously tedious if, in addition, it was a question of counting and quantifying a large number of objects found in a microscopic image.

However, the breakthrough that allowed the transition from a qualitative estimation to a reliable quantitative determination was the development of different software that, using a computer, makes it possible to quickly, objectively and accurately process images from a wide range of sources, once these images are digital or have been digitized.

Thus, we can define image analysis (IA) software as a programme specifically designed for capturing images, converting them into digital format and carrying out various manipulations on these images, including improvements to the quality and contrast of the image, as well as measurements of the different objects or elements present in the image [1].

Finally, once the microscopic images of our membranes have been acquired, from the point of view of the objective of this book, the next step would be to determine, from these images, the pore size and the corresponding PSD of our sample. This information must be obtained as accurately as possible and, moreover, as objectively as possible. Obviously, the rigorous and objective measurement (without the influence of the experimenter's judgement) of the pores present in a sample should be the ultimate goal of any porometry, but, as we will see in the following paragraphs, the analysis of microscopic images (whether SEM or AFM) is an effort that can easily give rise to a certain margin of arbitrariness that tarnishes the quality of the measurements. In any case, in order to carry out this task with maximum precision, we will have to make use of a suitable computerized IA software.

There are different softwares on the market suitable for this IA purpose, as will be mentioned in the next section, including those yet implemented in many microscopes (especially the AFM ones).

Whether it is commercial software, or software incorporated in the equipment, an adequate IA must be able to measure, as accurately as possible and with the least interference from the user, the size of the pores present in our sample. But to do this, we first have to tell the software what we consider to be a pore and what we do not. Generally, the IA software, once the contrast and definition of the original image have been improved as much as possible, converts the image into shades of grey before calculating the size. So, the software will work with a digitized image in which each pixel will be a grey level depending on the depth of the pixel in the image, the roughness of the photographed surface and the illumination. This digitization will assign a value between 0 and 255 to each pixel, where 0 will be black and 255 will be white. On this digitized image, we must indicate to the software which grey level we consider as a pore and which one not. Basically, pixels will be lighter if they are part of the surface while they will tend to be blacker if they are part of a pore. The software will ask us to assign a colour limit value (a threshold level) below which (darker colours) the software will consider such pixel as being part of a pore.

We can summarize the following steps that an IA software must fulfil to perform such an analysis in an optimal way (obviously in each software the steps will have their own way of being performed):

– Digitization of the image and conversion to grey levels: generally, the original image coming from SEM, STM or AFM equipment will be already a digitized file. The file (especially in the case of AFM) may be digitized in colour. It should be noted, however, that these colours are false colours as none of these techniques are able to determine the real colour of the sample. So, the first step should be to convert the colour image in a black and white one before being treated.

– Improvement of image quality: generally, the software allows us to apply various filters to the image that improve the contrast or definition of the image. This may be necessary when the image has pores with poorly defined edges. In that sense, the use of AFM phase contrast images could be a good alternative. Finally, the low pore border definition is a consequence of convolution of the membrane surface with the size of the tip. Since totally acute tips are not possible to be manufactured, there will always be some interaction of signals coming from adjacent atoms of the tip's sharp end. However, it is not advisable to abuse this type of filters that, really, are distorting the original image so that in the end, the analysed image may be clearly different from the one supplied by the equipment. A common step in this image enhancement is to represent the histogram of grey levels and maximize this histogram, so that spurious peaks are eliminated.

– A convenient step prior to working with our image is to perform the process known as brightfield equalization, whose objective is to eliminate any artefacts in the grey levels due to non-uniform illumination. It should be clear that by illumination we do not refer to light, which plays no role in these microphotographs, but to the signal intensity coming from the sample, which may have slight fluctuations between close points not attributable to true depth changes in the sample.

Since perfect brightfield equalization requires comparison with a white, non-porous reference (which is not feasible in practice), pseudo-brightfield equalization is usually performed by partitioning the original image into a convenient number of rectangles. We can then assign all pixels an intensity such that 95% of the original pixels have a lower intensity. Finally, these rectangles are placed together by a linear interpolation from rectangle to rectangle and subtracted from the original image [18].

– Next, we need to calibrate the software's measurement tool by using the scale included in the original images. SEM images used to include in the resulting image a white bar labelled with the actual size of such bar, while AFM pictures present its label as an x–y coordinate axes. By pointing the cursor on both ends of the given scale and measuring the length in pixels of this scale, we will assign to this length its real value according to the value indicated on the scale. The system is now ready to determine x and y distances on the resulting image.

– The next and possibly the most important step to obtain a reliable result is the definition of a grey threshold value from which we consider a pixel as belonging or not to a pore. In an ideal image of a membrane with well-defined pores and obtained after uniform illumination, the histogram of grey levels should represent a bimodal distribution, with one peak clearly in the light-coloured area (near 255) corresponding to the surface pixels and the other peak centred in the blackest area (grey level near 0), which we would consider to be part of pores. In between the two peaks, there would be a clearly differentiated flat area so that the grey threshold would be easy to choose [19]. In fact, this situation almost never occurs (among other reasons) because the pores in many cases will be interconnected, so that the image obtained will present a gradation of greys from the lightest colours on the surface to the darkest in depth, with a certain continuity. In any case, the software will present us with the resulting pore assignment of our choice, and this allows us to get an idea of the appropriateness of this choice.

– The result is that we have gone from an image with 256 grey levels to a binary image (0 and 1). The resulting binary image is sharpened by removing isolated pixels, and finally, the edges of the pores are smoothed to reduce the influence of a finite pixel size and low definition.

The software will automatically count the number of pixels in each pore and, based on the reassigned scale, determine its size. The result will be a list, usually exportable to Excel, with all the pores found in the image, as well as various parameters related to their measurements. Some are direct, such as the area of a pore (sum of all the 1's in contact with each other, multiplied by the area of an individual pixel) or its perimeter (sum of the length of all the 1's that are surrounded by at least one 0) [20, 21].

Other parameters must be defined appropriately. In particular, the one we will be most interested in is the pore diameter. But since the given pores will generally be irregular, we cannot measure the diameter as if they were perfect circles. Thus, the diameter can be defined in two different ways:

– Equivalent diameter is the diameter of a perfectly circular pore that would have the same area as our real pore. Thus,

$$d_{eq} = 2\sqrt{\frac{A_p}{\pi}}$$

(2.3)

where A_p is the area of the pore in question.

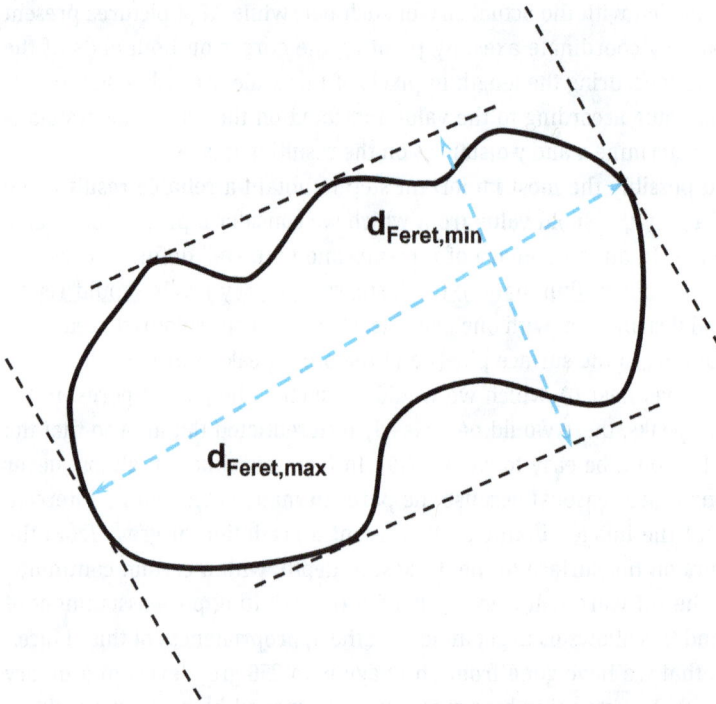

Fig. 2.9: Maximum and minimum Feret's diameters for an irregular shape particle.

– Feret's diameter: it is based on geometrical considerations. It can be defined as the distance between the two parallel planes that constrain the object perpendicular to that direction. Obviously, the Feret diameter will depend on the direction considered (see Fig. 2.9), so it is usually presented as the average of the Feret diameters in various directions. From Cauchy's theorem, it is shown that the average Feret diameter coincides with the diameter of a perfect circle having the same perimeter as the given object:

$$d_{eq} = \frac{P_p}{\pi} \tag{2.4}$$

with P_p being the perimeter of the given pore.

Other parameter that can be useful in the study of an image are:

– The pore shape factor is defined as

$$s_p = 4\pi \frac{A_p}{P_p^2} \tag{2.5}$$

This takes the value of 1 for perfectly circular pores, while any other geometry will give us a value lower than 1 (remember that the circle is the figure with the largest area for a given perimeter).

For each pore present in our microphotography, we can measure these parameters and some others of lesser interest. If the number of pores measured is sufficiently high, we can obtain the histograms corresponding to the size distributions of our membrane: pore area, perimeter, diameter and shape factor. A number of pores of no less than 200 is advisable in order to have a sufficiently continuous graph. This can be achieved with a single or several microphotographs of different areas of the sample taken at the same magnification and reasonably equal illumination. In that sense, SEM images can be easily formed by some hundreds of supposed pores, whereas AFM images can be scanned in a smaller area to get better definitions, which normally include much less pores, requiring some images to achieve a reliable statistical counting and measuring.

This procedure for determining PSDs can also be used for AFM images, if sufficient pores are observed in the scan field for a static determination. AFM images, contrary to SEM images, show a certain colour, but in both cases, this colour is false and is due to the colour palette used to mark the intensity of the electronic signal (SEM) or the strength detected by the tip at each point (AFM).

However, AFM images provide in-depth information that is not possible to obtain from SEM images (whatever their quality or definition). This is because, by its very nature, the AFM picks up the height information in its scan, which allows the pore entrances to be studied in greater detail. In this case, it is possible to use planar projections of the surface at different heights together with the level profiles on several reasonably chosen lines. The combination of planar images with depth profiles is of great interest in identifying the entrance of individual pores. However, the internal diameter of the pore detected in AFM is not reliable because the tip and the pore give rise to a certain convolution that affects the result. Also, these on-line analyses are too slow to allow the construction of a PSD from the analysis of individual pores.

Some attempts to use SEM images to obtain 3D information on pore sizing and structure have been made [22], but they must contend with the technical difficulty of obtaining very thin cut slices of our membrane without distortion of the actual pore structure.

Although SEM images have a certain depth of field that gives an idea of the presence of valleys, peaks and hollows in the scanned surface, it is only by means of AFM that these irregularities can be reasonably estimated. In the AFM topographic image (in any of the conventional working modes), for each point (x, y) of the scanned area, the information collected shows the height of the tip with respect to a baseline or reference, arbitrarily chosen but common to the whole image. This height, Z, allows us to define mean (Z_{avg}), median (Z_{med}) and maximum distances from which to determine the surface roughness of the analysed sample. Thus, we can define several parameters related to roughness that the AFM software (or our own IA) will provide us with:

- Thus, the average roughness is defined as

$$R_a = \frac{1}{m \cdot n} \sum_{j=1}^{m} \sum_{i=1}^{n} |Z_{ij} - Z_{avg}| \tag{2.6}$$

where n *and* m are the total number of points in the x, y directions, respectively, of the image matrix.
- Root mean square roughness, R_{ms}, is defined in a similar manner as

$$R_{ms} = \sqrt{\frac{1}{m \cdot n} \sum_{j=1}^{m} \sum_{i=1}^{n} |Z_{ij} - Z_{avg}|^2} \tag{2.7}$$

which can also be evaluated from the Fourier transform of the surface profile.

Roughness parameters here defined stand for the whole scanned surface. But similarly, we can focus on the roughness along a given line profile on the image (see Fig. 2.10).

In this case, the previous definitions modify to

$$R_a = \frac{1}{n} \sum_{i=1}^{n} |Z_{ij} - Z_{avg}| \tag{2.8}$$

and

$$R_{m,s} = \sqrt{\frac{1}{n} \sum_{i=1}^{n} |Z_{ij} - Z_{avg}|^2} \tag{2.9}$$

Although the roughness thus defined is a parameter commonly included in the characterization studies of membranes and other surfaces by AFM, the first thing to bear in mind is that these parameters are not univocal, that is, they are not scaling invariant. So, if we change the size of the surface scanned with our AFM equipment, the roughness values obtained will also change, in fact they grow as we increase the size of the scanned area [23]. This creates problems when comparing roughness results obtained by different groups with different equipment, even for the same membrane and the same scanned area.

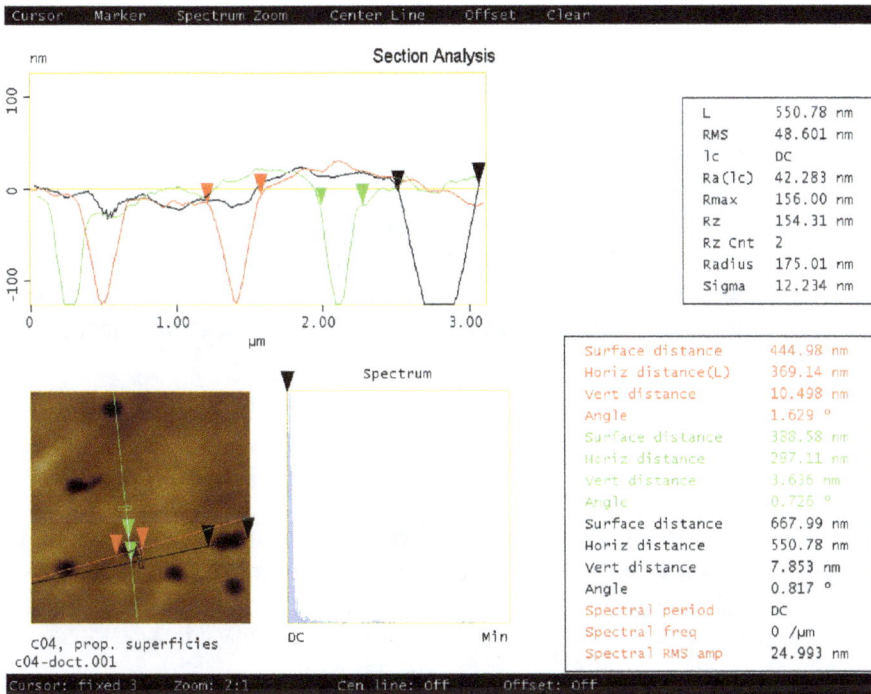

Fig. 2.10: Line profile and roughness calculations for an AFM image of a Cyclopore membrane surface.

Thus, we can say that the roughness of a sample (not only in AFM images as this fact is intrinsic to the definition of roughness itself) has a fractal behaviour with the scanned area.

It can be shown that the relationship between the measured roughness and the working scale is given by the following expression:

$$R_{ms} = a \cdot L_0^H \tag{2.10}$$

where L_0 is the scale length, a is the constant to be determined and H is the so-called roughness exponent (also called the Hurst exponent).

Taking into account the potential form of the above equation, the exponent H can be obtained by a double logarithmic representation of the various roughness values obtained for the same surface but using various scanned areas (i.e. various scale lengths) [18]. Thus [23]:

$$H = \frac{d \log R_{ms}}{d \log L_0} \tag{2.11}$$

but as mentioned above, this exponent is related to the fractality of the surface through the fractal dimension D_{fr} [24]:

$$H = n - D_{\text{fr}} \qquad\qquad (2.12)$$

where $n = 3$ in the case of a surface. Finally, D_{fr} can be obtained by fitting the frequency spectrum for a given AFM image [23], to obtain the roughness exponent from the fractal dimension.

Surfaces have values of the fractal dimension between 2 and 3, where a value close to 2 indicates a very flat surface, while a surface with high roughness would have a dimension close to 3, more typical of a volume than a real surface [18].

In any case, proper determination of the fractal dimension of the membrane surface could be useful to model its roughness [25] or to analyse the changes induced on membrane roughness due to fouling [26].

2.3.1 Image analysis software

Currently, various IA computer programmes can be found on the market or free of charge (open-source code), with more or less advanced characteristics, including, in the most developed cases, deep learning based on digital neural networks. In this way, increasingly sophisticated systems that make use of artificial intelligence to improve their performance are continuously emerging [1].

Among the best-known IA software are: ScanPro by Jandel, ImageJ (probably the most popular due to its simplicity and open-source nature), Ilastik, Gwyddion (developed specifically for SPM but can also be used for SEM analysis), INCA (by Oxford Instruments), MIPAR (recent software but very easy to use and very intuitive, while being also very powerful as includes AI features) and Image ProPlus. Most of those software are oriented towards the visualization and analysis of biomedical images and their use in pathological research or in the study of viruses or cells. In addition, many electron microscopes or AFMs include, as an accessory (sometimes optional), their own IA software, specifically designed to optimize the possibilities of the equipment in which they are installed.

2.3.2 Image analysis reliability considerations

It has already been mentioned at some point in this section that microscopic images of a membrane, although they are very visual and convenient to get an idea of the type of membrane we have, their analysis may be subject to a certain degree of subjectivity that must be taken into account when properly considering the results obtained.

Obviously, we must start with images that are as defined and precise as possible; otherwise, the quality of our analysis will leave much to be desired.

Therefore, we can list some points that can clarify this subjectivity and can guide us in obtaining reliable porosimetric data that can be compared with other more indirect techniques.

– First of all, we should remember that microscopic images only give us information on the surface of the membrane. Thus, to know the interior of the pores, we should be guided by images obtained from cross sections of the filters (cuts made with great care, generally using the cryogenic fracture technique). Another possibility is to take images of the membrane from both sides so that we are supposed to be able to analyse the behaviour of the active layer and the support separately. Readers should be aware that this approach would not be very useful in the case of most polymeric asymmetric membranes (e.g., those obtained from phase inversion), where the transition between the active layer and the support is continuous and not discrete. Thus, we can summarize that the IA will only be truly representative of the actual membrane structure and pore size in the case of membranes whose pores are reasonably uniform across the filter (which is only fully the case for track-etched membranes and some other membranes obtained by anodic deposition).

– Secondly, as previously discussed, the number of pores present in the image must be sufficiently representative of the actual behaviour of the existing pore distribution. This means that by using images containing a minimum amount, we can reasonably set at about 200 pores. It is obvious that with fewer pores analysed, a PSD can also be obtained but it will be less accurate and more subject to error. This means that we have two possibilities: either lower the resolution of our image so that more than a hundred possible pores appear in the analysed field or make several images of different areas, with a higher degree of resolution but with fewer pores in each image. Each option has its drawbacks. Thus, a large field can contain many pores in a single photo, but the low-resolution image will show problems in choosing the threshold that defines the pore entrance and, consequently, the pores obtained will be less defined and possibly more irregular. While decreasing the scanning area by increasing the resolution of the image will force us to take several images in similar conditions so that we can ensure comparable measurements and accumulate them in a single final PSD.

– Related to the previous point is the subjectivity inherent in the choice of the grey-level threshold necessary to consider a pixel as part of a pore or not. This subjectivity can be somewhat reduced by pre-treatment of the image (equalization, smoothing, contrast and brightness enhancement) but at the cost of slightly changing the final dimensions of the pores obtained. Although this is not serious in very close images (with few visible but well-defined pores), it can be a problem in more distant images. In any case, it is advisable to test several thresholds and visually check that we do not leave important parts of the pores unanalysed or that we do not include as pores simple darkening of the surface due to poor illumination.

This description will be clearer if we look at Fig. 2.11. Figure 2.11 shows an SEM image of an anodic alumina membrane (Anopore-02). These types of membranes present pores that are quite regular, which cross with almost no tortuosity through all the membrane thickness. In that sense, they can be used as models for membrane characterization (similarly to the track-etched ones, which also have the advantage of an almost circular pore entrance). The original image (Fig. 2.11a) has been treated with ImageJ software, and different thresholds have been applied (Fig. 2.11b–d). It seems complicated to decide which one is the most precise or better one, which leads to a better estimation of the actual pore sizes. Even in so regular and sharp images, a subjective selection of the colour threshold could give differences in PSD by as much as 50%.

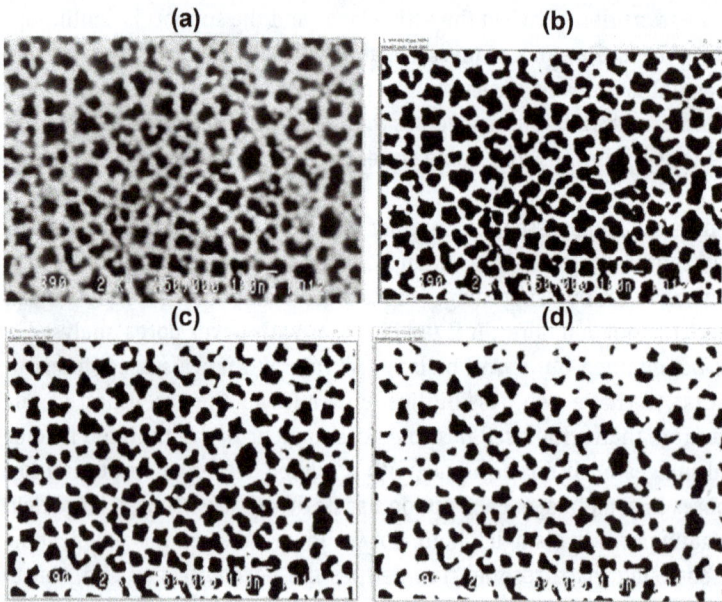

Fig. 2.11: SEM image of an Anopore (0.2 μm) membrane (a), along with different ImageJ® resulting images (b—d) after applying several grey-level threshold values.

–   On the other hand, only microscopic IA can give us information since it is impossible to obtain it from other types of porosimetry. Thus, the shape distribution of the pores (closely related to their manufacturing method) or the roughness of the membrane can only be determined from suitable microscopic images [27].
–   Generally, to minimize such operational errors associated with IA, the result of this analysis (especially when applied to the determination of pore sizes in membranes) is usually compared with other porosimetric techniques in order to ensure its reliability [27—31].

- Finally, we can conclude about the necessity of having a common and universally accepted standard about the use of IA for membrane characterization, not available actually [27].

References

[1] Sada O. Morphological and chemical characterization of nanoparticulate matter by SEM/EDX analytical techniques. PhD Thesis, Universitá di Genova, (2023), Genoa (Italy)

[2] Peinador R.I., Calvo J.I., ToVinh K., Thom V., Prádanos P., Hernández A. Liquid–liquid displacement porosimetry for the characterization of virus retentive membranes. J Membr Sci, 372 (2011) 366–372. https://doi.org/10.1016/j.memsci.2011.02.022.

[3] Williams D.B., Carter C.B. Transmission Electron Microscopy: A Textbook for Materials Science, 2nd. ed, Springer Science+Business Media, LLC, New York, USA, (2009).

[4] Palacio L. Caracterización estructural y superficial de membranas microporosas. PhD Thesis, Universidad de Valladolid, (1999), Valladolid (Spain). ISBN: 84-7762-944-7

[5] Girão A.V., Caputo G., Ferro M.C. Application of Scanning Electron Microscopy–Energy Dispersive X-ray Spectroscopy (SEM–EDS). In: T. Rocha-Santos, A. Duarte (Eds.), Characterization and Analysis of Microplastics. Elsevier, (2017), https://doi.org/10.1016/0376-7388(92)80207-Z10.1016/bs.coac.2016. 10.002.

[6] Mutalib M.A., Rahman M.A., Othman M.H.D., Ismail A.F., Jaafar J. Scanning Electron Microscopy (SEM) and Energy-Dispersive X-Ray (EDX) Spectroscopy. In: N. Hilal, A.F. Ismail, T. Matsuura, D. Oatley-Radcliffe (Eds.), Membrane Characterization. Elsevier, The Netherlands, (2017). 978-0-444-63776-5.

[7] Goldstein J.I., Newbury D.E., Michael J.R., Ritchie N.W.M., Scott J.H.J., Joy D.C. Scanning Electron Microscopy and X-Ray Microanalysis, 4th. ed., Springer Science+Business Media, LLC, New York, USA, (2018).

[8] Binning G., Quate C.F., Gerber C. Atomic force microscope. Phys Rev Lett, 56 (1986) 930–933. https://link.aps.org/doi/10.1103/PhysRevLett.56.93.

[9] Binning G., Rohrer H., Gerber Ch, Weibel E. Surface studies by scanning tunneling microscopy. Phys Rev Lett, 49 (1982) 57–61. https://link.aps.org/doi/10.1103/PhysRevLett.49.57.

[10] Chahboun A., Coratger R., Ajustron F., Beauvillain J., Aimar P., Sanchez V. First investigations on the use of scanning tunnelling microscopy (STM) for the characterisation of porous membranes. J Membrane Sci, 67 (1992) 295–300. https://doi.org/10.1016/0376-7388(92)80033-G.

[11] Booth T.J., Blake P., Nair R.R., Jiang D., Hill E.W., Bangert U., Bleloch A., Gass M., Novoselov K.S., Katsnelson M.I., Geim A.K. Macroscopic graphene membranes and their extraordinary stiffness. Nano Letters, 8(8) (2008) 2442–2446. https://doi.org/10.1021/nl801412y.

[12] Li X., Tao Y., Li F., Huang M. Efficient preparation and characterization of functional graphene with versatile applicability. J Harbin Inst Technol, 23(3) (2016) 1–29. https://doi.org/10.11916/j.issn.1005-9113.2016.03.001.

[13] Guo W., Mahurin S.M., Wang S., Meyer H.M., Luo H., Hu X., Jiang D., Dai S. Ion-gated carbon molecular sieve gas separation membranes. J Membrane Sci, 604 (2020) 118013. https://doi.org/10.1016/j.memsci.2020.118013.

[14] Huang S., Li S., Hus K.-j., Villalobos L.F., Agrawal K.V. Systematic design of millisecond gasification reactor for the incorporation of gas-sieving nanopores in single-layer graphene. J Membrane Sci, 637(2021) 119628. https://doi.org/10.1016/j.memsci.2021.119628.

[15] Voigtländer B. Atomic Force Microscopy. NanoScience and Technology. Springer, Berlin, Germany, (2019), http://dx.doi.org/10.1007/978-3-030-13654-3. ISBN 978-3-030-13653-6. S2CID 199490753.

[16]   Johnson D., Oatley-Radcliffe D.L., Hilal N. Atomic Force Microscopy (AFM). In: A.F. Ismail,
       T. Matsuura, D. Oatley-Radcliffe (Eds.), Membrane Characterization. Nidal Hilal. Elsevier, The
       Netherlands, (2017). 978-0-444-63776-5.

[17]   Brieven A.D. The Collected Letters. Volume 3, Brief No. 46. 15 Mei 1679., Alle de brieven. Deel 3:
       1679–1683, Anthoni van Leeuwenhoek. (n.d.). Retrieved November 27, 2022, from
       https://lensonleeuwenhoek.net/content/alle-de-brieven-collected-letters-volume-3

[18]   Calvo J.I., Bottino A., Prádanos P., Palacio L., Hernández A. Porosity. In: E.M.V. Hoek, V.V. Tarabara
       (Eds.), Encyclopedia of Membrane Science and Technology. Wiley Intersci. Pub, New York, USA,
       (2013), 1062–1086. ISBN: 978-0-470-90687-3.

[19]   Zeman L., Denault L. Characterization of microfiltration membranes by image analysis of electron
       micrographs. Part. I. Method development. J Membr Sci, 71 (1992) 221–231. https://doi.org/10.1016/
       0376-7388(92)80207-Z.

[20]   Swenson R.A., Attle J.R. Counting, measuring, and classifying with image analysis. Am Lab, 11 (1979)
       50–58.

[21]   Calvo J.I., Hernández A., Caruana G., Martínez L. Pore size distributions in microporous membranes.
       I. Surface study of track-etched filters by image analysis. J Colloid Interface Sci, 175 (1995) 138–150.
       https://doi.org/10.1006/jcis.1995.1439.

[22]   Chwojnowski A., Przytulska M., Wierzbicka D., Kulikowski J., Wojciechowski C. Membranes' porosity
       evaluation by computer-aided Analysis of SEM images – a preliminary study. Biocybern Biomed Eng,
       32 (2012) 65–75. https://doi.org/10.1016/S0208-5216(12)70050-5.

[23]   Johnson D., Hilal N. Polymer membranes – fractal characteristics and determination of roughness
       scaling exponents. J Membrane Sci, 570–571 (2019) 9–22. https://doi.org/10.1016/j.memsci.2018.10.024.

[24]   Mandelbrot B. The Fractal Geometry of Nature. W.H. Freeman and co, San Francisco, USA, (1982).

[25]   Zhang M., Chen J., Ma Y., Shen L., He Y., Lin H. Fractal reconstruction of rough membrane surface
       related with membrane fouling in a membrane bioreactor. Bioresour Technol, 216 (2016) 817–823.
       https://doi.org/10.1016/j.biortech.2016.06.034.

[26]   Feng S., Yu G., Cai X., Eulade M., Lin H., Chen J., Liu Y., Liao B.-.Q. Effects of fractal roughness of
       membrane surfaces on interfacial interactions associated with membrane fouling in a membrane
       bioreactor. Bioresour Technol, 244 (2017) 560–568. https://doi.org/10.1016/j.biortech.2017.07.160.

[27]   AlMarzooqi F.A., Bilad M.R., Mansoor B., Arafat H.A. A comparative study of image analysis and
       porometry techniques for characterization of porous membranes. J Mater Sci, 51 (2016) 2017–2032.
       https://doi.org/10.1007/s10853-015-9512-0.

[28]   Calvo J.I., Bottino A., Capannelli G., Hernandez A. Comparison of liquid–liquid displacement
       porosimetry and scanning electron microscopy image analysis to characterise ultrafiltration track-
       etched membranes. J Membr Sci, 239 (2004) 189–197. https://doi.org/10.1016/j.memsci.2004.02.038.

[29]   Calvo J.I., Peinador R.I., Prádanos P., Bottino A., Comite A., Firpo R., Hernández A. Porosimetric
       characterization of polysulfone ultrafiltration membranes by image analysis and liquid–liquid
       displacement technique. Desalination, 357 (2015) 84–92. http://dx.doi.org/10.1016/j.desal.2014.11.012.

[30]   Agarwal C., Pandey A.K., Das S., Sharma M.K., Pattyn D., Ares P., Goswami A. Neck–size distributions
       of through-pores in polymer membranes. J Membrane Sci, 415–416 (2012) 608–615. http://dx.doi.
       org/10.1016/j.memsci.2012.05.055.

[31]   Ley A., Altschuh P., Thom V., Selzer M., Nestler B., Vana P. Characterization of a macro porous
       polymer membrane at micron-scale by confocal-laser-scanning microscopy and 3D image analysis.
       J Membrane Sci, 564 (2018) 543–551. https://doi.org/10.1016/j.memsci.2018.07.062.

Chapter 3
Spectroscopic methods

3.1 Positron annihilation spectroscopy

3.1.1 Introduction

The positron is the antiparticle of the electron. It has the same rest mass as the electron but opposite charge. In condensed matter, each positron annihilates with an electron yielding γ-rays in a very short time. Information about electron momentum distribution or electron density in materials can be deduced by measuring the positron annihilation lifetime. Applications of positron annihilation to the study of crystal lattice defects started around the 1970s, when it was realized that positron annihilation is particularly sensitive to vacancy-type defects and that annihilation properties manifest the nature of each specific type of defect.

In fact, when some positrons enter a given material, they experience a process of thermalization during which a high-energetic positron reaches the thermal equilibrium state with the surrounding molecules in the material. Positrons emanate from the radioactive source with a continuous energy spectrum increasing up to 0.54 MeV. After the introduction of the positron into the sample, positrons lose energy by collisions with molecules and the subsequent dropping of energy by ionization, excitation, phonon scattering and so on. Eventually, molecule vibration and phonon scattering consume the energy of the inelastic collisions of positrons and allow them to come into thermal equilibrium with the material. After thermalization, the positrons with environmental thermal energy randomly travel through the sample till they annihilate or meet a hole and fall into a localized state by the diffusion process.

When positrons are finally trapped in holes where there are no or few electrons, they form a bound state linking a positron with a relatively free electron at the interface forming a system called positronium (Ps). Para-positronium (p-Ps) is formed if the spin states of the electron and positron are parallel, while ortho-positronium (o-Ps) is formed when the spin states of the electron and positron are antiparallel. Ps annihilates in characteristic lifetimes that can be measured, allowing the evaluation of the size of the trapping holes or voids at a molecular scale.

The determination of lifetimes and intensities of positron species in molecular media [1–3] is the basis of positron annihilation lifetime spectroscopy (PALS). The technique allows the determination of free or void volume corresponding to the very small holes, or pores.

Due to the small size of the Ps probe (1.59 Å), PALS is very sensitive to tiny voids, on the order of angstroms in radius, constituting what is called free volume. This can be accomplished without noticeable interferences from the bulk.

https://doi.org/10.1515/9783110792195-003

3.1.2 Positron source

Positrons can be produced in many ways [4], including radioactive sources and also from either linear accelerators or nuclear reactors. Highly energetic positron beams are moderated through the use of semi-crystalline metal foils or cryogenic noble gases [5]. The low-energy positron beam can then be transported by the convenient combination of electrostatic and/or magnetic fields. Beams based on a radioactive source are compact lab-scale devices that can deliver $<10^7$ positrons per second. Facility-scale beams using pair production from linear accelerators or nuclear reactors have the potential for higher positron rates approaching 10^{10} positrons per second [5].

The corresponding radioactive generation of positrons from $_{11}^{22}$Na (usually as $_{11}^{22}$NaCl) leading to the stable $_{10}^{22}$Ne is schematized in Fig. 3.1. Note that 90.4% of disappearing protons generate positrons, e^+, through a β^+ decay, leading to an excited $_{10}^{22}$Ne* that emits a 1,274 keV hard gamma photon after only 3.7 ps, in accordance with

$$_{10}^{22}\text{Ne}^* \rightarrow {}_{10}^{22}\text{Ne} + \gamma \tag{3.1}$$

It is worth mentioning that 9.5% of the number of protons in $_{11}^{22}$Na decreases by the capture of an electron from electronic shells of the radioactive atom. Electrons are usually captured from the inner K layer, leaving "holes" behind them. The resulting atom with a gap in its electronic structure rearranges itself immediately, emitting X-rays in the process or Auger electrons. When the hole is filled with an outer electron, an X-ray of an energy equal to the difference between the two electron shells is emitted. In the Auger effect, the energy absorbed when the outer electron replaces the inner electron is transferred to an outer electron. The outer electron, Auger electron, is ejected from the atom leaving a positive ion.

Both positron emission and electron capture result in reduction in the number of protons from Z to $Z-1$ and the production of a neutrino, according to the reactions:

$$\left.\begin{array}{l} _{11}^{22}\text{Na} \rightarrow {}_{10}^{22}\text{Ne}^* + {}_1^0 e^+ + {}_0^0 \nu_e \\[4pt] _{11}^{22}\text{Na} + {}_1^0 e^- \rightarrow {}_{10}^{22}\text{Ne}^* + {}_0^0 \nu_e \end{array}\right\} \tag{3.2}$$

$_{10}^{22}$Ne* corresponds to an excited state of $_{10}^{22}$Ne. The positron observed in the final stage of the beta decay is a new particle requiring an energy of 511 keV, corresponding to its rest mass energy, to be created. In the process, a proton is transformed into a neutron:

$$p^+ \rightarrow n + e^+ + \nu_e \tag{3.3}$$

In the case of electron capture, 511 keV is the energy corresponding to the disappearance of an electron:

$$p^+ + e^- \rightarrow n + \nu_e \tag{3.4}$$

In both cases, practically all the energy released is carried by the light particles.

$$^{22}_{11}Na$$

2.602 y

| Electron Capture 9.5% | β^+ $E_{\beta^+} = 545\ keV$ 90.4% | β^+ $E_{\beta^+} = 1830\ keV$ 0.1% |

$$^{22}_{10}Ne\ ^*$$

3.7 ps

$E_\gamma = 1274\ keV$

Stable

$$^{22}_{10}Ne$$

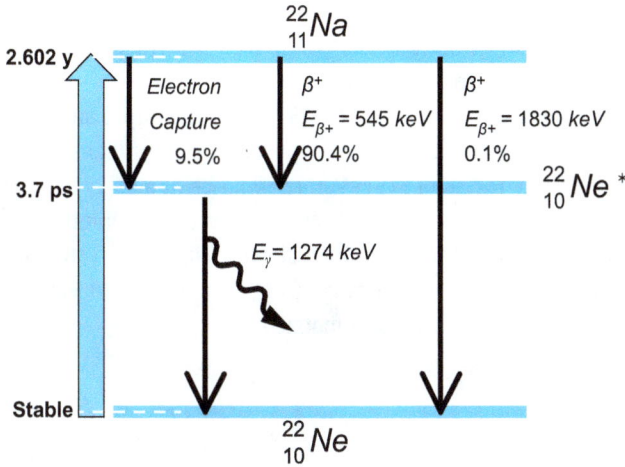

Fig. 3.1: Decay of $_{11}{}^{22}$Na to $_{10}{}^{22}$Ne and positron production.

The insertion of positrons from radioactive sources directly in contact with the sample is a direct technique. This line of attack is nevertheless problematic for thin films, given the high energies and broad energy range of the emitted positrons. Although they are more modular and autonomous, for example, $_{11}{}^{22}$Na has a half-life of 2.602 years.

A unique strength associated with controlled positron beams is the possibility to the depth profile by controlling the positron energy and the chance of resolving laterally by focusing the beam on a small spot [6]. In effect, positron beams [4, 5] can deliver focused, monoenergetic positrons onto the sample. Broadly speaking, a positron beam takes the high-energy positrons from β+ decay or pair production and, via moderation, cools them to electron volt energies, accelerates and transports them to the target, and focuses them onto the sample at a defined energy. Besides enabling the study of thin films, positron beams also offer the important capability of depth profiling the sample by varying the implantation energies of the incident positrons. One not minor disadvantage of using a beam is that the sample needs to be in a vacuum.

3.1.3 Positronium

For insulators, positrons need to find a loosely bound valence electron in order to annihilate. This favours the formation of Ps, which is an atom like hydrogen; thus, when positrons enter a polymeric porous material, Ps is eventually generated. Positrons collide with atoms and electrons in the solid and will slow down to some electron volt within picoseconds, leading to some 10–50% of these electrons capturing the bound molecular electrons or being paired with electrons already freed by prior ionizing collisions with positrons [7, 8]. A positron–electron pair is called Ps. p-Ps is formed if the spin states of the electron and positron are parallel, while o-Ps is formed when the spin states of the electron and positron are antiparallel.

The self-annihilation lifetime of *p-Ps* is very short, 125 ps, and this rapid annihilation occurs with the emission of two equal photons in opposite directions (conserving momentum and energy) in the γ-ray range, of 511 keV, that could be named 2γ annihilation. However, *o-Ps*, in vacuum, annihilates into three photons, 3γ annihilation, to preserve the angular momentum. This process is much slower with a characteristic lifetime of 142 ns, and its energies are within the range from 0 to 511 keV. In Fig. 3.2, the corresponding Ps annihilation processes are schematized.

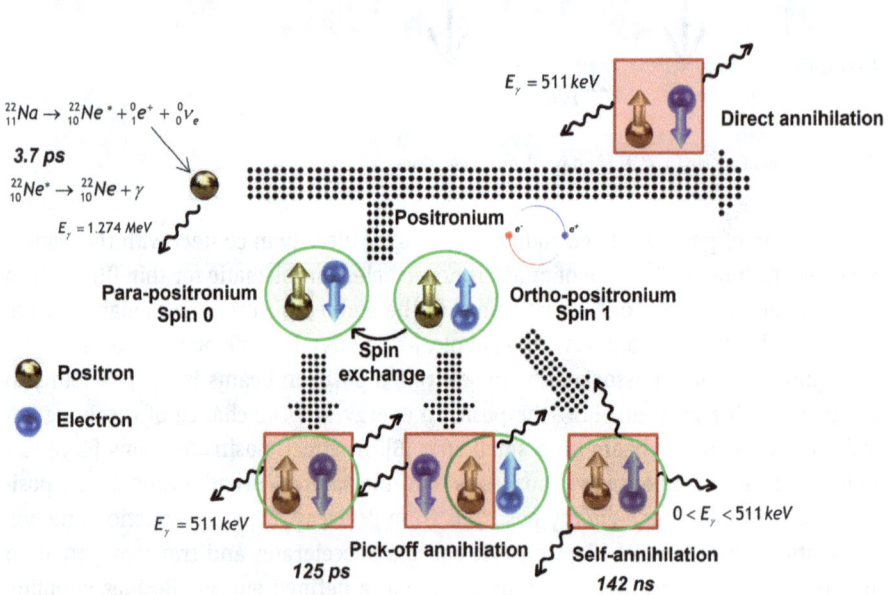

Fig. 3.2: Formation and decay of positronium. The source is supposed to be $_{11}^{22}$Na.

The 2γ and 3γ reactions are described as follows:

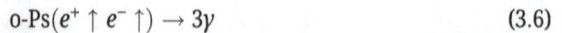

$$\text{p-Ps}(e^+ \uparrow\, e^- \downarrow) \rightarrow 2\gamma \tag{3.5}$$

$$\text{o-Ps}(e^+ \uparrow\, e^- \uparrow) \rightarrow 3\gamma \tag{3.6}$$

Actually, *p-Ps* can also decay by the emission of four γ-rays, but a branching ratio of 1.4×10^{-6} makes that 4γ decay insignificant. Also, *o-Ps* can decay by five γ-rays, but only in approximately 1.0×10^{-6} occasions [9].

o-Ps can experience a pick-off annihilation when the positron of Ps annihilates with a foreign electron in a molecule (water, for example):

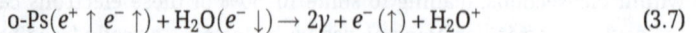

$$\text{o-Ps}(e^+ \uparrow\, e^- \uparrow) + \text{H}_2\text{O}(e^- \downarrow) \rightarrow 2\gamma + e^-\,(\uparrow) + \text{H}_2\text{O}^+ \tag{3.7}$$

In certain cases, when the surrounding molecules possess some unpaired electrons (paramagnetic materials), a spin conversion, through spin exchange, from o-Ps to p-Ps could happen as

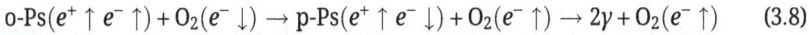

$$o\text{-}Ps(e^+ \uparrow e^- \uparrow) + O_2(e^- \downarrow) \rightarrow p\text{-}Ps(e^+ \uparrow e^- \downarrow) + O_2(e^- \uparrow) \rightarrow 2\gamma + O_2(e^- \uparrow) \quad (3.8)$$

The lifetime of *o-Ps* lifetime is reduced by both these reactions [10]. It is, nevertheless, clear that a significant fraction of Ps would have longer lifetimes anyway. Actually, the intrinsic *o-Ps* lifetime (142 ns) is typically shortened only in a few nanoseconds (110 ns) via 2γ annihilation processes in typical molecular substrates.

3.1.4 Positronium annihilation spectroscopy

Lifetime spectroscopy can easily distinguish the long-lived *o-Ps* state of Ps. Moreover, unlike the 2γ decays that are clustered in a narrow peak at 511 keV, the distribution in γ-ray energy of photons from 3γ decay of *o-Ps* is reasonably approximated by a linear increase from zero energy to a maximum cut-off at 511 keV. Hence, γ-ray energy spectroscopy can also easily distinguish these different decay modes. o-Ps plays a key role in probing porous materials, although its annihilation can be strongly perturbed by its interactions with surrounding electrons that measurably reduce the *o-Ps* lifetime and the 3γ/2γratio while it is within the pores [5].

Typically, PALS is performed by surrounding a positron source (e.g., ^{22}NaCl) with the sample of interest and placing it between two detectors. The detectors identify the photons produced when the positron forms, and the photons that are emitted when the Ps annihilates. The signals from these detectors can be processed and sorted to determine the time elapsed between the formation and annihilation of each positron generated. This information can be further processed and analysed to determine the average lifetime of each positron species that forms and annihilates in the sample.

For measuring *o-Ps* lifetime, two signals per annihilation event are required, that is, start and stop signals. The start signal informs on the moment when a positron is emitted, and the stop signal informs on the positron/Ps annihilation. The latter is determined by monitoring the annihilation photons. The former is defined by monitoring the prompt γ-ray emitted immediately after the positron production from the isotope used.

The sample and beta source are surrounded by two plastic scintillators. The scintillators connect directly to a photomultiplier tube (PMT) and absorbed. The photomultiplier records a given signal into a flashing light when the scintillator detects any form of γ-rays. This makes it possible for the following electronics to identify and record the individual γ-rays produced by the beta source and the Ps annihilation. The signals from both the start (from the source) and stop (from the sample) PMT are then sent to a constant fraction discriminator (CFD) with a significant delay. The discriminators are there to differentiate each signal, and the delays ensure that correct positrons are related to each other. The discriminators also clear the irrelevant information when needed. For instance, if a positron's lifetime does not reach within the reasonable range, the readings are not recorded. From here, the discriminators lead to an adequate delay box to slow down the recording to fit the time-to-amplitude converter (TAC). TAC changes the time readings into a point on a time-by-count graph. The fast coincidence device is also from CFD. Both discriminators lead to a fast coincidence, which checks whether each of the readings from both the stop and start discriminators are correlated. This is also considered by the TAC, and finally reported as a histogram of all readings [11]. A scheme is shown in Fig. 3.3.

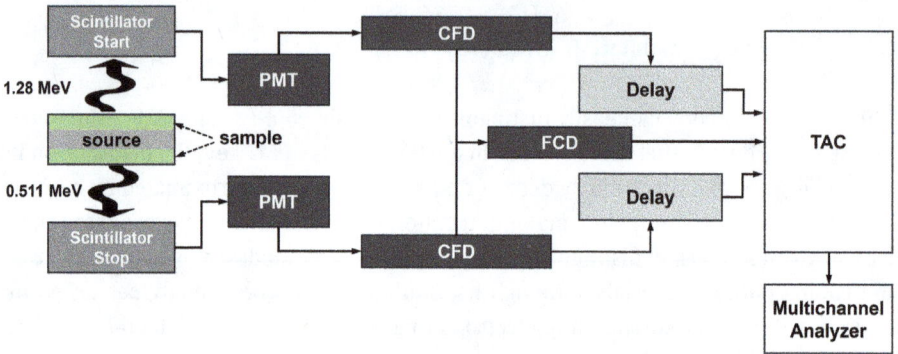

Fig. 3.3: Scheme of the typical experimental device to perform PALS.

The information conveyed would refer to the number of annihilations as a function of time $I_{Ps}(t)$. Actually, if there were only a characteristic lifetime we obtain

$$y(t) = R(t) \otimes I_{Ps}(t) = R(t) \otimes \left(\frac{A}{\tau_{Ps}} e^{-t/\tau_{Ps}} + B \right) \tag{3.9}$$

Here $\otimes$ stands for the convolution of a certain response function of the instrument and B corresponds to the baseline. But usually, we get a superposition of several exponentially decaying time dependences:

$$I_{Ps}(t) = \sum_{i=1}^{n} \frac{A_i}{\tau_{iPs}} e^{-t/\tau_{iPs}} + B \tag{3.10}$$

Usually, $\tau_1 = \tau_b$ (we drop, here on, the Ps subscript) corresponding to the bulk material without voids, while positron trapping, in open or free volumes, leads to longer lifetimes. A trapping rate can be defined as [9]

$$K_i = \frac{A_i}{A_1} \left(\frac{1}{\tau_1} - \frac{1}{\tau_i} \right) \tag{3.11}$$

Generally,

$$K_i = \mu_i C_i \tag{3.12}$$

μ_i is the trapping coefficient and C_i is the void concentration. Note that $\tau_1 < \tau_i$ and $A_1 > A_i$ for all i.

For a zero lifetime, the response would be approximately Gaussian [9]:

$$G(t) = \frac{1}{\sigma\sqrt{2\pi}} e^{\left[-\frac{(t-t_0)^2}{2\sigma^2}\right]} \tag{3.13}$$

A typical width at half maximum height of $2\sigma\sqrt{2\ln 2}$ usually ranges from 180 to 190 ps. Assuming that $R(t) = G(t)$, eq. (3.9) would be

$$y(t) = G(t) \otimes I(t) = \int_{-\infty}^{\infty} G(t-t')\, I(t')\, dt' \tag{3.14}$$

The response function distorts the beginning of the spectrum, $t = 0$, but has little effect at longer times. If the time resolution were infinitesimal, the spectrum should rise sharply from zero to the maximum, then followed by one or several exponential decays.

Exponentials would appear as straight lines in a plot with a logarithmic ordinate axis for $I(t)$ versus linear abscissas for time. Assuming a single lifetime, τ_i, and once the baseline is discounted, eq. (3.10), for long enough times, leads to

$$\ln I(t) = \ln\left(\frac{A_1}{\tau_i}\right) - \left(\frac{t}{\tau_i}\right) \tag{3.15}$$

This identifies the slope as negative, and being $-1/\tau_1$ allows an easy identification of τ_1. Concerning the logarithm of the response Gaussian, eq. (3.13), it decreases very fast with time, specifically linearly with $-(t-t_0)^2/\sigma^2$ [9]. In Figure 3.4 a typical PALS spectrum, with three characteristic lifetimes, is shown.

In general, it should be assumed that there is dispersion of hole sizes in the sample material, especially for polymers. In this case, eq. (10), with $B = 0$, should be substituted by [12]

$$I_{Ps}(t) = \sum_{i=1}^{n} A_i \left[\int_0^{\infty} a_i(\lambda)\lambda e^{-\lambda t} d\lambda\right] \tag{3.16}$$

Here we introduced a continuous spectrum of annihilation rates $\lambda = 1/\tau$ with a probability density function $a_i(\lambda)$. In this case, there are several approaches to obtain information on $a_i(\lambda)$ such as a numerical Laplace inversion [13] and a maximum entropy method [14, 15]. Nevertheless, usually it can be accepted that the probability distribution would be lognormal, which can be eventually fitted, term by term or altogether with

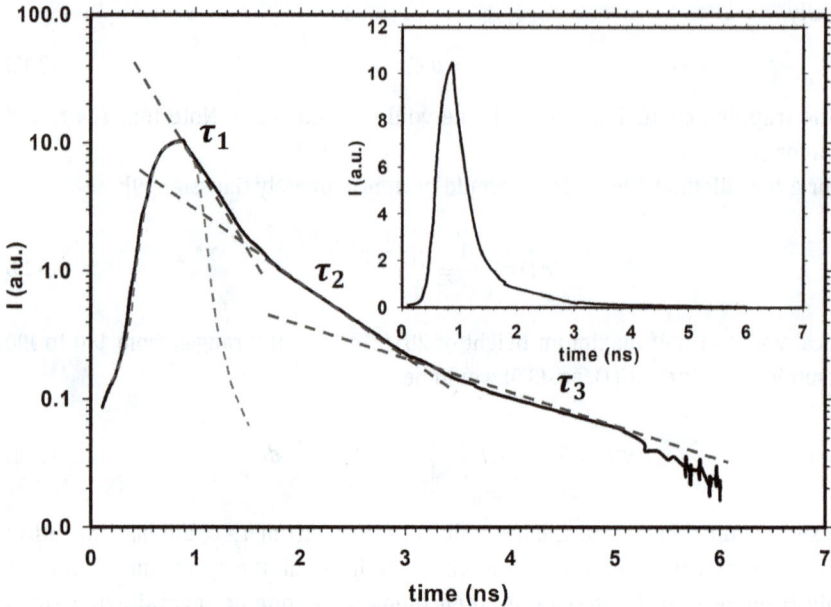

Fig. 3.4: A typical PALS spectrum showing three characteristic lifetimes. In the inset, the double linear corresponding plot is shown.

$$\alpha_i(\lambda)\lambda d\lambda = \frac{1}{\bar{\sigma}_i\sqrt{2\pi}} e^{-\frac{\left(\ln\frac{\lambda}{\lambda_{max}}\right)^2}{2\bar{\sigma}_i^2}} d\lambda \tag{3.17}$$

$$\left.\begin{array}{l} \sum_{i=1}^{n} A_i \\ \int_0^{\infty} \alpha_i(\lambda)d\lambda = 1 \end{array}\right\} \tag{3.18}$$

Here λ_{max} is the most probable value and $\bar{\sigma}_i$ is the standard deviation.

3.1.5 Pore radii

According to the Ps-free volume theory, the lifetime of o-Ps is determined by the reciprocal of the integral between the positron and the electron densities in the free volumes of molecular systems. Therefore, the o-Ps lifetime is expected to correlate directly with the dimensions where Ps is localized. A large hole, which contains a low mean electron density, results in a long o-Ps lifetime. A simple quantum mechanical model, where o-Ps resides in a spherical infinite potential barrier of radius R_0 with a homogeneous electron layer R in the region between the hole radius r_p and R_0 ($R_0 = r_p + R$) was proposed to correlate r_p with the annihilation time for o – Ps, τ_{o-Ps} [2, 12, 16–18].

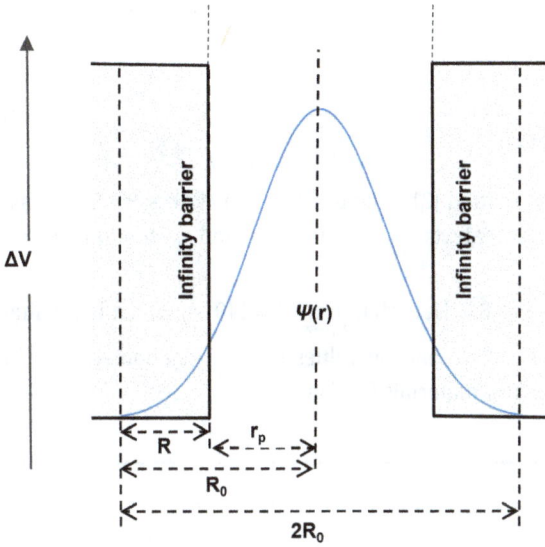

Fig. 3.5: Spherical barrier inside a void volume showing the electron-rich shell on the spherical walls.

In effect, using quantum mechanics for a particle inside a hard spherical box, we obtain the wave function for the ground state:

$$\psi(r) = \sqrt{\frac{1}{2\pi(r_p + R)}}\, \sin\left[\frac{1}{r}\left(\frac{\pi r}{r_p + R}\right)\right] \tag{3.19}$$

The probability $P(R)$ of finding the Ps inside the electron layer (from r_p to $r_p + R$), see Fig. 3.5, is

$$P(r) = 1 - \frac{\int_0^{r_p} |\psi|^2 r^2 dr}{\int_0^{r_p + R} |\psi|^2 r^2 dr} \tag{3.20}$$

The average annihilation rate (inverse of time life) of both the possible spin configurations in vacuum is [12]

$$\frac{1}{\tau_{ps}^v} = \frac{\left[\left(\frac{1}{\tau_{p-ps}^v}\right) + 3\left(\frac{1}{\tau_{o-ps}^v}\right)\right]}{4} \tag{3.21}$$

where it has been assumed that the formation ratio of p–Ps/o–Ps in vacuum is 1/3, and their respective values are $\tau_{p-ps}^v = 0.125$ ns and $\tau_{o-ps}^v = 142$ ns. According to eq. (3.21), we obtain $\tau_{ps}^v = 0.5$ ns.

The integration of eq. (3.20) with eq. (3.19) multiplied by $1/\tau_{ps}^v$ gives $1/\tau_{ps}^{rp}$, which corresponds to the average lifetime per annihilation event inside the electron-rich shell on the inner surfaces of the assumed spherical pores, leading to the Tao–Eldrup equation [19, 20]:

$$\tau_{pS}^{r_p} = \frac{1}{\left[2 - 2\frac{r_p}{r_p + R} + \frac{1}{\pi}\sin\left(2\pi\frac{r_p}{r_p + R}\right)\right]}$$ (3.22)

Here $\tau_{pS}^{r_p}$ would be in ns, while R is a fitted [21] electron layer thickness $R = 1.66$ Å, assumed to be equal irrespective of the material in question [22], and r_p should be given also in Å.

A plot of eq. (3.22) is shown in Fig. 3.6. Note that $\lim_{r_p \to \infty} \tau_{pS}^{r_p} = 140$ ns, which is the lifetime of *o-Ps* in vacuum. It has been tested that the values up to the nanometre are in accordance with the known data for test materials [23, 24].

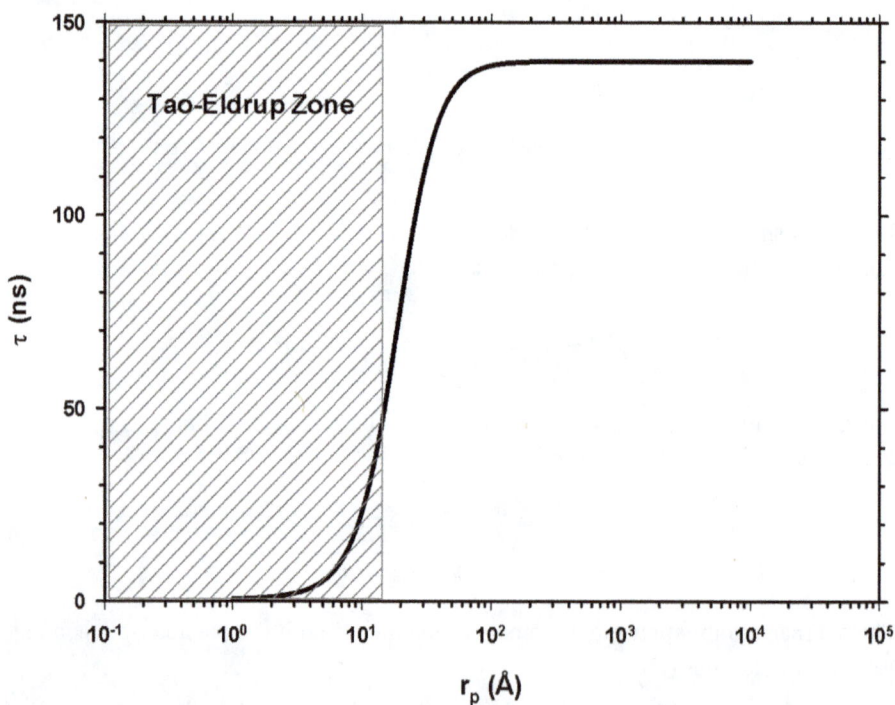

Fig. 3.6: A plot of the Tao–Eldrup equation.

To describe adequate lifetimes in free volumes with larger radii, the model should be somehow extended. The Tao–Eldrup model's simplicity is due to approximations like considering Ps as a particle without internal structure, using a spherical potential well of infinite depth and considering only the ground level of the particle [25].

To find simple solutions of the problem, Gidley et al. [26] changed the geometry of the voids from spherical to cubic. A new value of $R = 0.18$ nm was suggested to fit the obtained curve to the Tao–Eldrup model. Dull et al. [27] found that the use of this R value leads to a good agreement between the modified model and experimental data for $r_p > 1$ nm.

Goworek et al. [28–31] allowed excited levels of Ps as well as the ground one inside the potential well. Such a simple extension mainly changes the Tao–Eldrup model curve for $r_p > 1$ nm and explains the dependence of lifetime on the temperature for moderate temperatures. Temperature corrections to the Tao–Eldrup equation are shown for two temperatures in Figure 3.7 as function of pore radii.

The essential point in this model is the introduction of lifetime averaged over as many excited states as necessary. Unfortunately, frequently the resulting equations are more complicated and must be solved numerically, fitted or approximated [32, 33].

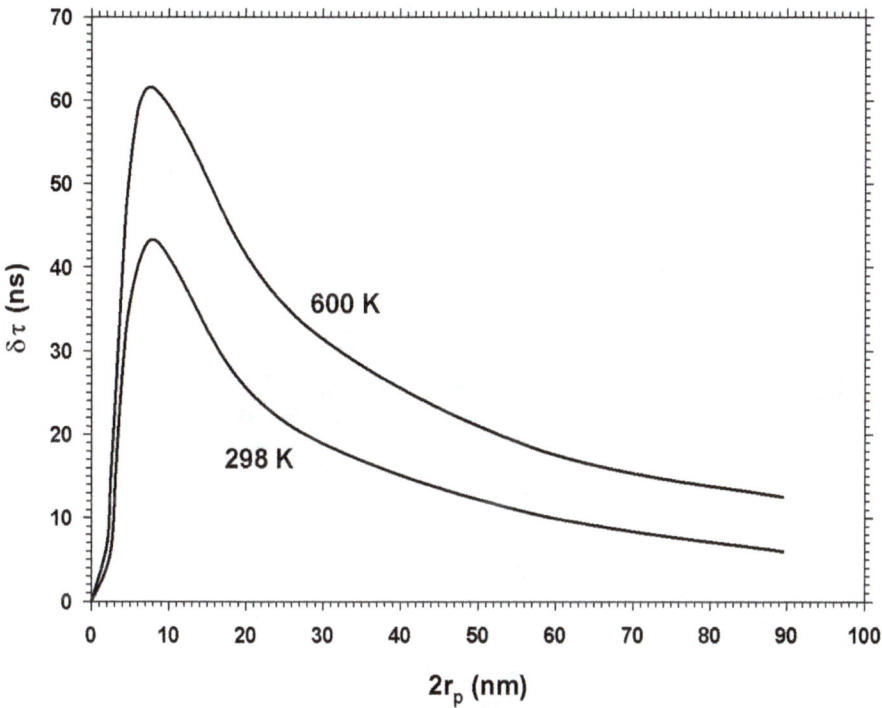

Fig. 3.7: Temperature corrections to the Tao–Eldrup equation. For a temperature of 0 K, the Tao–Eldrup equation would give correct results. Data taken from Gidley et al. [5] and Palacio Gómez [25].

In summary, PALS can quantitatively be used to determine free volume sizes, free volume size distribution and fractional free volume at dimensions ranging from 1 Å to approximately 100 nm [34, 35]. This information is specifically interesting for free volumes from 1 to 10 Å [3], which are particularly difficult to reveal by other methods and do not give any problem. This feature is useful for studying dense membranes for gas separations.

Note that free volume constituted by holes or voids in the sub-nanometric range plays an essential role in small molecular diffusion, and thus permeability, through otherwise dense membranes. This is the reason why PALS is mostly used in characterizing gas separation membranes.

3.1.6 Other positron annihilation techniques

Other techniques derived from PALS are 2D angular correlation of annihilation radiation (2D-ACAR) and Doppler broadening of positron annihilation radiation (DBAR), that is, angular correlation of the annihilation radiation and Doppler broadening of annihilation radiation. The fundamentals of both these techniques are shown in Figs. 3.8 and 3.9.

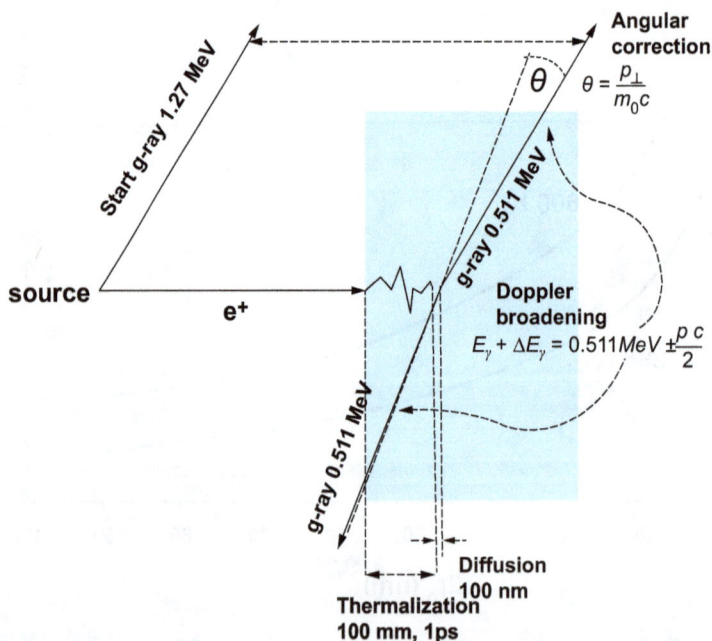

Fig. 3.8: Scheme of angular correction and Doppler broadening of photons produced by direct annihilation of positrons.

A basic set-up to perform 2D-ACAR, 2D measurement of the angular correlation of the positron annihilation radiation, would comprise two positions sensitive to 2D detectors. After entering into the sample, the positrons thermalize within a few picoseconds and thus lose their momentum. After diffusing through the sample, from 100 to 200 ps, the positron finally annihilates with an electron, direct annihilation in Fig. 3.2,

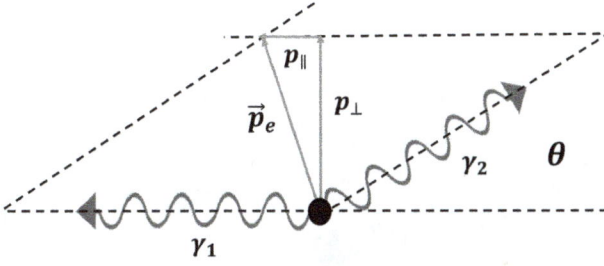

Fig. 3.9: Photons produced in a direct $e^+ + e^-$ annihilation. $\vec{p}_e$ corresponds to the momentum of the electron.

and emits two γ-particles. Due to the momentum of the electron and to the required conservation of momentum, the γ-rays are not emitted at exactly anti-parallel. For small angles, θ is proportional to the transversal momentum of the electron.

Moreover, due to the Doppler relativistic effect, the energy of each dispersed photon is modified, when measured in the laboratory frame, by an amount that is proportional to the longitudinal moment of the electron that is responsible for most of the movements of the system in the moment of collision. In effect, the relative change in energy, $\Delta E_\gamma / E_\gamma = (E_\gamma^{\text{lab}} - E_\gamma^{\text{CM}})/E_\gamma^{\text{CM}}$, is proportional to the change in the photon frequency, $\Delta v / v = (v^{\text{lab}} - v^{\text{CM}})/v^{\text{CM}}$:

$$\left| \frac{\Delta E_\gamma}{E_\gamma} \right| = \frac{\Delta v_\gamma}{v_\gamma} \tag{3.23}$$

According to the Doppler relativistic longitudinal shift:

$$v_\gamma^{\text{lab}} = v_\gamma^{\text{CM}} \sqrt{\frac{1 + v_\parallel/c}{1 - v_\parallel/c}} \approx v_\gamma^{\text{CM}} \left(1 + \frac{v_\parallel}{2c}\right)^2 \approx v_\gamma^{\text{CM}} \left(1 + \frac{v_\parallel}{c}\right) \tag{3.24}$$

where $v_\parallel$ is the longitudinal velocity of the centre of momenta as measured from the laboratory frame. The moment of the positron–electron system can be attributed mostly to the electron. It has been assumed that $v_\parallel \ll c, c$ is the speed of light. From eq. (3.24),

$$v_\gamma^{\text{lab}} \approx v_\gamma^{\text{CM}} + v_\gamma^{\text{CM}} \frac{v_\parallel}{c} \Rightarrow \frac{v_\gamma^{\text{lab}} - v_\gamma^{\text{CM}}}{v_\gamma^{\text{CM}}} = \frac{v_\parallel}{c} \tag{3.25}$$

Then

$$\left| \frac{\Delta E_\gamma}{E_\gamma} \right| = \frac{v_\parallel}{c} \tag{3.26}$$

But

$$v_{\parallel} = \frac{p_{\parallel}}{2m_e} \tag{3.27}$$

Thus,

$$\left|\Delta E_\gamma\right| = \left|\frac{p_{\parallel}}{2m_e c}\right| E_\gamma \tag{3.28}$$

But $E_\gamma = m_e\, c^2$, then

$$\left|\Delta E_\gamma\right| = \left|p_{\parallel}\right| \frac{c}{2} \tag{3.29}$$

Therefore, the energy of the photons, in the laboratory, is

$$E_\gamma^{\mathrm{lab}} = E_\gamma^{\mathrm{CM}} \pm \left|p_{\parallel}\right| \frac{c}{2} \tag{3.30}$$

where $p_{\parallel}$ is the longitudinal momentum of the electron. Equation (3.30) describes the so-called Doppler broadening.

On the other hand,

$$E_{\gamma 1}^{\mathrm{lab}} + E_{\gamma 2}^{\mathrm{lab}} = 2m_e c^2 \tag{3.31}$$

But for a photon $E_\gamma^{\mathrm{lab}} = p_\gamma c$. Now for the momenta and their projections (see Fig. 3.9):

$$\left. \begin{array}{c} \dfrac{E_{\gamma 1}^{\mathrm{lab}}}{c} - \dfrac{E_{\gamma 2}^{\mathrm{lab}}}{c} \cos\theta = p_{\parallel\gamma}^t \\[2mm] \dfrac{E_{\gamma 1}^{\mathrm{lab}}}{c} \sin 0 + \dfrac{E_{\gamma 2}^{\mathrm{lab}}}{c} \sin\theta = p_{\perp\gamma}^t \end{array} \right\} \tag{3.32}$$

Or, for small θ

$$\left. \begin{array}{c} \dfrac{E_{\gamma 1}^{\mathrm{lab}}}{c} - \dfrac{E_{\gamma 2}^{\mathrm{lab}}}{c} = p_{\parallel\gamma}^t \\[2mm] \dfrac{E_{\gamma 2}^{\mathrm{lab}}}{c} \theta = p_{\perp\gamma}^t \end{array} \right\} \tag{3.33}$$

p_γ^t is the total momentum in the laboratory frame, which is the moment of the electron $p_\gamma^t = p$ due to its conservation. Now from eqs. (3.33) (for the longitudinal component) and (3.31):

$$2E_{\gamma 2}^{\mathrm{lab}} + p_{\parallel}c = 2m_e c^2 \tag{3.34}$$

and

$$\left. \begin{array}{l} E_{\gamma 2}^{\mathrm{lab}} = m_e c^2 - \frac{p_\| c}{2} \\[2mm] E_{\gamma 1}^{\mathrm{lab}} = m_e c^2 + \frac{p_\| c}{2} \end{array} \right\} \tag{3.35}$$

That along with eq. (3.30) is the basis of Doppler broadening in the 2γ positron plus electron annihilation.

Now using the transversal part of eq. (3.33):

$$p_\perp = \frac{E_{\gamma 2}^{\mathrm{lab}}}{c} \theta \tag{3.36}$$

Using now eq. (3.35) we get

$$p_\perp = \theta \left(m_e c - \frac{p_\|}{2c} \right) \tag{3.37}$$

But given that $m_e c \gg p_\| / 2c$:

$$\theta = \frac{p_\perp}{m_e c} \tag{3.38}$$

This equation is the basis of the ACAR spectroscopy.

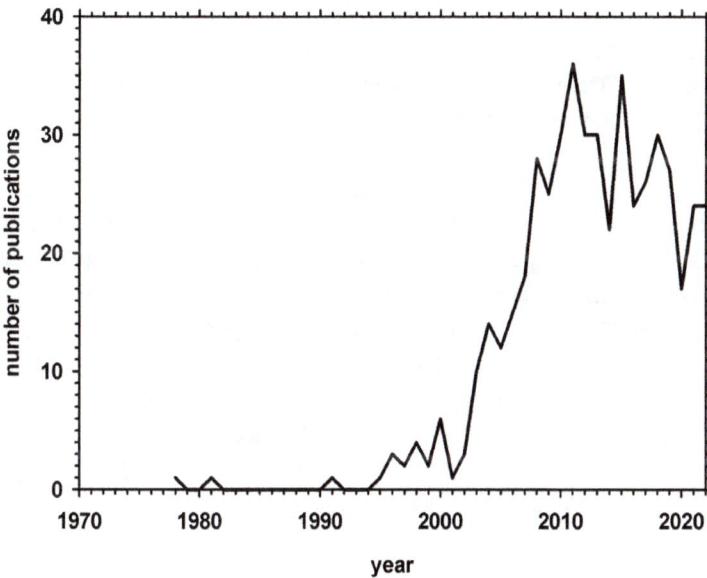

Fig. 3.10: Publications found in WOS, from Clarivate, on the topics such as membranes and positron annihilation lifetime spectroscopy.

ACAR and DBAR spectroscopies could give information on the movement of electrons in the bulk, which can be correlated with defects, vacancies or heterogeneities [36, 37]. Few studies have been dedicated to membranes yet [38, 39].

3.1.7 Applications

Most applications of PALS appear in the research and characterization of gas separation membranes [40–43]. In Fig. 3.10, the evolution of the publications on the application of PALS on membranes is depicted. Among them, the seminal works of Jordan and Koros [44] and Hill et al. [45] stand out.

3.2 Synchrotron radiation (SR)

A synchrotron is a device that allows charged particles to be accelerated to near-light speed. It is basically a particle accelerator using powerful magnetic fields. The difference between the synchrotron and other types of linear accelerators is that in the first case the particle trajectory is closed.

Synchrotrons were designed for the study of particle physics. In this sense, we have particle colliders, in which two beams of particles and their corresponding antiparticles are sent to collide after travelling through the synchrotron in such a way that the high energies involved in the collision allow matter to be broken down into its fundamental particles. The most spectacular example of such a collider is CERN's LHC, where the Higgs boson was discovered.

But over the years, the applications of synchrotrons outside the field of particle physics have been growing. In particular, in the field of materials research, it is synchrotron radiation (SR) that is used. This electromagnetic radiation is the consequence of the acceleration experienced by the particles present in the synchrotron (generally electrons, although they can also consist of positrons and their antiparticles) due to the curved trajectory that the magnets of the equipment force them to maintain.

Because of this curved trajectory, the electrons undergo a strong normal acceleration (in order to take them out of the expected straight line) whose excess energy is emitted in the form of electromagnetic radiation. It is obvious that the higher the velocity at which the electrons have been accelerated in the system, the higher the energy emitted in the form of radiation. Taking into account Planck's relation:

$$E = h\,v \tag{3.39}$$

The frequency of the wave associated with this energy loss will be higher and, consequently, the wavelength will be shorter (remember that in an electromagnetic wave, the relationship between frequency and wavelength is given by $\lambda \cdot v = c$, with c being the speed of light in the medium considered), but keep in mind that the energy of the electrons, and consequently the energy loss they experience, is controlled by the magnetic potential with which we have accelerated the particles in the synchrotron.

Depending on its frequency (which, as we have said, is related to the speed at which we have accelerated the electrons), this radiation can be in the range from X-rays to infrared. On the other hand, it is a very intense, concentrated radiation with high spatial coherence, which makes it particularly interesting for the study of the properties of materials.

This radiation, after passing through the membrane we want to analyse, is detected, and its intensity variations can be correlated with the number and size of the pores it has passed through [46]. All this allow for a real-time analysis of the sample [47], and provide a high-resolution images that can be used for pore sizing characterization purposes.

In the case of polymeric materials, and especially porous membranes, the study is generally carried out using X-rays; hence, synchrotrons built specifically for this purpose (different from those used in particle physics) called second- or third-generation synchrotrons are used.

Actually, the task of the synchrotron in many applications as a characterization technique is merely to provide an intense and coherent beam of X-rays, so any other X-ray source can be used to study these materials.

Once we have a suitable source of an X-ray beam, the small-angle X-ray scattering (SAXS) technique is applied, although wide-angle X-ray scattering can also be used.

Therefore, the technique has been used in the characterization of materials but also in a variety of fields such as micromechanics, biology, physics, geology, archaeology and medicine [48]. A complete paper by the authors and other colleagues accounts for a range of applications of X-ray diffraction and SAXS [46], for several types of materials, including of course membrane materials, especially in the last years.

Among these works we can mention, as more related to the structural characterization of membranes, the study by Lassinantti Gualtieri et al. [49] in which X-ray diffraction from a synchrotron, applied on zeolite powder samples on alumina support in MFI (Mordenite Framework Inverted) membranes, performed at high temperature, allowed visualizing and understanding the dynamics of crack formation inherent to the calcination process of the sample.

Also interesting, as an impressive application of SR to tomographic visualization of membranes, is the work of Remigy and Meireles [50, 51] on the three-dimensional (3D) structure of microfiltration hollow fibres based on polysulfone and poly(vinylidene-co-hexafluoropropylene fluoride). In this work, the SR technique was used to obtain a microtomographic view [52] that allows a 3D analysis of the pores inside the fibres.

Similarly Lee et al. [53] studied the 3D structure, porosity, pore connectivity and tortuosity of nanoporous polymeric membranes, based in images obtained by SR X-rays.

More recent works on the application of SR as an X-ray source can be found in the literature, as the work by Dutt et al. [54] focused on the study of the shape of nanopores present in track-etched polycarbonate membranes. Doudies et al. [55] studied the external membrane fouling during cross-flow ultrafiltration of milk protein (casein micelle) dispersions by in-situ SR-based SAXS. Guo et al. [56] studied the crystal orientation within zeolite membranes via nano-beam X-ray diffraction, then mapping the size of the grains within the zeolite and obtaining the pore size distribution (PSD).

Although there is not too many applications of SR to membrane characterization, the technique offers some advantages as wide spectrum range, high intensity, high polarization or small source, and moreover, it can be applied to nanometre-scale pores [46].

3.3 Ellipsometric porosimetry (ELLP)

3.3.1 Introduction

Ellipsometry is a powerful and versatile technique used in the field of materials science for characterizing the optical properties of thin films and surfaces [57–59]. It provides valuable insights into parameters such as film thickness, refractive index and surface roughness, making it an indispensable tool in both academic research and industrial applications. The fundamental principle of ellipsometry lies in the analysis of changes in the polarization state of light upon reflection or transmission through a material.

At its core, ellipsometry measures the change in polarization of light as it interacts with a sample. The incident light is typically linearly polarized, and upon reflection or transmission, its polarization state is altered [60]. Ellipsometry quantifies these changes by measuring the ellipticity and phase difference of the polarized light. The resulting data are then used to extract information about the optical properties of the material.

The key parameters in ellipsometry are the amplitude ratio (ψ) and the phase difference (Δ) between the p- and s-polarized components of the reflected or transmitted light. These parameters are highly sensitive to the properties of the material, allowing for precise determination of film characteristics. The ellipsometric data are usually represented in a plot known as an ellipsogram, providing a visual representation of the polarization changes.

To enhance the accuracy and versatility of ellipsometry, multiple wavelength ellipsometry is often employed. By using light at different wavelengths, researchers can obtain information about the material's optical properties at various depths, allowing for a more comprehensive analysis of complex structures. This approach is particularly valuable when dealing with multi-layered or anisotropic materials.

3.3.2 Reflection and transmission: the Fresnel equations

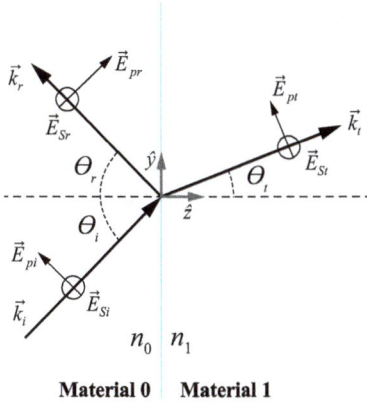

Fig. 3.11: Reflection and transmission. Components p are parallel to the incidence plane while the S components are perpendicular ("senkrecht" in German) to this plane.

In Figure 3.11, the reflection and transmission of the electric field are shown. The wave vectors for the rays are:

$$\left.\begin{array}{l} \vec{k}_i = k_i(\hat{y}\sin\theta_i + \hat{z}\cos\theta_i) \\ \vec{k}_r = k_r(\hat{y}\sin\theta_r - \hat{z}\cos\theta_r) \\ \vec{k}_t = k_t(\hat{y}\sin\theta_t + \hat{z}\cos\theta_t) \end{array}\right\} \quad (3.40)$$

The plane wave electrical fields are

$$\left.\begin{array}{l} \vec{E}_i = \left[E_{pi}(\hat{y}\cos\theta_i - \hat{z}\sin\theta_i) + E_{Si}\hat{x}\right]e^{i\left[k_i(y\sin\theta_i + z\cos\theta_i) - \omega_i t\right]} \\ \vec{E}_r = \left[E_{pr}(\hat{y}\cos\theta_r + \hat{z}\sin\theta_r) + E_{Sr}\hat{x}\right]e^{i\left[k_r(y\sin\theta_r - z\cos\theta_r) - \omega_r t\right]} \\ \vec{E}_t = \left[E_{pt}(\hat{y}\cos\theta_t - \hat{z}\sin\theta_t) + E_{St}\hat{x}\right]e^{i\left[k_t(y\sin\theta_t + z\cos\theta_t) - \omega_t t\right]} \end{array}\right\} \quad (3.41)$$

Because the total fields at both sides of the interface at $z = 0$ must be equal:

$$\left[E_{pi}\hat{y}\cos\theta_i + E_{Si}\hat{x}\right]e^{i\left[k_i y\sin\theta_i - \omega_i t\right]} + \left[E_{pr}\hat{y}\cos\theta_r + E_{Sr}\hat{x}\right]e^{i\left[k_r y\sin\theta_r - \omega_r t\right]} =$$
$$= \left[E_{pt}\hat{y}\cos\theta_t + E_{St}\hat{x}\right]e^{i\left[k_t y\sin\theta_t - \omega_t t\right]} \quad (3.42)$$

This identity must hold for all y and t, thus

$$\left.\begin{array}{l} \omega_i = \omega_r = \omega_t \equiv \omega \\ k_i\sin\theta_i = k_r\sin\theta_r = k_t\sin\theta_t \end{array}\right\} \quad (3.43)$$

On the other hand, we know that $k_i = k_r = \omega/(c/n_0)$ and $k_t = \omega/(c/n_1)$, introducing the refraction index; therefore

$$\left. \begin{array}{c} \theta_i = \theta_r \\ k_r \sin \theta_r = k_t \sin \theta_t \end{array} \right\} \left. \begin{array}{c} \theta_i = \theta_r \equiv \theta_0, \ \theta_t \equiv \theta_1 \\ n_0 \sin \theta_0 = n_1 \sin \theta_1 \end{array} \right. \tag{3.44}$$

These are the reflection and Snell's laws. Moreover, eq. (3.41) reduces to

$$\left. \begin{array}{c} (E_{pi} + E_{pr}) \cos \theta_0 = E_{pt} \cos \theta_t \\ E_{Si} + E_{Sr} = E_{St} \end{array} \right\} \tag{3.45}$$

Now for the magnetic field:

$$\vec{B} = \frac{\vec{k} \wedge \vec{E}}{\omega} = \frac{n}{c} \hat{k} \wedge \vec{E} \tag{3.46}$$

and

$$\left. \begin{array}{l} \vec{B}_i = \frac{n_0}{c} \left[-\hat{x} E_{pi} + E_{Si} (\hat{y} \cos \theta_0 - \hat{z} \sin \theta_0) \right] e^{i[k_i (y \sin \theta_0 + z \cos \theta_0) - \omega t]} \\ \vec{B}_r = \frac{n_0}{c} \left[\hat{x} E_{pr} + E_{Sr} (-\hat{y} \cos \theta_0 - \hat{z} \sin \theta_0) \right] e^{i[k_r (y \sin \theta_0 - z \cos \theta_0) - \omega t]} \\ \vec{B}_t = \frac{n_1}{c} \left[-\hat{x} E_{pt} + E_{St} (\hat{y} \cos \theta_1 - \hat{z} \sin \theta_1) \right] e^{i[k_1 (y \sin \theta_1 + z \cos \theta_1) - \omega_t t]} \end{array} \right\} \tag{3.47}$$

Because $\vec{B}_i + \vec{B}_r = \vec{B}_t$ for $z = 0$, if no magnetic effects appear at the interface, eq. (3.47) leads to

$$\left. \begin{array}{c} n_0 (E_{pi} - E_{pr}) = n_1 E_{pt} \\ n_0 (E_{Si} - E_{Sr}) \cos \theta_0 = n_1 E_{St} \cos \theta_1 \end{array} \right\} \tag{3.48}$$

Finally, from eqs. (3.45) and (3.48):

$$\frac{E_{Sr}}{E_{Si}} = \frac{n_0 \cos \theta_0 - n_1 \cos \theta_1}{n_0 \cos \theta_0 + n_1 \cos \theta_1} \tag{3.49}$$

$$\frac{E_{St}}{E_{Si}} = \frac{2 n_0 \cos \theta_0}{n_0 \cos \theta_0 + n_1 \cos \theta_1} \tag{3.50}$$

In addition, there are two analogous equations for the components of the field parallel to the plane of incidence:

$$\frac{E_{pr}}{E_{pi}} = \frac{n_1 \cos \theta_0 - n_0 \cos \theta_1}{n_0 \cos \theta_1 + n_1 \cos \theta_0} \tag{3.51}$$

$$\frac{E_{pt}}{E_{pi}} = \frac{2 n_0 \cos \theta_0}{n_0 \cos \theta_1 + n_1 \cos \theta_0} \tag{3.52}$$

These are the Fresnel's equations. They can be rewritten by using the Snell laws (eq. (3.44)) as follows:

$$\frac{E_{Sr}}{E_{Si}} = \frac{n_0 \cos\theta_0 - n_1 \cos\theta_1}{n_0 \cos\theta_0 + n_1 \cos\theta_1} = \frac{\sin\theta_1 \cos\theta_0 - \sin\theta_0 \cos\theta_1}{\sin\theta_1 \cos\theta_0 + \sin\theta_0 \cos\theta_1} = \frac{\sin(\theta_1 - \theta_0)}{\sin(\theta_1 + \theta_0)} \tag{3.53}$$

$$\frac{E_{St}}{E_{Si}} = \frac{2n_0 \cos\theta_0}{n_0 \cos\theta_0 + n_1 \cos\theta_1} = \frac{2\sin\theta_1 \cos\theta_0}{\sin\theta_1 \cos\theta_0 + \sin\theta_0 \cos\theta_1} = \frac{2\sin\theta_1 \cos\theta_0}{\sin(\theta_1 + \theta_0)} \tag{3.54}$$

$$\frac{E_{pr}}{E_{pi}} = \frac{n_1 \cos\theta_0 - n_0 \cos\theta_1}{n_0 \cos\theta_1 + n_1 \cos\theta_0} = \frac{\sin\theta_1 \cos\theta_1 - \sin\theta_0 \cos\theta_0}{\sin\theta_1 \cos\theta_1 + \sin\theta_0 \cos\theta_0} = \frac{\tan(\theta_1 - \theta_0)}{\tan(\theta_1 + \theta_0)} \tag{3.55}$$

$$\frac{E_{pt}}{E_{pi}} = \frac{2n_0 \cos\theta_0}{n_0 \cos\theta_1 + n_1 \cos\theta_0} = \frac{2\sin\theta_1 \cos\theta_0}{\sin\theta_1 \cos\theta_1 + \sin\theta_0 \cos\theta_0} =$$
$$= \frac{2\sin\theta_1 \cos\theta_0}{\sin(\theta_1 + \theta_0)\cos(\theta_1 - \theta_0)} \tag{3.56}$$

If we admit a complex refraction index, we can correlate it with the dielectric constant:

$$\left. \begin{array}{l} \tilde{n} = n + i\kappa \\ \tilde{n}^2 = \tilde{\varepsilon} \Rightarrow (n + i\kappa)^2 = (\varepsilon_{re} + i\varepsilon_{im}) \end{array} \right\} \quad \begin{array}{l} n^2 - \kappa^2 = \varepsilon_{re} \\ 2n\kappa = \varepsilon_{im} \end{array} \tag{3.57}$$

Note that E_{Sr}/E_{Si}, E_{St}/E_{Si}, E_{pr}/E_{pi} and E_{pt}/E_{pi} are real according to eqs. (3.53)–(3.55).
By using eq. (3.57):

$$\left. \begin{array}{l} n = \frac{1}{\sqrt{2}}\left(\sqrt{\varepsilon_{re}^2 + \varepsilon_{im}^2} + \varepsilon_{re}\right) \\ \kappa = \frac{1}{\sqrt{2}}\left(\sqrt{\varepsilon_{re}^2 + \varepsilon_{im}^2} - \varepsilon_{re}\right) \end{array} \right\} \tag{3.58}$$

Thus, we can correlate the imaginary part of the dielectric constant with conductivity:

$$2n\kappa = \varepsilon_{im} = \frac{\sigma}{\varepsilon_0 \omega} \tag{3.59}$$

and

$$\left. \begin{array}{l} n = \frac{1}{\sqrt{2}}\left(\sqrt{\varepsilon_{re}^2 + \left(\frac{\sigma}{\varepsilon_0 \omega}\right)^2} + \varepsilon_{re}\right) \\ \kappa = \frac{1}{\sqrt{2}}\left(\sqrt{\varepsilon_{re}^2 + \left(\frac{\sigma}{\varepsilon_0 \omega}\right)^2} - \varepsilon_{re}\right) \end{array} \right\} \tag{3.60}$$

In terms of such a complex refraction index, we obtain for the refracted electric field, for example:

$$\vec{E}_t = \left[E_{pt}(\hat{y}\cos\theta_t - \hat{z}\sin\theta_t) + E_{St}\hat{x}\right]e^{i\left[\frac{\tilde{n}_1}{c}(y\sin\theta_t + z\cos\theta_t) - \omega_t t\right]} \tag{3.61}$$

$$\vec{E}_t = E_t e^{i\left[\frac{n_0 + i\kappa}{c}(y \sin \theta_t + z \cos \theta_t) - \omega t\right]} =$$

$$= E_t e^{i\left[\frac{n_0}{c}(y \sin \theta_t + z \cos \theta_t) - \omega t\right]} e^{-\frac{\kappa}{c}(y \sin \theta_t + z \cos \theta_t)}$$

(3.62)

This means, assuming $\kappa \geq 0$, that there is attenuation along both the y and z axes (note that $0 \leq \theta_t \leq 1$). Consequently, κ is called the adsorption coefficient.

3.3.3 Light polarization

Comprehending and controlling the polarization of light have significant importance in various optical applications. While optical design often emphasizes the wavelength and intensity of light, the polarization aspect is frequently overlooked. Nevertheless, polarization is a critical characteristic influencing optical systems, even those not explicitly designed to measure it [61]. The polarization of light plays a role in laser beam focus, determines the cut-off wavelengths of filters and is crucial for preventing undesired back-reflections. It proves essential in metrology applications like stress analysis in glass or plastic, pharmaceutical ingredient analysis and biological microscopy. Additionally, materials may absorb different polarizations of light to varying degrees, a crucial property for LCD screens, 3D movies and glare-reducing sunglasses.

Light, being an electromagnetic wave, features an electric field oscillating perpendicularly to the direction of propagation and a magnetic field perpendicular to both the electric field and to the direction of propagation. Light is considered unpolarized when the electric field's direction fluctuates randomly over time. Common light sources like sunlight, halogen lighting, LED spotlights and incandescent bulbs emit unpolarized light. When the electric field direction is well-defined, the light is termed polarized. Most lasers emit polarized light.

Based on the orientation of the electric field, polarized light can be classified into three types:
1. Linear polarization: the electric field is confined to a single plane along the direction of propagation.
2. Circular polarization: the electric field comprises two linear components perpendicular to each other, equal in amplitude but with a phase difference of $\pi/2$. This results in an electric field that rotates in a circle around the direction of propagation, known as left- or right-hand circularly polarized light, depending on the rotation direction.
3. Elliptical polarization: the electric field traces an ellipse, arising from the combination of two linear components with different amplitudes and/or a phase difference not equal to $\pi/2$. This provides a more general description of polarized light, with circular and linear polarized light being specific cases of elliptically polarized light.

The elliptic polarization corresponds to two components of the electric field that oscillate with a dephasing γ.

If we take the appearance of phase changes in reflection into account, we can define

$$\left.\begin{array}{l} \tilde{R}_p = \frac{\vec{E}_{pr}}{\vec{E}_{pi}} = r_p e^{i\delta_p} \\[2mm] \tilde{R}_S = \frac{\vec{E}_{Sr}}{\vec{E}_{Si}} = r_S e^{i\delta_S} \end{array}\right\} \tag{3.63}$$

with (see eqs. (3.53) and (3.55))

$$r_S = \frac{E_{Sr}}{E_{Si}} = \frac{n_0 \cos\theta_0 - n_1 \cos\theta_1}{n_0 \cos\theta_0 + n_1 \cos\theta_1} \tag{3.64}$$

$$r_p = \frac{E_{pr}}{E_{pi}} = \frac{n_1 \cos\theta_0 - n_0 \cos\theta_1}{n_0 \cos\theta_1 + n_1 \cos\theta_0} \tag{3.65}$$

Here, the wavy tilde on $\tilde{R}_p$ and $\tilde{R}_S$ means that they are complex numbers, and

$$\frac{\tilde{R}_p}{\tilde{R}_S} = \frac{r_p}{r_p} e^{i(\delta_p - \delta_S)} = \tan\Psi e^{i\Delta} = \tilde{\rho} \tag{3.66}$$

Here $\tilde{\rho}$ is the complex reflectivity, and the following definitions have been applied:

$$\left.\begin{array}{l} \tan\Psi = \frac{r_p}{r_S} \\[2mm] \Delta = \delta_p - \delta_S \end{array}\right\} \tag{3.67}$$

with

$$\left.\begin{array}{l} 0 \le \Psi \le \pi/2 \\[2mm] 0 \le \Delta \le 2\pi \end{array}\right\} \tag{3.68}$$

Note that by using eqs. (3.53) and (3.55)

$$\begin{aligned} \tan\Psi &= \frac{r_p}{r_S} = \frac{E_{pr}/E_{pi}}{E_{Sr}/E_{Si}} = \frac{E_{Si}/E_{pi}}{E_{Sr}/E_{pr}} = \\[2mm] &= \frac{\dfrac{\tan(\theta_1 - \theta_0)}{\tan(\theta_1 + \theta_0)}}{\dfrac{\sin(\theta_1 - \theta_0)}{\sin(\theta_1 + \theta_0)}} = \frac{\cos(\theta_1 + \theta_0)}{\cos(\theta_1 - \theta_0)} \end{aligned} \tag{3.69}$$

$$\tan\Psi = \frac{r_p}{r_S} = \frac{E_{pr}/E_{pi}}{E_{Sr}/E_{Si}} = \frac{E_{pr}/E_{Sr}}{E_{pi}/E_{Si}} = \frac{\tan\gamma_r}{\tan\gamma_i} \tag{3.70}$$

Here the angle that the plane of polarization of the incident light makes with the plane of incidence is γ_i and the angle after reflection is γ_r. If the incident light is linearly polarized with $\gamma_i = \pi/4$, then $\tan \gamma_i = 1$ or $E_{pi} = E_{Si}$ and $\tan \Psi = \tan \gamma_r = E_{pr}/E_{Sr}$.

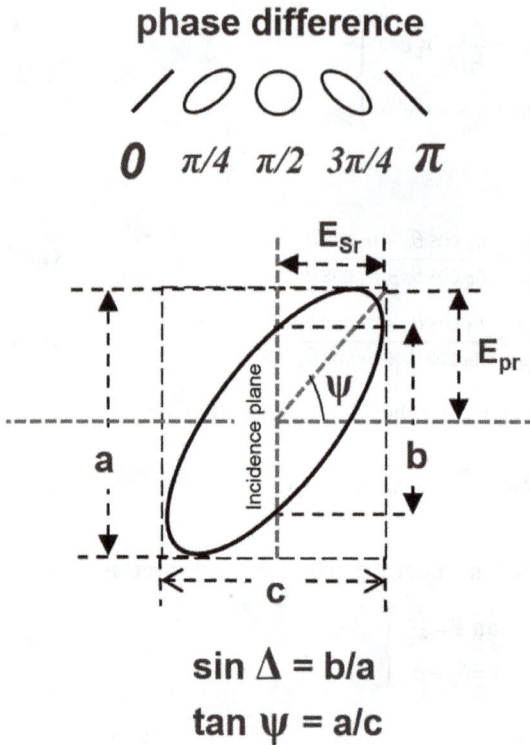

phase difference

0   π/4 π/2 3π/4 π

$$\sin \Delta = b/a$$
$$\tan \psi = a/c$$

Fig. 3.12: Dephasing between orthogonal components of electrical fields in an elliptically polarized light wave. It has been assumed that the incident light was linearly polarized with π/4 dephasing and that the electrical field has equal p and S amplitudes.

A scheme of a simple ellipsometric apparatus is shown in Fig. 3.13. A general description of the components involved in an ellipsometric apparatus include [62]:

1. Light source: ellipsometers typically use a light source that emits polarized light, often in the form of a laser. The light source generates a beam of light with a well-defined polarization state and a more or less wide range of wavelengths.
2. Polarizer: the initial polarizer in the system ensures that the light beam entering the ellipsometer is linearly polarized. This means that the light waves oscillate in a specific direction.
3. Sample chamber: the sample chamber holds the material under investigation, which may have thin films or coatings. The sample is positioned in the path of the polarized light.

4. Analyser: the analyser is another polarizing element that allows only a specific polarization component of the light to pass through. It is positioned after the sample.
5. Detector: the detector measures the intensity of the light that has passed through the sample. The ellipsometer detects changes in the polarization state caused by interactions with the sample.
6. Rotating stage: a rotating stage is often incorporated to change the angle of incidence of the polarized light on the sample. This helps in obtaining information at different angles, improving the accuracy of measurements.
7. Data analysis system: The raw data collected by the detector needs to be analysed to extract information about the optical properties and thickness of the sample. Advanced software is employed to fit the experimental data to theoretical models, allowing for the determination of parameters such as film thickness and refractive index.
8. Automated control system: Many modern ellipsometers feature automated control systems for precise adjustment of the instrument's components, making the measurements more efficient and accurate.

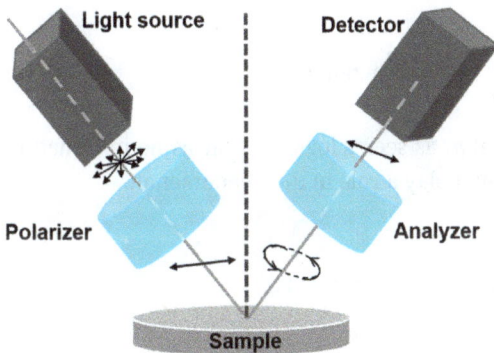

Fig. 3.13: Scheme of an ellipsometric apparatus.

Several different details in the apparatus configuration define different techniques as, for example, polarizer, compensator, sample, analyser) ellipsometry (as shown in Fig. 3.3); rotating polarizer and analyser ellipsometry; rotating polarizer and analyser ellipsometry; polarization modulation ellipsometry and multi-channel ellipsometry [62–64].

3.3.4 Single interface

According to Fresnel's equations (3.53) and (3.55):

$$\rho = \frac{r_p}{r_s} = \frac{E_{pr}/E_{pi}}{E_{sr}/E_{si}} = \tan \Psi =$$

$$= \frac{n_1 \cos \theta_0 - n_0 \cos \theta_1}{n_0 \cos \theta_1 + n_1 \cos \theta_0} \left(\frac{n_0 \cos \theta_0 + n_1 \cos \theta_1}{n_0 \cos \theta_0 - n_1 \cos \theta_1} \right) \tag{3.71}$$

For $n_0 = 1$ (air):

$$\tan \Psi = \frac{n_1 \cos \theta_0 - \cos \theta_1}{n_1 \cos \theta_0 + \cos \theta_1} \left(\frac{\cos \theta_0 + n_1 \cos \theta_1}{\cos \theta_0 - n_1 \cos \theta_1} \right) \tag{3.72}$$

$$\rho = \tan \Psi = \frac{n_1^2 \cos \theta_0 - \sqrt{n_1^2 - \sin^2 \theta_0}}{n_1^2 \cos \theta_1 + \sqrt{n_1^2 - \sin^2 \theta_0}} \left(\frac{\cos \theta_0 + \sqrt{n_1^2 - \sin^2 \theta_0}}{\cos \theta_0 - \sqrt{n_1^2 - \sin^2 \theta_0}} \right) \tag{3.73}$$

This equation can be solved for n_1:

$$n_1 = \sin \theta_0 \sqrt{1 + \left(\frac{1-\rho}{1+\rho} \right)^2 \tan^2 \theta_0} \tag{3.74}$$

Note that this relationship gives the real n_1 as seems logical as far as no information on the dissipation within the phase 1 could play a role in the phenomenon.

3.3.5 Three-layered materials (air–film–substrate systems)

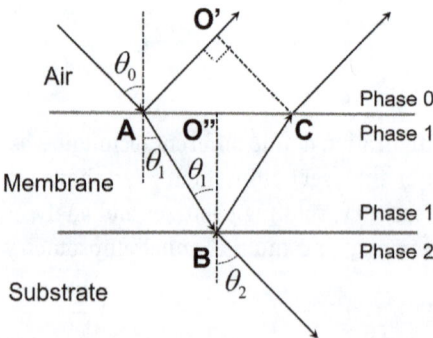

Fig. 3.14: Optical paths in reflection and transmission in a three-layered film.

The difference in the path for the two reflected rays in Fig. 3.14 (rays 1 and 2) is, admitting the possibility of adsorptive media,

$$\Delta x = \tilde{n}_1 \left(\overline{AB} + \overline{BC} \right) - \tilde{n}_0 \overline{AO'} \tag{3.75}$$

Making some trigonometry,

$$\cos \theta_1 = \frac{\overline{BO''}}{\overline{AB}} = \frac{\overline{BO''}}{\overline{BC}} \Rightarrow \frac{d}{\cos \theta_1} = \overline{AB} = \overline{BC} \tag{3.76}$$

where $\overline{BO''} = d$ is the thickness of the membrane

$$\sin \theta_0 = \frac{\overline{AO'}}{\overline{AC}} \Rightarrow \overline{AO'} = \overline{AC} \sin \theta_0 \tag{3.77}$$

$$\tan \theta_1 = \frac{\overline{AO''}}{\overline{BO''}} \Rightarrow \overline{AO''} = d \tan \theta_1 \Rightarrow \overline{AC} = 2\overline{AO''} = 2d \tan \theta_1 \tag{3.78}$$

Therefore,

$$\overline{AO'} = \overline{AC} \sin \theta_0 = 2d \tan \theta_1 \sin \theta_0 \tag{3.79}$$

Then

$$\Delta x = \tilde{n}_1 \left(\overline{AB} + \overline{BC} \right) - \tilde{n}_0 \overline{AO'} = \frac{2\tilde{n}_1 d}{\cos \theta_1} - 2\tilde{n}_0 d \sin \theta_0 \tan \theta_1 \tag{3.80}$$

$$\Delta x = \frac{2\tilde{n}_1 d}{\cos \theta_1} - 2\tilde{n}_1 d \sin \theta_1 \tan \theta_1 = 2\tilde{n}_1 d \cos \theta_1 \tag{3.81}$$

Consequently, the phase difference is

$$\tilde{\delta}' = \frac{2\pi}{\lambda} \Delta x = \frac{2\pi}{\lambda} \left(2\tilde{n}_1 d \cos \theta_1 \right) \tag{3.82}$$

and

$$\tilde{\delta} = \frac{\tilde{\delta}'}{2} = \frac{2\pi}{\lambda} \tilde{n}_1 d \cos \theta_1 \tag{3.83}$$

where $\tilde{\delta}$ is the dephasing per passage through phase 1 (see Fig. 3.15).

Then, successive reflected rays (in accordance with Fig. 3.15) have successive factors $\tilde{R}_0$, $\tilde{R}_1$, $\tilde{R}_2$, $\tilde{R}_3$, ... correlating the incident field with the reflected one. These factors are given by

$$\left. \begin{aligned} \tilde{R}_0 &= r_{01} e^{-0\tilde{\delta}t} \\ \tilde{R}_1 &= t_{01} t_{10} r_{12} e^{-2\tilde{\delta}t} \\ \tilde{R}_2 &= t_{01} t_{10} r_{10} r_{12}^2 e^{-4\tilde{\delta}t} \\ \tilde{R}_3 &= \cdots \end{aligned} \right\} \tag{3.84}$$

where t_{ij} and r_{ij} correspond to the transmission and reflection coefficients when going from phase i to phase j.

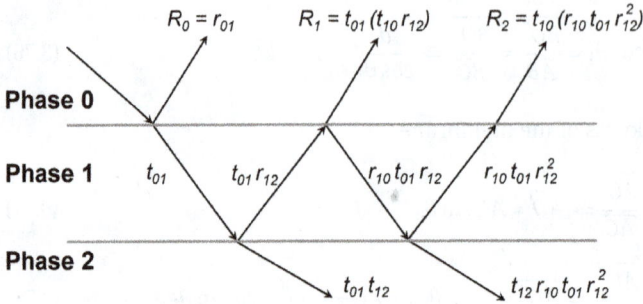

Fig. 3.15: Successive reflections and transmissions in a three-layered material.

Then the resulting reflected ray

$$\tilde{R} = \tilde{R}_0 + \tilde{R}_1 + \tilde{R}_2 + \cdots =$$
$$= r_{01}e^{-0\tilde{\delta}t} + t_{01}t_{10}r_{12}e^{-2\tilde{\delta}t} + t_{01}t_{10}r_{10}r_{12}^2e^{-4\tilde{\delta}t} + \cdots \tag{3.85}$$

$$\tilde{R} = r_{01} + t_{01}t_{10}r_{12}e^{-2\tilde{\delta}t}\left(1 + r_{10}r_{12}e^{-2\tilde{\delta}t} + \cdots\right) \tag{3.86}$$

But it is known that

$$\frac{1}{1-x} = 1 + x + x^2 + \cdots \tag{3.87}$$

Therefore,

$$\tilde{R} = r_{01} + \frac{t_{01}t_{10}r_{12}e^{-2t\tilde{\delta}}}{1 - r_{10}r_{12}e^{-2t\tilde{\delta}}} \tag{3.88}$$

In accordance with Fresnel's laws:

$$r_{01} = -r_{10} \tag{3.89}$$

$$t_{10} = \frac{1 - r_{01}^2}{t_{01}} \Rightarrow t_{01}t_{10} = 1 - r_{01}r_{10} = 1 - r_{01}^2 \tag{3.90}$$

Consequently,

$$\tilde{R} = \frac{r_{01} + r_{12}e^{-2t\tilde{\delta}}}{1 + r_{10}r_{12}e^{-2t\tilde{\delta}}} \tag{3.91}$$

This equation can be written, for both the parallel and perpendicular components to the incidence plane, as

$$
\left.
\begin{aligned}
\tilde{R}_{\mathrm{p}} &= \frac{\vec{E}_{\mathrm{pr}}}{\vec{E}_{\mathrm{pi}}} = \frac{r_{01}^{\mathrm{p}} + r_{12}^{\mathrm{p}} e^{-2t\tilde{\delta}}}{1 + r_{10}^{\mathrm{p}} r_{12}^{\mathrm{p}} e^{-2t\tilde{\delta}}} \\[2mm]
\tilde{R}_{\mathrm{S}} &= \frac{\vec{E}_{\mathrm{Sr}}}{\vec{E}_{\mathrm{Si}}} = \frac{r_{01}^{\mathrm{S}} + r_{12}^{\mathrm{S}} e^{-2t\tilde{\delta}}}{1 + r_{10}^{\mathrm{S}} r_{12}^{\mathrm{S}} e^{-2t\tilde{\delta}}}
\end{aligned}
\right\}
\tag{3.92}
$$

The complex reflectivity $\tilde{\rho} = \tilde{R}_{\mathrm{p}} / \tilde{R}_{\mathrm{S}} = \tan \Psi e^{i\Delta}$ is

$$
\tilde{\rho} = \frac{r_{01}^{\mathrm{p}} + r_{12}^{\mathrm{p}} e^{-2t\tilde{\delta}}}{1 + r_{10}^{\mathrm{p}} r_{12}^{\mathrm{p}} e^{-2t\tilde{\delta}}} \left(\frac{1 + r_{10}^{\mathrm{S}} r_{12}^{\mathrm{S}} e^{-2t\tilde{\delta}}}{r_{01}^{\mathrm{S}} + r_{12}^{\mathrm{S}} e^{-2t\tilde{\delta}}} \right)
\tag{3.93}
$$

Therefore, by taking into account the Fresnel equations (3.53) and (3.55), the Snell laws:

$$
\left.
\begin{aligned}
\frac{\sin \theta_0}{\tilde{n}_1} &= \sin \theta_1 & \frac{\sin \theta_0}{\tilde{n}_1} &= \sin \theta_1 \\[2mm]
\frac{\sin \theta_1}{\tilde{n}_2} &= \sin \theta_2 & \frac{\sin \theta_0}{\tilde{n}_1 \tilde{n}_2} &= \sin \theta_2
\end{aligned}
\right\}
\tag{3.94}
$$

and eq. (3.83), we reach

$$
\tilde{\rho} = f(\lambda, \theta_0; \tilde{n}_1, \tilde{n}_2, d)
\tag{3.95}
$$

For a given incidence angle, we can measure $\tilde{\rho}$ (i.e. $\tan \Psi$ and Δ) for different wavelengths, an ellipsometric spectroscopy, and perform a fitting procedure to obtain $\tilde{n}_1$, $\tilde{n}_2$ and d.

3.3.6 Simple dispersion models: the Lorentz and Drude models

The simplified dielectric constant versus wavelength dependence is shown in Fig. 3.16.

Many correlations between the dielectric constant and the wavelength of the radiation (dispersion relationships) have been proposed and applied for different materials and ranges of dispersion models [65]. A simple dispersion model in the electronic radiation range is the Lorentz oscillatory model [66]. This model assumes that electrons can only have radial oscillations because they are linked by a spring-like force (see Figure 3.17) to the nucleus in each atom and therefore subjected to a radial attractive electric field with a damping term, which is proportional to speed. Therefore, the motion equation is

$$
m_e \frac{d^2 r(t)}{dt^2} = -m_e \omega_0^2 r(t) + q_e E_r(t) - m_e \gamma \frac{dr(t)}{dt}
\tag{3.96}
$$

$$
\frac{d^2 r(t)}{dt^2} + \gamma \frac{dr(t)}{dt} + \omega_0^2 r(t) = \frac{q_e}{m_e} E_r(t)
\tag{3.97}
$$

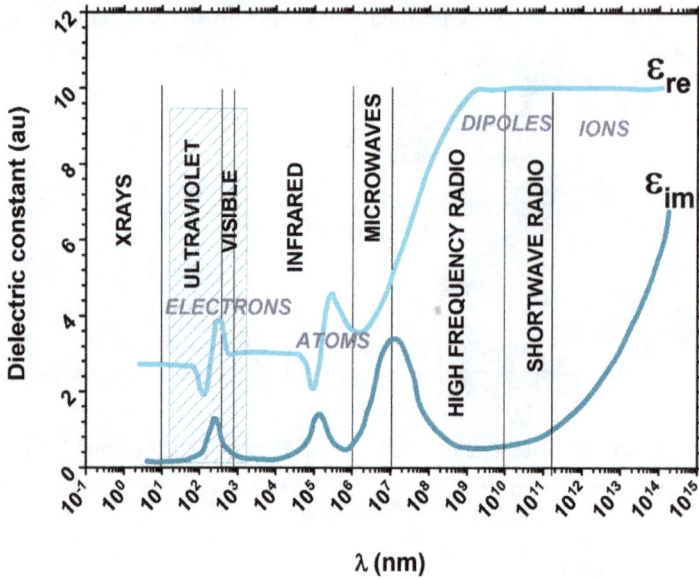

Fig. 3.16: Schematic dielectric constant for a wide range of wavelengths.

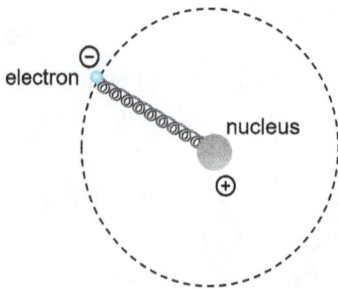

Fig. 3.17: Lorentz electronic oscillatory model.

We can assume that

$$\left. \begin{aligned} E_r(t) &= \mathrm{Re}\left(E_r e^{-i\omega t}\right) \\ r(t) &= \mathrm{Re}\left(r e^{-i\omega t}\right) \end{aligned} \right\} \tag{3.98}$$

Therefore,

$$-\omega^2 r - i\omega\gamma r + \omega_0^2 r = \frac{q_e}{m_e} E_r \tag{3.99}$$

And

$$r = \frac{q_e}{m_e}\left(\frac{1}{\omega_0^2 - \omega^2 - i\omega\gamma}\right)E_r \tag{3.100}$$

This expression for the position of the electron can be correlated with polarization by taking into account that

$$\vec{D} = \varepsilon_0\vec{E} + \vec{P} \tag{3.101}$$

With polarization comprising the sum of N dipolar moments:

$$P_r = Nq_e r \tag{3.102}$$

Then, eq. (3.97) would give

$$\left[\frac{d^2}{dt^2} + \gamma\frac{d}{dt} + \omega_0^2\right]P_r(t) = \frac{Nq_e^2}{m_e}E_r(t) \tag{3.103}$$

$$\left[\frac{d^2}{dt^2} + \gamma\frac{d}{dt} + \omega_0^2\right]P_r(t) = \varepsilon_0\omega_p^2 E_r(t) \tag{3.104}$$

With $\omega_p = \sqrt{Nq_e^2/\varepsilon_0 m_e}$. Now, using $P(t) = \mathrm{Re}(P \cdot e^{-i\omega t})$:

$$P_r = \left(\frac{\omega_p^2}{\omega_0^2 - \omega^2 - i\omega\gamma}\right)\varepsilon_0 E_r \tag{3.105}$$

and

$$D_r = \left[\varepsilon_0\left(1 + \frac{\omega_p^2}{\omega_0^2 - \omega^2 - i\omega\gamma}\right)\right]E_r \tag{3.106}$$

Or, because $\vec{D} = \varepsilon\varepsilon_0\vec{P}$:

$$\tilde{\varepsilon} = \left(1 + \frac{\omega_p^2}{\omega_0^2 - \omega^2 - i\omega\gamma}\right) \tag{3.107}$$

Finally, by using $\tilde{\varepsilon} = (\varepsilon_{re} + i\varepsilon_{im})$, eq. (3.107) is

$$\varepsilon_{re} = 1 + \frac{\omega_p^2(\omega_0^2 - \omega^2)}{(\omega_0^2 - \omega^2)^2 + \omega^2\gamma^2} \tag{3.109}$$

and

$$\varepsilon_{im} = \frac{\omega_p^2\omega\gamma}{(\omega_0^2 - \omega^2)^2 + \omega^2\gamma^2} \tag{3.110}$$

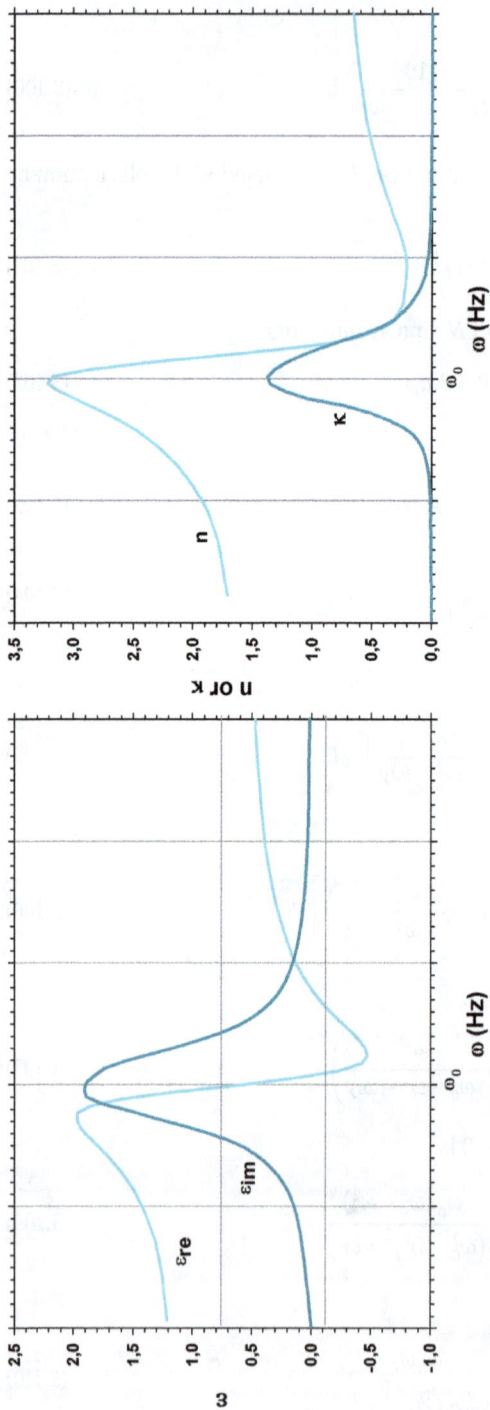

Fig. 3.18: Lorentz model for the dielectric constant (real and imaginary) and for the refraction index and the adsorption coefficient.

In Figure 3.18, the theoretical (according to Lorentz) dielectric constant and the refraction index along with the adsorption coefficient are shown. According to eq. (3.60), both n and κ could be evaluated. Note that

$$\left.\begin{aligned}
\lim_{\omega\to\infty} \tilde{\varepsilon} &= 1 \\[4pt]
\lim_{\omega\to 0} \tilde{\varepsilon} &= 1 + \frac{\omega_p^2}{\omega_0^2} \\[4pt]
\lim_{\omega\to\infty} \varepsilon_{im} &= 0 = \lim_{\omega\to 0} \varepsilon_{im}
\end{aligned}\right\} \tag{3.111}$$

In metals, the valence electrons are free to roam around the material rather than being fixed to their nuclei. Still, they behave similarly to the electrons in insulators, from which the Lorentz oscillator model applies. This is so in terms of how they respond to the electric field and undergo damping. We can simply make $\omega_0 = 0$ to indicate that there are no restoring forces. This could constitute the so-called Drude model [67], with

$$\tilde{\varepsilon} = \left(1 - \frac{\omega_p^2}{\omega^2 + i\omega\gamma}\right) \tag{3.112}$$

$$\varepsilon_{re} = 1 - \frac{\omega_p^2}{\omega^2 + \gamma^2} \tag{3.113}$$

and

$$\varepsilon_{im} = \frac{\omega_p^2 \gamma}{\omega\gamma^2 + \omega^3} \tag{3.114}$$

Note that now

$$\left.\begin{aligned}
\lim_{\omega\to 0} \tilde{\varepsilon}_{re} &= 1 - \frac{\omega_p^2}{\gamma^2} \\[4pt]
\lim_{\omega\to\infty} \tilde{\varepsilon}_{re} &= 1 \\[4pt]
\lim_{\omega\to 0} \varepsilon_{im} &= \infty \\[4pt]
\lim_{\omega\to\infty} \varepsilon_{im} &= 0
\end{aligned}\right\} \tag{3.115}$$

An empirical relationship between the refractive index and the wavelength of light for a many transparent materials known as Cauchy's transmission equation is both simple and useful. It was due to Augustin-Louis Cauchy, a mathematician who first described it in 1836. The Cauchy's model is, in fact, a purely empirical formulation that predates the Sellmeier equation. It is, however, conceivable to derive it from a binomial series expansion of a single Sellmeier term, written as

$$\left.\begin{array}{l} n = A + \dfrac{B}{\lambda^2} + \dfrac{C}{\lambda^4} \\[2mm] k = 0 \end{array}\right\}$$ (3.116)

3.3.7 Mixed phases

In his 1850 publication, the Italian physicist Ottaviano-Fabrizio Mossotti investigated the correlation between the dielectric constants of two different media. The explicit formula for the Clausius–Mossotti relation was later provided by the German physicist Rudolf Clausius in 1879, although it was based on refractive indices rather than dielectric constants [68–71]. This formula, known as Maxwell's formula, remains applicable in the context of conductivity as well. When applied within the realm of refractivity, it is referred to as the Lorentz–Lorenz equation. This equation is utilized when polarization remains unaffected by permanent electric dipole moments, either due to the presence of solely nonpolar molecules or when the frequency of the applied field is high.

The Lorentz–Lorenz equation, also known as the Clausius–Mossotti equation, is a fundamental relationship in physical chemistry and optics that describes the relationship between the refractive index of a substance and its molar refractivity. Proposed independently by physicists Hendrik Lorentz and Ludwig Lorenz in the late nineteenth century, the equation provides insight into the optical properties of materials, particularly in the context of electromagnetic wave propagation. The equation states that the difference between the refractive index of a substance and the refractive index of a vacuum is proportional to the molar concentration of the substance, its molar refractivity and a constant known as the polarizability of the substance.

Mathematically [72, 73], the Lorentz–Lorenz equation is expressed as

$$\frac{n^2 - 1}{n^2 + 1} = \frac{4}{3}\pi a N$$ (3.117)

Here n represents the refractive index, N stands for the number density of molecules and "a" denotes the polarizability of the substance.

The refractive index of a mixture of two-component phases can be written as

$$n^2 - 1 = \phi\left(n_1^2 - 1\right) + (1 - \phi)\left(n_2^2 - 1\right)$$ (3.118)

$$n^2 - 1 = \phi f_1\left(n_1^2 - 1\right) + (1 - \phi)f_2\left(n_2^2 - 1\right)$$ (3.119)

$$f_i = \frac{n_i^2 + 2}{3}$$ (3.120)

3.3.8 Ellipsometry and pore size distribution

Ellipsometry shows its potential in the characterization of composition, roughness, thickness, conductivity and other associated properties in thin films of various materials. Furthermore, for the purpose of this subject, the ellipsometric technique can also be used for the determination of porosity and PSD of porous materials. The ellipsometric porosimetry (ELLP) technique was proposed by Baklanov and co-workers in several articles published in the late 1990s, as well as in a 1998 patent application [74–77].

Basically, ELLP consists of using an ellipsometer to determine the adsorption of a suitable adsorbate (usually a gas) on the material to be analysed. As we will see in Section 5.1, the gas adsorption–desorption (GAD) technique, especially using nitrogen as adsorbate, is one of the most widely used techniques in the porosimetric analysis of membranes or other types of porous materials. The equipment used for this analysis uses various types of techniques to determine the amount of adsorbed gas: volumetric, chromatographic or gravimetric. What they all have in common is the difficulty of determining as accurately as possible the adsorbed volume as a function of vapour pressure.

In this context, the use of ellipsometry can mean a higher accuracy in the determination of the amount of adsorbate and, consequently, a possible improvement in the determination of the PSD.

Thus, in the case of ELLP, we will need to combine an ellipsometer with a vacuum chamber with appropriate pressure controls and a source of adsorbate vapour.

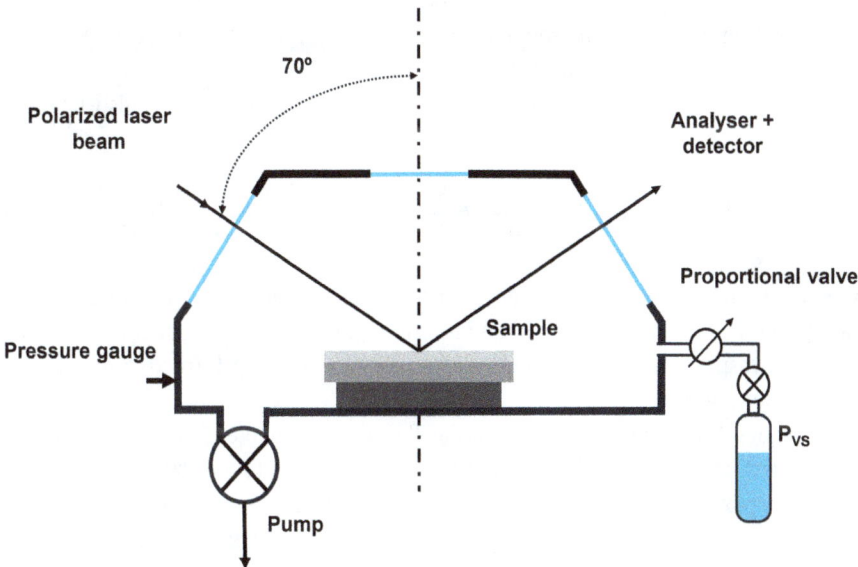

Fig. 3.19: Scheme of an ELLP device constructed by coupling an ellipsometer to a gas adsorption and pressure control system (adapted from [78]).

Figure 3.19 shows a suitable set-up for ELLP analysis, adapted from the work of Rouessac et al. [78].

Two quartz windows in the vacuum chamber allow the reflection of the laser beam on the sample to be analysed, while a third window at the top serves to ensure proper alignment of the laser beam.

The vacuum chamber in which the sample is placed must be properly sealed, since the complete determination of the adsorption isotherm will require several hours of measurement and any type of loss would falsify the results. For this reason, it is advisable to ensure an adequate vacuum of the chamber before starting the measurements and to check the stability of the vacuum for a sufficient time period.

By means of this or a similar device (generally mounted in the research laboratory itself), the optical properties of the adsorption layer deposited on the sample are determined by means of ellipsometry. These optical properties must be related to the composition and quantity of adsorbed material, for which the Lorentz–Lorenz equation (3.117) is used.

From this we can derive another relationship that allows us to determine the porosity, ρ, of the adsorbed layer, which is usually called the Clausius–Mossotti equation [79]:

$$\frac{n_p^2 - 1}{n_p^2 + 2} = (1 - \rho)\left(\frac{n_{np}^2 - 1}{n_{np}^2 + 2}\right) \tag{3.121}$$

where n_p and n_{np} are the refractive indices of the porous and non-porous layers, respectively, previously determined by ellipsometry.

As mentioned in previous sections, ellipsometry allows us to determine the refractive index, n, and the thickness of the material, d. Both values will change during the physisorption process on the material, so that we can determine the adsorbed volume as follows:

$$V_{ads} = (P - P_0)V_m\left(\frac{d_1 - d}{a_{ads}}\right) = \frac{3V_m(d_1 - d)}{4\pi a_{ads}}\left(\frac{n_1^2 - 1}{n_1^2 + 2} - \frac{n^2 - 1}{n^2 + 2}\right) \tag{3.122}$$

where V_m is the volume of liquid adsorbed in the pores and a_{ads} is the polarizability of an individual molecule. Finally, d and d_1 are the layer thicknesses before and after adsorption, respectively, while n and n_1 are the corresponding refractive indices before and after adsorption.

Subsequently, Kelvin's equation, as we will discuss in Chapter 5, allows us to relate the relative vapour pressure during the adsorption isotherm to the size of the pores in which adsorption takes place, in order to finally determine the relationship between adsorbed volume and pore size, which is the basis for the construction of the PSD.

The different ways of applying Kelvin's equation to obtain PSD from the adsorption isotherm data (adsorbed volume vs relative pressure) will be detailed in Chapter 5. It should be noted that the problem must be approached differently depending on whether the pores present in the sample are only mesopores (between 2 and

50 nm) or micropores (smaller than 2 nm). In the first case, Kelvin's equation is perfectly valid, and its application allows us to quickly obtain PSD from the adsorption isotherm (or desorption isotherm, since both isotherms allow us to obtain PSD with sufficient agreement between both results), whereas, in the case of the presence of micropores, Kelvin's equation loses validity. In this case, the use of methods such as the Dubinin–Radushkevich method (based on the Dubinin equation) or others such as the micropore method (MP method, based on the t-plot) will allow PSD to be obtained.

Finally, we should mention that this porosimetric method is no longer valid in the presence of macropores (pores larger than 50 nm) and, in this case, other porosimetric methods must be used.

Following Baklanov et al. [76], we can consider the following advantages of the use of ELLP, with respect to conventional nitrogen GAD:

a) Measurements are performed directly on the surface of the porous film, which must be previously deposited on a smooth solid substrate. This feature is very convenient for the analysis of different thin layers deposited on a support material, which is perfectly applicable to composite membranes. However, the method does not have such a clear application for asymmetric polymeric membranes obtained by the phase inversion method, where it is not easy to have such smooth support. Also, for symmetric membranes it will generally be difficult to have a sufficiently thin membrane thickness for ellipsometric analysis to access the entire membrane.

b) Really small surface areas can be analysed, due to the use of a particularly precise laser beam.

c) The use of organic solvents such as toluene, heptane and CCl_4 [75] allows to be carried out at room temperature without the requirements associated with the low temperatures involved in using nitrogen as an adsorbent.

Although the ELLP technique is basically a modification of GAD in terms of the acquisition of information leading to the determination of the adsorption isotherm, several studies have been using it in the porosimetric analysis of various materials.

Thus, Baklanov and Dultsev's groups have used this method for the analysis of porosity in thin films (silicon wafers, dielectric films, etc.) [75, 76, 79, 80], while others have used the technique for the analysis of thin layers of mesoporous anatase [78] or Nafion nanomembranes obtained for use in fuel cells [81].

3.3.9 Commercial equipment

The increasing use of ellipsometry on the analysis of thin-layer structure has led to a number of companies manufacturing more than 30 models of ellipsometers with different features. Among those companies, we can mention, for example:

– J.A. Woollam Co., Inc.
– Bruker Corporation

- Semilab Semiconductor Physics Laboratory Co. Ltd.
- Angstrom Sun Technologies Inc.
- Horiba, Ltd.
- Axometrics, Inc.
- Film Sense LLC
- Sentech Instruments GmbH

None of the ellipsometers marketed by these companies are designed solely for their use in ELLP. As mentioned earlier, ELLP needs coupling a conventional ellipsometer with an isolated chamber able to implement controlled pressure of adsorbing gas onto the sample, while avoiding any leakage during the operation. Normally, the researchers using ELLP for their studies decide to build their own set-ups.

References

[1]   Schrader D.M., Jean Y.C. Studies in Physical and Theoretical Chemistry 57: Positron and Positronium Chemistry. Elsevier, New York, (1988). ISBN 0-444-43009-1.
[2]   Pethrick R.A. Positron annihilation - A probe for nanoscale voids and free volume?. Prog Polym Sci, 22 (1997) 1–47. https://doi.org/10.1016/S0079-6700(96)00023-8.
[3]   Jean Y.C., Mallon P.E., Schrader D.M. Principles and Applications of Positron and Positronium Chemistry. World Scientific, London, UK, (2003). ISBN: 978-981-238-144-6.
[4]   Coleman P.G. Positron Beams and Their Applications. World Sci. Publ. Co., Singapore, (2000). ISBN 978-981-4496-38-4.
[5]   Gidley D.W., Peng H.G., Vallery R.S. Positron Annihilation as a method to Characterize Porous Materials. Annu Rev Mater Res, 36 (2006) 49–79. https://doi.org/10.1146/annurev.matsci.36.111904.135144.
[6]   Rowe B.W., Pas S.J., Hill A.J., Suzuki R., Freeman B.D., Paul D.R. A variable energy positron annihilation lifetime spectroscopy study of physical aging in thin glassy polymer films. Polymer, 50 (2009) 6149–6156. https://doi.org/10.1016/j.polymer.2009.10.045.
[7]   Charlton M., Humberston J.W. Positron Physics. Cambridge Univ. Press, Cambridge, U.K, (2001). ISBN 0 521 41550 0.
[8]   Mogensen O.E. Positronium formation in condensed matter and high-density gases. In: P.G. Coleman, S.C. Sharma, L.M. Diana (Eds.), Positron Annihilation. North Holland, Amsterdam, The Netherlands, (1982), ISBN 0 444 86534 9.
[9]   ORTEC–AMETEK "Experiment 27: Positron annihilation lifetime spectroscopy", Tech. Rep https://www.ortec-online.com/-/media/ametekortec/third-edition-experiments/27-positron-annihilation-lifetime-spectrometry.pdf?la=en&revision=f7ac0fd5-8ee1-4502-ae4f-a25ba8c56088&hash=C51B933CD0411388608EFC0318B89AEE
[10]   Shibuya K., Saito H., Nishikido F., Takahashi M., Yamaya T. Oxygen sensing ability of positronium atom for tumor hypoxia imaging. Comm. Phys, 3 (2020) 173. https://doi.org/10.1038/s42005-020-00440-z.
[11]   Liebow E. Electron–Positron Annihilation Lifetime Spectroscopy of MgO and Aluminum-Doped MgO. Union College – Schenectady, NY, (2022), Honor thesis
[12]   Yu Y. Positron Annihilation Lifetime Spectroscopy Studies of Amorphous and Crystalline Molecular Materials. Ph. D. Thesis, Martin-Luther-Universität Halle-Wittenberg, Germany, (2011).

[13] Gregory R.B. Free-volume and pore size distributions determined by numerical Laplace inversion of positron annihilation lifetime data. J Appl Phys, 70(9) (1991) 4665–4670. https://doi.org/10.1063/1.349057.

[14] Hoffmann L., Shukla A., Peter M., Barbiellini B., Manuel A.A. Linear and non-linear approaches to solve the inverse problem: Applications to positron annihilation experiments. Nuclear Instruments and Methods in Physics Research Section A: Accelerators, Spectrometers, Detectors and Associated Equipment, 335(1–2) (1993) 276–287. https://doi.org/10.1016/0168-9002(93)90282-M.

[15] Shukla A., Peter M., Hoffmann L. Analysis of positron lifetime spectra using quantified maximum entropy and a general linear filter. Nuclear Instruments and Methods in Physics Research Section A: Accelerators, Spectrometers, Detectors and Associated Equipment, 335(1–2) (1993) 310–317. https://doi.org/10.1016/0168-9002(93)90286-Q.

[16] Jean Y.C. Positron annihilation spectroscopy for chemical analysis: A novel probe for microstructural analysis of polymers. Microchem J, 42–1 (1990) 72–102. https://doi.org/10.1016/0026-265X(90)90027-3.

[17] Liao K.S., Chen H., Awad S., Yuan J.P., Hung W.S., Lee K.R., Lai J.Y., Hu C.C., Jean Y.C. Determination of free-volume properties in polymers without orthopositronium components in positron annihilation lifetime spectroscopy. Macromolecules, 44 (2011) 6818–6826. https://doi.org/10.1021/ma201324k.

[18] Tanzi-Marlotti G., Theory of positronium interactions with porous materials. PhD Thesis, Universita degli Studi di Milano, Italy, (2018).

[19] Tao S.J. Positronium annihilation in molecular substances. J Chem Phys, 56 (1972) 5499–5510. https://doi.org/10.1063/1.1677067.

[20] Eldrup M., Lightbody D., Sherwood J.N. The temperature dependence of positron lifetimes in solid pivalic acid. Chem Phys, 63 (1981) 51–58. https://doi.org/10.1016/0301-0104(81)80307-2.

[21] Wang Y.Y., Nakanishi H., Jean Y.C., Sandreczki T.C. Positron annihilation in amine-cured epoxy polymers – pressure dependence. J Polym Sci B Polym Phys, 28 (1990) 1431. https://doi.org/10.1002/polb.1990.090280902.

[22] Nakanishi H., Wang S.J., Jean Y.C. In Positron Annihilation Studies of Fluids. S.C. Sharma (Eds.), World Scientific, Singapore, (1988).

[23] Dutta D., Chatergee S., Pillai K.T., Pujari P.K., Ganguly B.N. Pore structure of silica gel: A comparative study through BET and PALS. Chem Phys, 312 (2005) 319–324. https://doi.org/10.1016/j.chemphys.2004.12.008.

[24] Ito K., Nakanishi H., Ujihira Y. Extension of the equation for the annihilation lifetime of ortho-positronium at a cavity larger than 1 nm in radius. J Phys Chem B, 103 (1999) 4555–4558. https://doi.org/10.1021/jp9831841.

[25] Palacio Gómez C.A. Some effects on polymers of low-energy implanted positrons. PhD Thesis, University of Ghent, Belgium, (2008).

[26] Gidley D.W., Frieze W.E., Dull T.L., Yee A.F., Ryan E.T., Ho H.M. Positronium annihilation in mesoporous thin films. Phys Rev B, 60(8) (1999) R5157. https://doi.org/10.1103/PhysRevB.60.R5157.

[27] Dull T.L., Frieze W.E., Gidley D.W., Sun J.N., Yee A.F. Determination of pore size in mesoporous thin films from the annihilation lifetime of positronium. J Phys Chem B, 105(20) (2001) 4657–4662. https://doi.org/10.1021/jp004182v.

[28] Goworek T., Ciesielski K., Jasińska B., Wawryszczuk J. Positronium in large voids silicagel. Chem Phys Let, 272 (1997) 91–95. https://doi.org/10.1016/S0009-2614(97)00504-6.

[29] Goworek T., Ciesielski K., Jasińska B., Wawryszczuk J. Lifetimes of o-Ps in the pores of silica gel. In: Positron annihilation, ICPA-11: Proceedings of the 11th International Conference on Positron Annihilation, Kansas City, Missouri, USA, (May 1997). ISBN: 9780878497799.

[30] Goworek T., Ciesielski K., Jasińska B., Wawryszczuk J. Positronium states in the pores of silica gel. Chem Phys, 230 (1998) 305–315. https://doi.org/10.1016/S0301-0104(98)00068-8.

[31] Goworek T., Ciesielski K., Jasińska B., Wawryszczuk J. Mesopore characterization by PALS. Radiat Phys Chem, 68 (2003) 331–337. https://doi.org/10.1016/S0969-806X(03)00180-4.

[32] Thanh N.D., Dung T.Q., Tuven L.A., Tuan K.T. Semi-empirical formula for large pore-size estimation from the o-Ps annihilation lifetime. Int J Nucl Energ Sci & Tech, 4(2) (2008) 81–87. https://doi.org/10.1504/IJNEST.2008.020528.

[33] Wada K., Hyodo T. A simple shape-free model for pore-size estimation with positron annihilation lifetime spectroscopy. J Phys: Conf Ser, 443 (2013) 012003. http://dx.doi.org/10.1088/1742-6596/443/1/012003.

[34] Hill A.J., Freeman B.D., Jaffe M., Merkel T.C., Pinnau I. Tailoring nanospace. J Mol Struct, 739 (2005) 173–178. https://doi.org/10.1016/j.molstruc.2004.05.041.

[35] Tung K.L., Chang K.S., Wu T.T., Lin N.J., Lee K.R., Lai J.Y. Recent advances in the characterization of membrane morphology. Curr Opin Chem Eng, 4 (2014) 121–12. http://dx.doi.org/10.1016/j.coche.2014.03.002.

[36] Eijt S.W.H., van Veen T., Schut H., Mijnarends P.E., Denison A.B., Barbiellini B., Bansil A. Study of colloidal quantum-dot surfaces using an innovative thin-film positron 2D-ACAR method. Nat Mat, 5 (2006) 23–26. https://doi.org/10.1038/nmat1550.

[37] Wu Z., de Krom T., Colombi G., Chaykina D., van Hattem G., Schut H., Dickmann M., Egger W., Hugenschmidt C., Brück E., Dam B., Eijt S.W.H. Formation of vacancies and metallic-like domains in photochromic rare-earth oxyhydride thin films studied by in-situ illumination positron annihilation spectroscopy. Phys Rev Mater, 6 (2022) 065201. https://doi.org/10.1103/PhysRevMaterials.6.065201.

[38] Sodaye H.S., Pujari P.K., Goswami A., Manohar S.B. Probing the microstructure of Nafion-117 using positron annihilation spectroscopy. J Polym Sci Part B: Polym Phys, 35(5) (1997) 771–776. https://doi.org/doi:10.1002/(sici)1099-0488(19970415)35:5<771::aid-polb5>3.0.co;2-p.

[39] Mohamed H.F.M., Kobayashi Y., Kuroda S., Ohira A. Positron trapping and possible presence of SO$_3$H clusters in dry fluorinated polymer electrolyte membranes. Chem Phys Lett, 544 (2012) 49–52. https://doi.org/10.1016/j.cplett.2012.06.060.

[40] Yampolski Y., Pinnau I., Freeman B. Materials Science of Membranes for Gas and Vapor Separation. John Wiley & Sons, Chichester, UK, (2006). ISBN:9780470853450.

[41] Bernardo P., Drioli E., Golemme G. Membrane Gas Separation: A Review/State of the Art. Ind Eng Chem Res, 48 (2009) 4638–4663. https://doi.org/10.1021/ie8019032.

[42] Freeman B., Yampolski Y. Membrane Gas Separation. John Wiley & Sons, Chichester, UK, (2010). ISBN: 978-0-470-74621-9.

[43] Jean Y.C., van Horn J.D., Hung W.S., Lee K.R. Perspective of Positron Annihilation Spectroscopy in Polymers. Macromolecules, 46(18) (2013) 7133–7145. https://doi.org/10.1021/ma401309x.

[44] Jordan S.S., Koros W.J. A Free Volume Distribution Model of Gas Sorption and Dilation in Glassy Polymers. Macromolecules, 28(7) (1995) 2228–2235. https://doi.org/10.1021/ma00111a017.

[45] Hill A.J., Weinhold S., Stack G.M., Tant M.R. Effect of copolymer composition on free volume and gas permeability in poly(ethylene terephthalate)–poly(1,4 cyclohexylenedimethylene terephthalate) copolyesters. Eur Polym J, 32 (1996) 843–849. https://doi.org/10.1016/0014-3057(95)00204-9.

[46] Tanis-Kanbur M.B., Peinador R.I., Calvo J.I., Hernández A., Wei Chew J. Porosimetric membrane characterization techniques: A review. J Membrane Sci, 619 (2021) 118750. https://doi.org/10.1016/j.memsci.2020.118750.

[47] Adams F.C. X-ray absorption and diffraction – overview. In: P. Worsfold, C. Poole, A. Townshend, M. Miró (Eds.), Encyclopedia of Analytical Science, 3rd ed., Academic Press, Oxford, UK, (2019) 391–403.

[48] Streli C., Wobrauschek P., Kregsamer P. X-ray fluorescence spectroscopy, applications. In: J.C. Lindon (Eds.), Encyclopedia of Spectroscopy and Spectrometry. Elsevier, Oxford, UK, (1999) 2478–2487.

[49] Lassinantti-Gualtieri M., Andersson C., Jareman F., Hedlund J., Gualtieri A.F., Leoni M., Meneghini C. Crack formation in α-alumina supported MFI zeolite membranes studied by in situ high

temperature synchrotron powder diffraction. J Membrane Sci, 290(1) (2007) 95–104. https://doi.org/10.1016/j.memsci.2006.12.018.

[50]   Remigy J.-.C., Meireles M. Assessment of pore geometry and 3-D architecture of filtration membranes by synchrotron radiation computed microtomography. Desalination, 199(1) (2006) 501–503. https://doi.org/10.1016/j.desal.2006.03.193.

[51]   Remigy J.C., Meireles M., Thibault X. Morphological characterization of a polymeric microfiltration membrane by synchrotron radiation computed microtomography. J Membrane Sci, 305 (2007) 27–35. https://10.1016/j.memsci.2007.06.059.

[52]   Baruchel J., Buffière J.-.Y., Maire E., Merle P., Peix G. X-ray Tomography in Material Science. Hermes Science Publications, Paris, France, (2000).

[53]   Lee S.-.H., Chang W.-.S., Han S.-.M., Kim D.-.H., Kim J.-.K. Synchrotron X-ray nanotomography and three-dimensional nanoscale imaging analysis of pore structure–function in nanoporous polymeric membranes. J Membrane Sci, 535 (2017) 28–34. https://doi.org/10.1016/j.memsci.2017.04.024.

[54]   Dutt S., Apel P., Lizunov N., Notthoff C., Wen Q., Trautmann C., Mota-Santiago P., Kirby N., Kluth P. Shape of nanopores in track-etched polycarbonate membranes. J Membrane Sci, 638 (2021) 119681. https://doi.org/10.1016/j.memsci.2021.119681.

[55]   Doudiès F., Loginov M., Hengl N., Karrouch M., Leconte N., Garnier-Lambrouin F., Pérez J., Pignon F., Gésan-Guiziou G. Build-up and relaxation of membrane fouling deposits produced during crossflow ultrafiltration of casein micelle dispersions at 12 °C and 42 °C probed by in situ SAXS. J Membrane Sci, 618 (2021) 118700. https://doi.org/10.1016/j.memsci.2020.118700.

[56]   Guo J-C., Zou Ch., Chiang Ch-Y., Chang T-A., Chen J-J., Lin L-Ch., Kang D-Y. NaP1 zeolite membranes with high selectivity for water–alcohol pervaporation. J Membrane Sci, 639 (2021) 119762. https://doi.org/10.1016/j.memsci.2021.119762.

[57]   Rosow U. A Brief History of Ellipsometry. Phys Status Solidi B, 256(2) (2018) 1800307. https://doi.org/10.1002/pssb.201800307.

[58]   Stenzel O., Ohlídal M. Optical Characterization of Thin Solid Films: 64 (Springer Series in Surface Sciences). Springer, Berlin, Germany (2018). ISBN: 9783319753249.

[59]   Wee A.T.S., Yin X., Tang C.S. Introduction to Spectroscopic Ellipsometry of Thin Film Materials: Instrumentation, Data Analysis, and Applications. Wiley-VCH, Weinheim, Germany (2022). ISBN: 978-3-527-34951-7.

[60]   Kliger D.S., Lewis J.W. Polarized Light in Optics and Spectroscopy. Elsevier, Amsterdam, The Netherlands, (1990). ISBN: 9780080571041. https://doi.org/10.1016/C2009-0-22282-3.

[61]   Singh J. Optical Properties of Materials and Their Applications. John Wiley & Sons, Hoboken, New Jersey, USA, (2020). ISBN 978-1119506317.

[62]   Tompkins H.G., Irene E.A. Handbook of Ellipsometry, W. Andrew Pub, Norwich, New York, USA, (2005). ISBN: 978-081551747-4.

[63]   Tompkins H.G. A User's Guide to Ellipsometry. Dover Pub, Mineola, New York, USA, (2013). ISBN 9781306366076.

[64]   Wahaia F. Ellipsometry – Principles and Techniques for Materials Characterization. IntechOpen, London, UK, (2017). ISBN: 978-953-51-3624-8. https://doi.org/10.5772/65558.

[65]   Hilfiker J.N., Tiwald T. Dielectric Function Modeling. In: H. Fujiwara, R. Collins (Eds.), Spectroscopic Ellipsometry for Photovoltaics. Springer Series in Optical Sciences, Vol. 212. Springer, Cham, Switzerland, (2018), https://doi.org/10.1007/978-3-319-75377-5_5.

[66]   Moliton A. Basic Electromagnetism and Materials. Springer Science, New York, USA, (2006). ISBN978-0-387-30284-3.

[67]   Drude P. The Theory of Optics. Longmans Green and Co., New York, (1902). Reprinted by Kessinger Publishing (June 2, 2008), Whitefish, Montana, USA. ISBN: 978-0548998816.

[68]   Oughstun K.E., Natalie A., Cartwright N.A. On the Lorentz-Lorenz formula and the Lorentz model of dielectric dispersion. Opt Express, 11 (2003) 1541–1546. https://doi.org/10.1364/OE.11.001541.

[69]   Silaghi M.A. Dielectric Material. InTech, (2012). https://doi.org/10.5772/2781.

[70]   Kragh H. The Lorenz–Lorentz Formula: Origin and Early History. Substantia, 2(2) (2018) 7–18. https://doi.org/10.13128/substantia-56.

[71]   Mayerhöfer T.G., Popp J. Beyond Beer's Law: Revisiting the Lorentz-Lorenz Equation. Chem Phys Chem, 21 (2020) 1218–1223. https://doi.org/10.1002/cphc.202000301.

[72]   Born M., Wolf E., Bhatia A.B. Principles of Optics: Electromagnetic Theory of Propagation, Interference and Diffraction of Light. Cambridge University Press, Cambridge, U.K, (2000). ISBN: 978-0521642224.

[73]   Losurdo M., Hingerl K. Ellipsometry at the Nanoscale. Springer, Heidelberg, Germany, (2013). ISBN 978-3-642-33955-4.

[74]   Baklanov M.R., Vasilyeva L.L., Gavrilova T.A., Dultsev F.N., Mogilnikov K.P., Nenasheva L.A. Porous structure of SiO2 films synthesized at low temperature and pressure. Thin Solid Films, 171(1) (1989) 43–52.

[75]   Mogilnikov K.P., Polovinkin V.G., Dultsev F.N., Baklanov M.R. Calculation of Pore Size Distribution in the Ellipsometric Porosimetry: Method and Reliability. MRS Proceedings. 565 (1999) 81. https://doi.org/10.1557/PROC-565-8

[76]   Baklanov M.R., Mogilnikov K.P., Polovinkin V.G., Dultsev F.N. Determination of pore size distribution in thin films by ellipsometric porosimetry. J Vac Sci Technol, B18c (2000) 1385–1391.

[77]   Baklanov M.R., Dultsev F.N., Mogil'nikov K.P., Maex K., US patent Atty. Ref. No. 98, (1998), 540.

[78]   Rouessac V., Coustel R., Bosc F., Durand J., Ayral A. Characterisation of mesostructured TiO2 thin layers by ellipsometric porosimetry. Thin Solid Films, 495 (2006) 232–236. https://doi.org/10.1016/j.tsf.2005.08.334.

[79]   Dultsev F.N., Solowjev A.P. Synthesis and ellipsometric characterization of insulating low permittivity SiO2 layers by remote-PECVD using radio-frequency glow discharge. Thin Solid Films, 419 (2002) 27–32. https://doi.org/10.1016/S0040-6090(02)00760-5.

[80]   Dultsev F.N., Baklanov M.R. Nondestructive Determination of Pore Size Distribution in Thin Films Deposited on Solid Substrates. Electrochem Solid State Lett, 2 (1999) 192. https://doi.org/10.1149/1.1390780.

[81]   Abuin G.C., Fuertes M.C., Corti H.R. Substrate effect on the swelling and wáter sorption of Nafion nanomembranes. J Membr Sci, 428 (2013) 507–515. http://dx.doi.org/10.1016/j.memsci.2012.10.060.

Chapter 4
Thermodynamic basis of indirect methods

4.1 Introduction

Among the indirect porosimetric methods that will be discussed in the following chapters, each has its own basis or fundamental equations for relating various measurable properties to the pore size of a sample. The fact that pores and pore size (in the methods presented here) are not directly measurable, but must be correlated with other measurable properties, is what makes them indirect characterization methods.

However, the scientific fundamentals on which these techniques are based differ between them, although most of the methods relate to the behaviour of interfaces between fluids, depending on whether the interface is flat or curved (as occurs inside the pores). Thus, we can group these indirect porosimetric techniques whether they are based on three different equations: Young–Laplace, Kelvin and Gibbs–Thomson:

- The Young–Laplace equation accounts for the relationship between the curvature of an interface, the surface tension between the different fluid phases in contact and, finally, the pressure difference between the two fluid phases. Since the curvature of the interface will be related to the size of the pores in which the interface is formed, the Young–Laplace equation will allow us to determine this size from the existing pressures. Among the porosimetric techniques that will find their basis in the Young–Laplace equation are liquid displacement porosimetry (both gas–liquid displacement porosimetry and liquid–liquid displacement porosimetry), liquid intrusion/extrusion porosimetry or mercury intrusion porosimetry (HgP).
- The Kelvin equation, on the other hand, describes the variation that occurs in vapour pressure as a consequence of the existence of a curved liquid–vapour interface, such as the surface of a droplet. Again, the behaviour of an experimental phenomenon, in this case, the vapour pressure of a given fluid, is different based on whether it is a convex curved surface or a flat surface. The porosimetric techniques based on the Kelvin equation are: the recently developed evapoporometry, the permporometry and the gas adsorption–desorption technique. Also, ELLP could be included in this group, when the Kelvin equation is used to interpret the adsorption isotherms obtained from ellipsometry.
- Finally, we must consider the Gibbs–Thomson equation, which, in its most general form, relates the change in total free energy caused by the curvature of an interfacial surface under stress. The Gibbs–Thomson equation like the Kelvin equation is a particular case of the more general Gibbs equation. Thus, the Kelvin equation applies when the process is at constant temperature, while the Gibbs–Thomson equation refers to situations at constant pressure. In any case, the Gibbs–Thomson equation will be the basis of several techniques related to the freezing point of flu-

https://doi.org/10.1515/9783110792195-004

ids inside pores. There are several ways to do this: one possibility is to determine the heat fluxes associated with these phase transitions, using a differential scanning calorimeter , which gives rise to the technique usually known as thermoporometry. We can also determine the amount of liquid remaining during the phase change, either by means of nuclear magnetic resonance(NMR, thus we will have the NMR cryoporometry (NCMR)) or by measuring the amplitude of neutron scattering in the crystalline or liquid phases embedded in the sample (in this latter case, we will speak of ND cryoporometry). In this section we will discuss the thermoporometric technique, which is older in time and has been more widely used in the characterization of membranes, leaving the relatively new NCMR for the next section.

Here we will obtain the thermodynamic relationships concerning to the solid–gas–liquid interfaces inside pores. These equations can be studied in common Thermodynamics textbooks [1—4] or more specific publications [5—8]. Here we will follow the summary that can be encountered in the works of Landry [9].

In the following sections, we will discuss the origin, basis, assumptions and implications of these three equations, so that when we further analyse the porosimetric techniques to which they give rise to, we can easily understand the underlying physical basis.

4.2 Pressure at an interface: the Young–Laplace equation

Suppose that a gas–liquid interface has only volume expansion and surface extension interchanges of energy according to the Gibbs equation at constant temperature, for the equilibrium in an interface between gas and liquid phases we have

$$(p_g - p_\ell)dV_g = \gamma_{g\ell}d\Sigma_{g\ell} \tag{4.1}$$

where p_g and p_ℓ are the pressures in the gas and liquid phases, respectively, dV_g is the variation of volume of the gas, $\gamma_{g\ell}$ is the surface tension between gas and liquid and $d\Sigma_{g\ell}$ is the extension of the interface.

The overpressure on the interface (p_g-p_ℓ) can be termed as p_p and thus eq. (4.1) becomes

$$p_p dV_g = \gamma_{g\ell}d\Sigma_{g\ell} \tag{4.2}$$

or

$$p_g - p_\ell = \gamma_{g\ell}\frac{d\Sigma_{g\ell}}{dV_g} \tag{4.3}$$

But in accordance to Fig. 4.1:

$$\Sigma_{g\ell} = \gamma_{g\ell} r_1 \beta r_2 \tag{4.4}$$

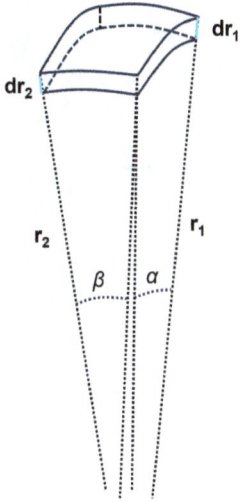

Fig. 4.1: Principal curvature radii and expansion.

Then:

$$\Sigma_{g\ell} + d\Sigma_{g\ell} = \alpha\beta(r_1 + dr_1)(r_2 + dr_2) \tag{4.5}$$

Of course, both α and β are small:

$$dr_1 \approx dr_2 \equiv dr \tag{4.6}$$

Therefore,

$$d\Sigma_{g\ell} = \alpha\beta(r_1 + dr)(r_2 + dr) - \alpha\beta r_1 r_2 = \alpha\beta(r_1 + r_2)dr \tag{4.7}$$

$$dV_g = \Sigma_{g\ell} dr = \alpha\beta r_1 r_2 dr \tag{4.8}$$

Then eq. (4.3) is

$$p_g - p_\ell = \gamma_{g\ell} \frac{d\Sigma_{g\ell}}{dV_g} = \gamma_{g\ell} \frac{\alpha\beta(r_1 + r_2)dr}{\alpha\beta r_1 r_2 dr} = \gamma_{g\ell} \frac{r_1 + r_2}{r_1 r_2} = \gamma_{g\ell}\left(\frac{1}{r_2} + \frac{1}{r_1}\right) \tag{4.9}$$

For a hemispherical interface $r_1 = r_2 \equiv r$, then

$$p_g - p_\ell = \frac{2\gamma_{g\ell}}{r} \tag{4.10}$$

But, according to Fig. 4.2:

$$r_p = r \cos \theta_{g\ell} \tag{4.11}$$

$$p_g - p_\ell = \frac{2\gamma_{g\ell}}{r_p} \cos \theta_{g\ell} \tag{4.12}$$

This equation is known as the Young–Laplace equation.

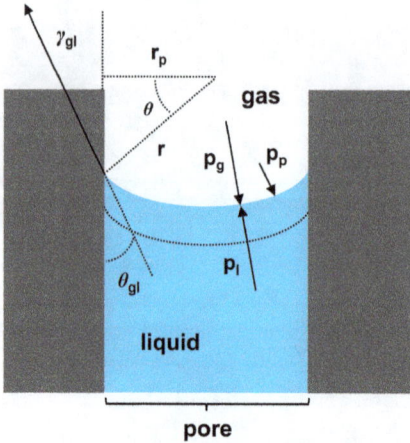

Fig. 4.2: Gas–liquid interface within a pore.

In the case of a solid–liquid interface, the same Fig. 4.2 would be accurate but only the gas phase should be substituted by a solid phase and the interfacial tension would be $\gamma_{s\ell}$. The contact angle θ would be different from that of a gas–liquid interface as well. Then the Laplace equation would be

$$p_s - p_\ell = \frac{2\gamma_{s\ell}}{r_p} \cos \theta_{s\ell} \tag{4.13}$$

4.3 Pressure at a gas–liquid interface: the Kelvin equation

The fundamental equation of thermodynamics for both the liquid and the gas is

$$dG = dU + pdV - TdS - \mu dn \tag{4.14}$$

But the Gibbs thermodynamic potential is

$$\left. \begin{array}{l} G = U + pV - TS - \mu n \\ dG = dU + pdV + Vdp - TdS - SdT - \mu dn - nd\mu \end{array} \right\} \tag{4.15}$$

When eq. (4.15) is compared with eq. (4.14), we can conclude that

$$0 = Vdp - SdT - nd\mu \tag{4.16}$$

which is the Gibbs–Duhem equation. According to this equation for the equilibrium in isothermal conditions with interchange of mass through a constant area surface:

$$\left.\begin{array}{l} -V_\ell dp_\ell + n_\ell d\mu_\ell = 0 \\ -V_g dp_g + n_g d\mu_g = 0 \end{array}\right\} \tag{4.17}$$

where $V_{g,\ell}$, $p_{g,\ell}$, $n_{g,\ell}$ and $\mu_{g,\ell}$ are the volume, pressure, number of moles and chemical potential in the gas/liquid phases, respectively.

As there is a material equilibrium between gas and liquid:

$$d\mu_\ell = d\mu_g \tag{4.18}$$

This is equivalent to

$$\frac{V_\ell}{n_\ell} dp_\ell = \frac{V_g}{n_g} dp_g \tag{4.19}$$

$$dp_\ell = \frac{V_g n_\ell}{V_\ell n_g} dp_g \tag{4.20}$$

Or, in terms of molar volumes:

$$dp_\ell = \frac{v_g}{v_\ell} dp_g \tag{4.21}$$

Now by using eqs. (4.10) and (4.21):

$$dp_g \left(1 - \frac{v_g}{v_\ell}\right) = d\left(\frac{2\gamma_{g\ell}}{r}\right) \tag{4.22}$$

If the gas phase behaves ideally ($v_g = RT/p_g$) and $v_\ell \ll v_g \Leftrightarrow v_g/v_\ell \gg 1$:

$$d\left(\frac{2\gamma_{g\ell}}{r}\right) = \left(-\frac{v_g}{v_\ell}\right) dp_g = \left(-\frac{RT}{v_\ell}\right) \frac{dp_g}{p_g} \tag{4.23}$$

$$\left(-\frac{2\gamma_{g\ell} v_\ell}{RT}\right) d\left(\frac{1}{r}\right) = \frac{dp_g}{p_g} \tag{4.24}$$

$$-\frac{2\gamma_{g\ell} v_\ell}{rRT} = \ln\frac{p_g}{p_g^0} \tag{4.25}$$

where p_g^0 is the pressure of the gas in the bulk phase: $p_g^0 = \lim_{r \to \infty} p_g$. Equation (4.25) is the Kelvin equation, which describes the change in vapour pressure due to a curved liquid–vapour interface. Finally, if we introduce the contact angle, we obtain

$$-\frac{2\gamma_{g\ell}v_{\ell}}{r_{p}RT}\cos\theta_{g\ell} = \ln\frac{p_{g}}{p_{g}^{0}} \tag{4.26}$$

4.4 Melting temperature in a pore: the Gibbs–Thompson equation

In this case, we will also quit from the Gibbs–Duhem equation for non-isothermal processes, but considering three phases such as the gas and liquid phases in contact and also the solid phase corresponding to the pore in which the gas and liquid are contained. Then the Gibbs–Duhem equations are

$$\left.\begin{array}{l} S_{s}dT - V_{s}dp_{s} + n_{s}d\mu_{s} = 0 \\ S_{\ell}dT - V_{\ell}dp_{\ell} + n_{\ell}d\mu_{\ell} = 0 \\ S_{g}dT - V_{g}dp_{g} + n_{g}d\mu_{g} = 0 \end{array}\right\} \tag{4.27}$$

where $S_{s,\ell,g}$ is the entropy of each phase and T is the temperature. By introducing molar entropy and volumes, we obtain

$$\left.\begin{array}{l} -s_{s}dT + v_{s}dp_{s} = d\mu_{s} \\ -s_{\ell}dT + v_{\ell}dp_{\ell} = d\mu_{\ell} \\ -s_{g}dT + v_{g}dp_{g} = d\mu_{g} \end{array}\right\} \tag{4.28}$$

Subtracting the Gibbs–Duhem equation for the liquid and the solid from that of the gas, we obtain

$$\left.\begin{array}{l} (s_{\ell} - s_{g})dT + v_{g}dp_{g} - v_{\ell}dp_{\ell} = d\mu_{g} - d\mu_{\ell} = 0 \\ (s_{s} - s_{g})dT + v_{g}dp_{g} - v_{s}dp_{s} = d\mu_{g} - d\mu_{s} = 0 \end{array}\right\} \tag{4.29}$$

Or

$$\left.\begin{array}{l} \left(\dfrac{s_{g} - s_{\ell}}{v_{g} - v_{\ell}}\right)dT = \dfrac{v_{g}}{v_{g} - v_{\ell}}dp_{g} - \dfrac{v_{\ell}}{v_{g} - v_{\ell}}dp_{\ell} \\ \left(\dfrac{s_{g} - s_{s}}{v_{g} - v_{s}}\right)dT = \dfrac{v_{g}}{v_{g} - v_{s}}dp_{g} - \dfrac{v_{s}}{v_{g} - v_{s}}dp_{s} \end{array}\right\} \tag{4.30}$$

And subtracting these two equations:

$$\begin{aligned} \left(\frac{s_{g} - s_{s}}{v_{g} - v_{s}} - \frac{s_{g} - s_{\ell}}{v_{g} - v_{\ell}}\right)dT = {} & \frac{v_{g}}{v_{g} - v_{s}}dp_{g} - \frac{v_{s}}{v_{g} - v_{s}}dp_{s} - \\ & - \frac{v_{g}}{v_{g} - v_{\ell}}dp_{g} + \frac{v_{\ell}}{v_{g} - v_{\ell}}dp_{\ell} \end{aligned} \tag{4.31}$$

Then we can write, as done for eq. (4.3):

$$p_\ell - p_g = \gamma_{\ell g} \frac{d\Sigma_{\ell g}}{dV_\ell} \tag{4.32}$$

$$p_g - p_s = \gamma_{g\ell} \frac{d\Sigma_{gs}}{dV_g} \tag{4.33}$$

$$p_s - p_\ell = \gamma_{s\ell} \frac{d\Sigma_{s\ell}}{dV_s} \tag{4.34}$$

These equations are valid provided that the changes in volume and interfacial surface do not involve changes in the temperature. Of course, these interfaces are not independent as far as two of them are known and the other is totally determined. We can choose gas–liquid and gas–solid interfaces, for example, represented by eqs. (4.32) and (4.33), which after differentiation give

$$dp_\ell - dp_g = d\left(\gamma_{\ell g} \frac{d\Sigma_{\ell g}}{dV_\ell} \right) \tag{4.35}$$

$$dp_g - dp_s = d\left(\gamma_{gs} \frac{d\Sigma_{gs}}{dV_g} \right) \tag{4.36}$$

Now eq. (4.31) can be reordered as

$$\left[\left(\frac{S_g - S_s}{V_g - V_s} \right) - \left(\frac{S_g - S_\ell}{V_g - V_\ell} \right) \right] dT =$$
$$= \left(\frac{V_g}{V_g - V_s} - \frac{V_g}{V_g - V_\ell} \right) dp_g + \frac{V_\ell}{V_g - V_\ell} dp_\ell - \frac{V_s}{V_g - V_s} dp_s \tag{4.37}$$

But

$$\frac{V_g}{V_g - V_s} - \frac{V_g}{V_g - V_\ell} = \frac{\cancel{V_g^2} - V_g V_\ell - \cancel{V_g^2} + V_g V_s}{(V_g - V_s)(V_g - V_\ell)} =$$
$$= \frac{V_s V_g - \cancel{V_s V_\ell} - V_\ell V_g + \cancel{V_s V_\ell}}{(V_g - V_s)(V_g - V_\ell)} = \frac{V_s}{V_g - V_s} - \frac{V_\ell}{V_g - V_\ell} \tag{4.38}$$

Then eq. (4.37) transforms to

$$\left[\left(\frac{S_g - S_s}{V_g - V_s} \right) - \left(\frac{S_g - S_\ell}{V_g - V_\ell} \right) \right] dT =$$
$$= \left(\frac{V_s}{V_g - V_s} - \frac{V_\ell}{V_g - V_\ell} \right) dp_g + \frac{V_\ell}{V_g - V_\ell} dp_\ell - \frac{V_s}{V_g - V_s} dp_s \tag{4.39}$$

Or according to eqs. (4.35) and (4.36):

$$\left[\left(\frac{s_g - s_s}{v_g - v_s}\right) - \left(\frac{s_g - s_\ell}{v_g - v_\ell}\right)\right] dT =$$

$$= \frac{v_s}{v_g - v_s}(dp_g - dp_s) + \frac{v_\ell}{v_g - v_\ell}(dp_\ell - dp_g) = \tag{4.40}$$

$$= \frac{v_s}{v_g - v_s} d\left(\gamma_{g\ell}\frac{d\Sigma_{gs}}{dV_g}\right) + \frac{v_\ell}{v_g - v_\ell} d\left(\gamma_{\ell g}\frac{d\Sigma_{\ell g}}{dV_\ell}\right)$$

If we assume that $v_g \gg v_s$ and $v_g \gg v_\ell$:

$$\left[\left(\frac{s_g - s_s}{v_g}\right) - \left(\frac{s_g - s_\ell}{v_g}\right)\right] dT = \frac{v_s}{v_g} d\left(\gamma_{g\ell}\frac{d\Sigma_{gs}}{dV_g}\right) + \frac{v_\ell}{v_g} d\left(\gamma_{\ell g}\frac{d\Sigma_{\ell g}}{dV_\ell}\right) \tag{4.41}$$

$$(s_\ell - s_s)dT = v_s d\left(\gamma_{gs}\frac{d\Sigma_{gs}}{dV_g}\right) + v_\ell d\left(\gamma_{\ell g}\frac{d\Sigma_{\ell g}}{dV_\ell}\right) \tag{4.42}$$

Moreover, if the pores are assumed to be cylinders with hemispherical interfaces between phases α and β with radii $r_{\alpha\beta}$, then according to eq. (4.10):

$$\frac{d\Sigma_{\alpha\beta}}{dV_\alpha} = \pm \frac{2}{r_{\alpha\beta}} \tag{4.43}$$

With a "+" sign, if the interface is concave to phase β, and a "–" sign if the interface is convex to phase β

$$(s_\ell - s_s)dT = 2\left[\pm v_s d\left(\frac{\gamma_{gs}}{r_{gs}}\right) \pm v_\ell d\left(\frac{\gamma_{\ell g}}{r_{\ell g}}\right)\right] \tag{4.44}$$

Because the molar enthalpy of melting is $(s_\ell - s_s)\,T = \Delta h_f$, which comes from the Legendre transform to get the Gibbs potential from enthalpy equation $\Delta g_f = \Delta h_f - T\,\Delta s_f$ with $\Delta g_f = 0$, we obtain

$$\frac{dT}{T} = \frac{2}{\Delta h_f}\left[\pm v_s d\left(\frac{\gamma_{gs}}{r_{gs}}\right) \pm v_\ell d\left(\frac{\gamma_{\ell g}}{r_{\ell g}}\right)\right] \tag{4.45}$$

After integration, this gives

$$\ln\frac{T}{T^0} = \frac{2}{\Delta h_f}\left[\pm v_s \frac{\gamma_{gs}}{r_{gs}} \pm v_\ell \frac{\gamma_{\ell g}}{r_{\ell g}}\right] \tag{4.46}$$

Here $T_0 = \lim_{r_{gs}\to\infty} T = \lim_{r_{g\ell}\to\infty} T$ is the melting temperature in the bulk. In each term within square brackets, we should use a "+" sign if the interface is concave to phase s (first term) or g (second term), and a "–" sign otherwise.

In the case of a porous material saturated with the liquid phase, it is convenient to pay attention to solid–liquid and gas–solid interfaces. To that end, we should substitute eq (4.34) by

$$\left[\left(\frac{S_s - S_g}{v_s - v_g}\right) - \left(\frac{S_\ell - S_s}{v_\ell - v_s}\right)\right] dT =$$

$$= -\frac{v_g}{v_s - v_g} d\left(\gamma_{gs}\frac{d\Sigma_{gs}}{dV_g}\right) + \frac{v_\ell}{v_\ell - v_s} d\left(\gamma_{s\ell}\frac{d\Sigma_{s\ell}}{dV_s}\right) \tag{4.47}$$

Assuming again that $v_g \gg v_s$ and $v_g \gg v_\ell$:

$$-\left(\frac{S_\ell - S_s}{v_\ell - v_s}\right) dT = \pm d\left(\gamma_{gs}\frac{2}{r_{gs}}\right) \pm \frac{v_\ell}{v_\ell - v_s} d\left(\gamma_{s\ell}\frac{2}{r_{s\ell}}\right) \tag{4.48}$$

But now the gas–solid interface would be planar and $r_{gs} = \infty$; thus,

$$-\left(\frac{S_\ell - S_s}{v_\ell - v_s}\right) dT = \pm \frac{v_\ell}{v_\ell - v_s} d\left(\gamma_{s\ell}\frac{2}{r_{s\ell}}\right) \tag{4.49}$$

$$(S_\ell - S_s) dT = \mp v_\ell d\left(\gamma_{s\ell}\frac{2}{r_{s\ell}}\right) \tag{4.50}$$

$$(s_\ell - s_s) dT = \mp v_\ell d\left(\gamma_{s\ell}\frac{2}{r_{s\ell}}\right) \tag{4.51}$$

$$\frac{dT}{T} = \mp \frac{v_\ell}{\Delta h_f} d\left(\gamma_{s\ell}\frac{2}{r_{s\ell}}\right) \tag{4.52}$$

Here, the minus sign corresponds to the interface being concave to the solid phase, and there would be a plus sign otherwise.

Assuming that $\gamma_{s\ell}$, Δh_f and v_ℓ are constants, integrating them leads to

$$\ln\frac{T}{T^0} = \mp \frac{2\gamma_{s\ell}}{\Delta h_f} \frac{v_\ell}{r_{s\ell}} \tag{4.53}$$

where we introduce T^0 as the bulk melting temperature.

If we introduce now the contact angle:

$$\ln\frac{T}{T^0} = \mp \frac{2\gamma_{s\ell}}{\Delta h_f} \frac{v_\ell}{r_p} \cos\theta_{s\ell} \tag{4.54}$$

If we define $\Delta T = T_0 - T$, the logarithm can be expanded to

$$\ln\frac{T}{T^0} = \ln\left(\frac{T_0 - \Delta T}{T_0}\right) = \ln\left(1 - \frac{\Delta T}{T_0}\right) \simeq -\frac{\Delta T}{T_0} \tag{4.55}$$

Then

$$-\Delta T = \mp \frac{2T_0 v_\ell \gamma_{s\ell}}{\rho_\ell \Delta h_f} \frac{\cos \theta_{s\ell}}{r_p} \tag{4.56}$$

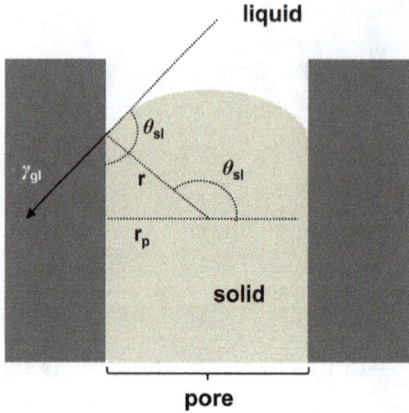

Fig. 4.3: Solid–liquid interface within a pore.

If, as shown in Fig. 4.3, the liquid–solid interface is concave to the solid side:

$$\Delta T = \frac{2T_0 (\gamma_{s\ell} \cos \theta_{s\ell})}{\rho_\ell \Delta h_f} \left(\frac{v_\ell}{r_p}\right) \tag{4.57}$$

This is the Gibbs–Thompson equation.

$$T = T_0 - \frac{2T_0 (\gamma_{s\ell} \cos \theta_{s\ell})}{\rho_\ell \Delta h_f} \left(\frac{v_\ell}{r_p}\right) \tag{4.58}$$

Now we can consider that $(s_\ell - s_s)\, T_0 = \Delta h_f$ or $T_0 \Delta s_f = \Delta h_f$, then

$$T = T_0 - \frac{2(\gamma_{s\ell} \cos \theta_{s\ell})}{\rho_\ell \Delta s_f} \left(\frac{v_\ell}{r_p}\right) \tag{4.59}$$

$$T = T_0 - \frac{2T_0 \gamma_{s\ell} \cos \theta_{s\ell}}{\rho_\ell \Delta H_f} \frac{1}{r_p} \tag{4.60}$$

with ΔH_f being the heat of fusion per unit of mass.

References

[1]   Adamson A.W., Gast A.P. Physical Chemistry of Surfaces, 6th ed., John Wiley & Sons, New York, USA, (1997). ISBN 0-471-14873-3.

[2]   Meier G.H. Thermodynamics of Surfaces and Interfaces Concepts in Inorganic Materials. MRS, Cambridge Univ. Press, Cambridge, United Kingdom, (2014). ISBN 978-0-521-87908-8.

[3]   Gaskell D.R., Laughlin D.E. Introduction to the Thermodynamics of Materials, 6th ed., CRC Press, Taylor & Francis Group, Boca Raton, FL, USA, (2018). ISBN 978-1-4987-5700-3.

[4]   Cantor B. The Equations of Materials. Oxford, United Kingdom, (2020). ISBN 978-0-19-885187-5.

[5]   Johnson C.A. Generalization of the Gibbs–Thomson equation. Surf Sci, 3 (1965) 429–444. https://doi.org/10.1016/0039-6028(65)90024-5.

[6]   Melrose J.C. Thermodynamic Aspects of Capillarity. Ind Eng Chem, 60-3 (1968) 53-70. https://doi.org/10.1021/ie50699a008.

[7]   Soustelle M. Thermodynamics of Surfaces and Capillary Systems. John Wiley & sons, Hoboken, NJ, USA, (2016). ISBN 978-1-84821-870-3.

[8]   Fultz B. Phase Transitions in Materials, Cambridge Univ. Press, Cambridge, United Kingdom, (2020). ISBN 9781108485784.

[9]   Landry M.R. Thermoporometry by differential scanning calorimetry: Experimental considerations and applications. Thermochim Acta, 433 (2005) 27–50. https://doi.org/10.1016/j.tca.2005.02.015.

Chapter 5
Kelvin equation-based techniques

5.1 Gas adsorption–desorption (GAD)

The gas adsorption–desorption (GAD) method is one of the most recognized and reliable methods for the study of porous samples of all kinds. It has therefore become a standard procedure for the characterization of porous materials in many different industrial fields. It is used as a first-choice method to determine the surface area of ceramics, carbons or catalytic beds. It has also been used frequently in membranes, although perhaps not as frequently and successfully as in other materials with a higher surface area. Likewise, in the case of membranes, as important as the surface area (and probably more so) is the determination of the size of the pores in the membrane. This determination (as well as the corresponding pore size distribution (PSD)) is very sensitive, in this case, to the porosity of the sample, which can make it more difficult to obtain significant results in the case of membranes (with low porosity or at least in their active layer) compared to other porous materials. Even the fact that most polymeric membranes (and many other composites) have two layers differentiated by their properties and/or pore size makes the interpretation of the results obtained in the technique more complicated. However, in the scientific literature, several applications of GAD to membrane characterization can be found as early as the 1950s [1–5].

In any case, the adsorption of gases on porous solids has been known for several centuries. Thus, Scheele [6] describes the increase in the volume of carbon due to adsorption of air, adsorption that can be reversed if the carbon is heated. The name "adsorption" seems to have been introduced, however, rather later by Kaiser [7] to distinguish the condensation of gas on a solid surface from simple gaseous adsorption, where gas molecules simply penetrate into the voids or pores of a solid. This adsorption of gas molecules to the surface of the solid (and their subsequent condensation) is a phenomenon in which various types of forces come into play. Thus, we will speak of physical adsorption or physisorption when the predominant interactions are van der Waals-type forces, whereas we will speak of chemical adsorption or chemisorption when the forces involve the formation of more or less complex chemical bonds.

Both processes are (usually) exothermic, although they differ in the values of the energies involved. Therefore, we will have enthalpies of –(20–40) kJ/mol for physisorption, while in the case of chemisorption, where the forces involved are more intense, we find enthalpies of –(100–500) kJ/mol.

Brunauer et al. [8] have proposed various names to distinguish between these phenomena. Thus, adsorption would be when the gas is adsorbed on a solid surface, absorption would indicate the entry of gaseous molecules into the porous solid and

https://doi.org/10.1515/9783110792195-005

finally capillary condensation would be reserved for adsorption inside the pores. However, it seems that this nomenclature has not taken root.

In order to establish concepts, we will call the solid on which (or on the walls of its pores) adsorption takes place the adsorbent, whereas the gas that has been effectively adsorbed by the adsorbent is called adsorbate (see Fig. 5.1).

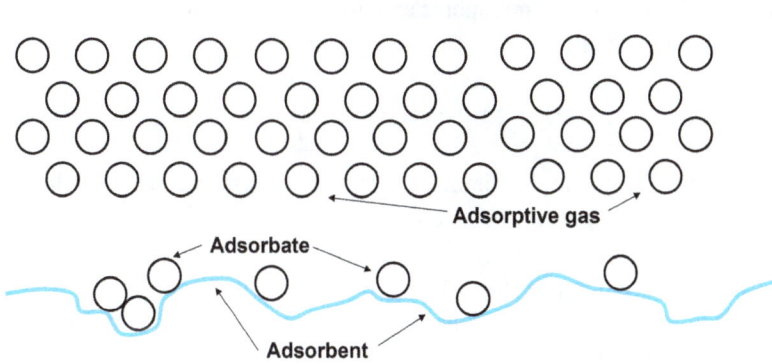

Fig. 5.1: Basic scheme of the adsorption process and the elements involved.

In any case, the basis of the technique consists of experimentally determining the so-called adsorption isotherm. Let us suppose that, in an isolated container, we bring a solid pore into contact with a gas in such a way that we control the pressure of the gas and the temperature inside the system. We will observe (either by the volume of gas entering from the external gas supply vessel or by the change in weight of the sample placed on a balance) that the solid has begun to adsorb gas, while the pressure of the gas decreases. After some time, an equilibrium situation will be obtained, where the pressure is constant and the weight of the sample (as well as the volume of gas adsorbed) is stabilized. If we now vary the gas pressure, the amount adsorbed will be different and a new equilibrium pressure will be reached. If all these measurements are carried out at constant temperature, the set of measurements will form what is called an adsorption isotherm (if we start from a minimum pressure and gradually increase the pressure of the gas, the quantity adsorbed increases). Once the process of maximum gas adsorption has been completed, this adsorption isotherm can be followed by an equally gradual process in which the applied pressure is slowly decreased. This new isotherm is called desorption isotherm and the volume of gas adsorbed on the solid decreases.

What is important is that when both curves (adsorption and desorption isotherms) are represented, the shape of the resulting curves is very informative about the existence of pores inside the solid being analysed as well as the shape of these pores. Thus, it is considered that all adsorption isotherms can be classified into a set of six ideal isotherms.

The IUPAC [9] proposed the classification of isotherms shown in Fig. 5.2. The reason for the differences between the different isotherms lies in the mechanism of pore filling in the sample, and to interpret this classification properly, we must remember how pores are classified according to their size.

According to the definition proposed by Dubinin and officially adopted by IUPAC [10], we will speak of macropores, mesopores and micropores:

- Micropores: Pores with sizes smaller than 2 nm (0.002 μm).
- Mesopores: Pores with sizes between 2 and 50 nm (0.002–0.05 μm).
- Macropores: Pores with sizes larger than 50 nm (0.05 μm).

Very briefly, the main factor to discriminate between pores in this classification is given by the underlying physical process during gas adsorption in them. Thus, it is considered that:

- Micropores: They are filled by direct interaction of the solid and the gas, which corresponds to the first adsorption layers. In this case, the effects of the interaction potential of the walls are important.
- Mesopores: After the first layers, pores of these sizes continue to fill by capillary condensation. This process results in the appearance of a meniscus between the liquid and the gas.
- Macropores: Finally, in macropores, we speak more of multilayer adsorption, with values of the relative pressure, p_r, very close to 1. In this case, the Kelvin equation can no longer be applied to model pore filling, so that this type of pore cannot be characterized experimentally using GAD, whereas they could be measured with other intrusive techniques such as Hg porosimetry or gas–liquid displacement porosimetry (GLDP).

Clearly, strict boundaries between categories in this classification are difficult to ensure, so some authors consider 15–16 Å as a more suitable boundary between mesopores and micropores (the two types of pores for which GAD is a suitable technique), which is relevant when characterizing UF (ultrafiltration) or NF (nanofiltration) membranes (with pore diameters between 1 and 100 nm) [11, 12].

In practice, the GAD technique determines the amount of gas adsorbed on the surface under study as a function of gas pressure. This is done at constant temperature and the result is the so-called adsorption isotherm. The resulting curve is fundamental for interpreting the results obtained, since its shape is closely related to the internal structure of the adsorbent material.

To this end, it is useful to discuss the different types of isotherms that can be found experimentally, so that we can know from the outset whether there are pores inside the material, as well as the type of pores that we hope to analyse.

The IUPAC classifies the different adsorption isotherms into six main groups [13–15], which are shown in Fig. 5.2.

We will briefly comment on the characteristics of each type outlined in Fig. 5.2:

–   Isotherm type I: This is also called Langmuir isotherm [16], who modelled it in 1918. It is the type of isotherm that occurs when the sample contains only micropores in its interior.

–   Type II and III isotherms: Both types of isotherms correspond either to nonporous solids or to solids that only have macropores, so that, as we have already mentioned, their filling would be produced by a multilayer adsorption process (not analysable on the basis of the Kelvin equation). The difference between the two types lies in the relative weakness or strength of the solid–gas interaction. In type II isotherm (slightly stronger interaction), there is an inflection point (marked as B in the figure) which corresponds to the filling of the initial monolayer. Whereas isotherm type III occurs when the solid–gas interaction is particularly weak.

–   Type IV and V isotherms: The main characteristic of these isotherms is the presence of hysteresis between the adsorption and desorption curves [17]. This hysteresis is attributed to the presence of mesopores, with mesoporous solids being those in which the Kelvin equation is most easily applicable, leading to the most reliable results. Finally, we can indicate that the difference between types IV and V is based on the existence of weaker interactions in the case of the latter type of isotherm.

–   Type VI isotherm: This last case is characterized by a multi-layer adsorption step on a non-porous surface. In this isotherm, each step corresponds to an adsorption layer or adsorption on different faces of a crystalline solid. An example of this type of adsorption isotherm occurs when argon is adsorbed on carbon black at the temperature of liquid nitrogen (77 K).

It is evident that the six types of adsorption isotherms described in the figure are ideal cases, so that in a certain adsorption experiment, we can find several intermediate possibilities and the interpretation of the underlying mechanisms, as well as the analysis of the results is not always so clear.

In any case, for the study of the GAD technique (especially applied to membrane porosimetry), only types I (micropores) as well as IV and V (mesopores) will be of practical interest.

On the other hand, the shape of the hysteresis cycles found in types IV and V is also a relevant indicator of the type of pores found in the solid under analysis. Thus, the IUPAC, in these cases, proposes a new classification, in which the hysteresis loop is classified into five types, which correspond to different situations, and these are shown in Fig. 5.3:

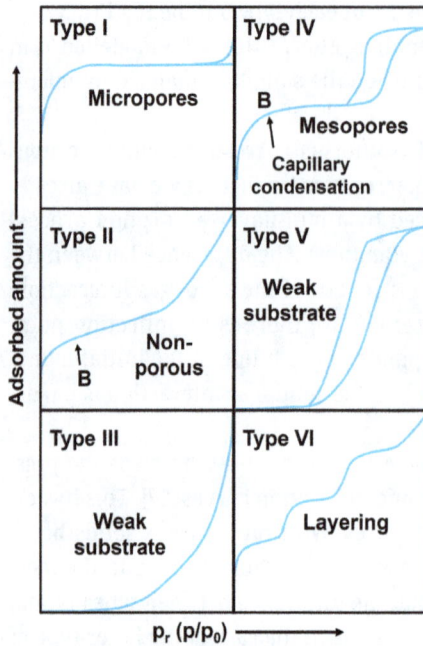

Fig. 5.2: Types of adsorption isotherms according to the IUPAC classification.

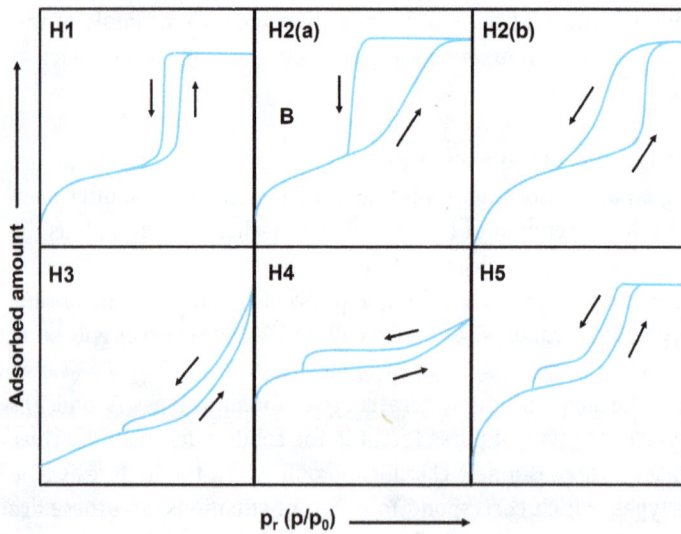

Fig. 5.3: IUPAC classification of the types of hysteresis found in adsorption isotherms corresponding to mesopores.

- H1 hysteresis, where we see adsorption and desorption curves that are fairly vertical and parallel to each other, corresponds to materials containing open or interconnected cylindrical mesopores with relatively narrow size distributions. This is the type of hysteresis found in many porous membranes, and also, for example, in Mobil crystalline materials (MCM-41) or in agglomerates of particles with reasonably uniform size distributions.
- H2 hysteresis is observed in complex porous structures, for example, containing ink-bottle pores (pores in which a narrow external access neck gives way to a significantly wider cavity). Thus, there is a delay between the evaporation of the narrow neck of the pore and the body remaining full until lower pressures are reached, resulting in hysteresis. Depending on the relative size of the two distributions (pore neck and pore body), there are two types of hysteresis (H2(a) or H2(b)). This type of hysteresis is found, for example, in inorganic oxides (silica gels) as well as in micro/mesoporous zeolites, clays or activated carbons, being the most frequent type of hysteresis.
- H3 hysteresis is assigned to the presence of wedge-shaped pores obtained due to the accumulation or stacking of platelets or flaky particles. It is frequently observed with clay aggregates (pillared clays) and in some macroporous systems.
- H4 hysteresis is similar to the previous case due to the presence of slit-like pores due to the presence of parallel platelets. It is also a hysteresis characteristic of some types of activated carbons, provided that the PSD is in the micropore range.
- Finally, the H5 type of hysteresis is rare and is usually found in materials with open or partially open mesopore systems [18].

Thus, as we have mentioned, the shape of the hysteresis cycle (and, in fact, the shape of the adsorption isotherm itself) gives us relevant information about the geometry and size of the pores [19].

The fact is that the nitrogen adsorption–desorption technique has become a widespread method of PSD in the mesopore range. Thus, there are several publications that standardise the necessary procedure for the adequate characterization of this type of materials [20].

Likewise, the GAD technique is very effective for the determination of the surface area of a porous sample, which is why it is also the reference technique for this type of determination. Therefore, when analysing the data obtained by GAD, we will start by obtaining the surface area, and then we will detail the different ways of obtaining the PSD in the sample.

5.1.1 Determination of surface area

When dealing with porous materials, the specific surface area of the material is defined as the interstitial surface area of all existing pores and voids, either per unit mass (S) or per unit volume (S_V) of the sample. This value coincides with the sample area in which adsorption can occur and is a very important parameter in the analysis of porous samples. The specific surface area can be determined from the adsorption isotherm of a gas on the sample, using the BET method, after the name of the authors (Brunauer, Emmett and Teller) on which this theory is based [21], a method considered standard for the determination of S_V.

According to the BET model, the kinetics of adsorption on the sample is carried out by fulfilling the following hypotheses:

1. In all the layers, except the first one, the molar enthalpy of adsorption corresponds to the enthalpy of condensation, L.
2. In all layers except the first one, the condensation–evaporation conditions are equal.
3. At saturation pressure (i.e. when $p_r = 1$), all gases condense on the surface of the solid, tending to an infinite number of adsorption layers.

With these simple assumptions, we arrive at the following expression which relates the adsorbed volume to the gas pressure:

$$\frac{p}{V(p_0 - p)} = \frac{1}{V_m K} + \frac{K-1}{V_m C}\frac{p}{p_0} \tag{5.1}$$

where V_m is the total volume adsorbed in the first adsorption monolayer (per unit mass), V is the volume adsorbed per unit mass (in the rest of the layers) and K is the constant to be determined depending on the type of experimental isotherm obtained.

Equation (5.1) can be converted into a linear regression, if we plot $p/[V\cdot(p_0-p)]$ against p_r (= p/p_0). In this way, the fit of the experimental data to the equation allows us to evaluate K and V_m from the slope and ordinate of the fit.

However, the BET theory is still an ideal case and, therefore, it loses its validity as more layers are adsorbed on the solid. For this reason, it is advisable to use only experimental data in the range 0.05–0.3 of relative pressures for the adjustment.

Once V_m is obtained, and taking into account the adsorbed gas (whose cross section of the molecule, A_m, is assumed to be known), we can determine S_V as follows:

$$S_V = \left(\frac{V_m}{v_g}\right) N_A\, A_m \tag{5.2}$$

where v_g is the molar volume of the gas, measured at STP conditions, and N_A is the Avogadro number [9].

5.1.2 Determination of pore size

The PSD of a porous sample is obtained from the analysis of the adsorption–desorption isotherm, for which we must distinguish whether we are dealing with mesopores or micropores (where the Kelvin equation is no longer valid). In the following pages, we will discuss the analysis of the experimental data in both cases, commenting on the main treatment methods depending on the type of pores present.

5.1.2.1 Mesoporous analysis

In samples containing mesopores, several mathematical methods have been developed to obtain the PSD from the adsorption isotherms. The best known and most widely used method was proposed by Barrett, Joyner and Halenda (BJH method) [22]. Subsequently, Dollimore and Heal [23] simplified and systematized the BJH calculation. In both cases, as well as in other proposed methods, use is made of the Kelvin equation, which, in its simplest form, can be written as

$$r_K = -\frac{2\gamma V_m}{RT\ln(p_r)} \tag{5.3}$$

where r_K is the radius of the equivalent hemispherical meniscus (the Kelvin radius) at which gas condensation occurs at a relative pressure, p_r, γ is the surface tension and V_m is the molar volume of the condensed liquid.

A first clarification must be made before describing the use of the Kelvin equation. As we know, when capillary condensation of the gas begins, this condensation is not directly on the walls of the solid, since these walls are already covered by a pre-adsorbed layer of thickness, t (this is the so-called t-layer). We can therefore say that capillary condensation takes place in the core of the pore. This must be taken into account when analysing the results obtained. Thus, the real pore radius would not be the one obtained by direct application of Kelvin's equation, but the sum of this plus the thickness of the adsorbed layer, t:

$$r_p = r_K + t \tag{5.4}$$

as shown in Fig. 5.4 for a cylindrical pore. Whereas for a parallel-sided slit (a geometrical model that is reasonably well suited to many types of membranes), the diameter of the slit will be given by

$$d_{sl} = r_K + 2 \cdot t \tag{5.5}$$

The equivalent radius r_K and the radius of the meniscus formed during desorption, r_m, are related through the contact angle, θ, between the adsorbed gas and the solid. This contact angle is usually not known, but it is known that as the number of adsorbed layers increases, the value of $\theta \to 0$, so it is generally considered to be zero for practical purposes.

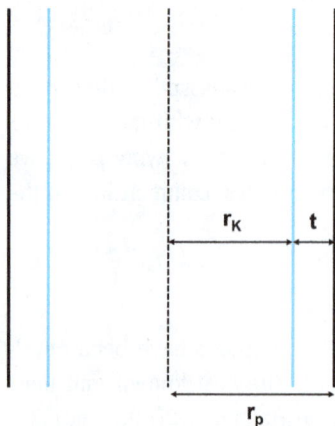

Fig. 5.4: Cross section of a cylindrical pore, showing the relationship between the pore radius, r_p, the equivalent Kelvin radius, r_K, and the thickness of the adsorbed layer, t.

Including a correct value of t in our calculations is a critical point in order to arrive at reliable results. However, such a value is usually not known, since strictly speaking it should have been measured for a completely flat surface of the same material. Instead, some empirical correlation is used, based on measurements on a large set of measurements on a variety of flat materials. Among the most used correlations is that of Halsey [9, 24], which gives the adsorbed layer thickness, measured in angstroms, as

$$t = 0.354 \left[\frac{5}{\ln\left(\frac{p_0}{p}\right)} \right]^{1/3} \tag{5.6}$$

Another empirical correlation that allows estimating the value of t is the one proposed by DeBoer et al. [25], also known as the Harkins and Jura method [26], which is most frequently used in the micropore analysis. In this case, the expression for t is given as follows:

$$t = \left[\frac{13.99}{\ln\left(\frac{p_0}{p}\right) + 0.034} \right]^{1/2} \tag{5.7}$$

In any case, we must start, as original data, from the various volumes of gas adsorbed (or desorbed) at each of the relative pressures, p_r.

Barrett, Joyner and Halenda analysed each step of the desorption isotherm stage by stage. Thus, at a given stage of pressure decrease, the volume of the desorbed gas will not only be a consequence of the emptying of the pores of the corresponding size but the decrease in the remaining adsorption layer t on the material, t_i, must also be taken into account. From these considerations, the authors arrive at the following expression for the pore volume at the n-th desorption step:

$$V_{p,n} = R_n \left[\Delta V_n - \Delta t_n \sum_{j=1}^{n-1} c_j \cdot A_{p,j} \right] \tag{5.8}$$

The factor R_n is given by the following expression:

$$R_n = \left(\frac{r_{p,n}}{r_{k,n} + \Delta t_n} \right)^2 \tag{5.9}$$

Regarding the pore area, A_p, considering cylindrical capillary pores, we have

$$A_p = \frac{2 V_p}{r_p} \tag{5.10}$$

Finally, the factor c, at each stage, is related to the quotient between the average value of the pore radius and the Kelvin radius at that stage:

$$c_j = \left(\frac{\bar{r}_{k,j}}{\bar{r}_{p,j}} \right) = \left(\frac{\bar{r}_{p,j} - t_{\bar{r},j}}{\bar{r}_{p,j}} \right) \tag{5.11}$$

Obviously, eq. (5.8) must be combined with the Kelvin equation to relate the pore radius to the relative pressure of each stage of the isotherm.

Dollimore and Heal revised the BJH calculations to simplify them. Thus, they substitute the right-hand side term in the subtraction of eq. (5.8) by the following expression:

$$V_{\Delta t,n} = \Delta t_n \left(\sum_{j=1}^{n-1} A_{p,n} - 2\pi t_n \sum_{j=1}^{n-1} L_{p,n} \right) \tag{5.12}$$

where, also for cylindrical pore geometry, the area is given by (5.10) while the length is expressed as

$$l_p = \frac{A_p}{2\pi r_p} \tag{5.13}$$

The calculation procedure previously described is similar in any of the other methods proposed to analyse the GAD curves in mesopores. The main differences between the methods are in the way the adsorbed layer thickness is calculated, as well as the correction factor for the mean pore radius.

Operationally, the above calculations are applied starting from a point on the isotherm plate, that is, corresponding to a relative pressure value close to unity. Usually, a value of $p_r = 0.95$ is considered as a starting point, which corresponds, according to Kelvin's equation, to a maximum pore radius of 20 nm (pore diameter of 40 nm), assuming a cylindrical geometry. Although the membranes analysed may contain pores of larger sizes, the truth is that, as we approach the range of microfiltration (MF) pores, the validity of the assumptions on which the application of the Kelvin equation is based becomes less evident, so that the technique is no longer useful in this range (remember that from 50 nm in size, the IUPAC already considers them to be macropores, not being analysable by GAD).

Although the reasoning followed in the explanation of the BJH method seems to indicate that the data used correspond to the desorption isotherm, the fact is that conceptually, both branches of the isotherm can be applied indistinctly to the aforementioned calculation or any of those proposed by other authors. Although it has been customary to use the desorption branch of the isotherm, it is true that there are various factors to be taken into account when analysing the experimental results, which can give rise to results that are not completely reliable in the case of the desorption isotherm. Particularly important is the phenomenon of pore blockage due to neighbouring pores that may impede the gas path during desorption, a phenomenon that does not occur in the case of adsorption. Therefore, it can be said that it is preferable to use the adsorption isotherm in the calculation of the PSD except when pore blocking is not relevant [19].

Finally, it should be noted that all proposed calculation methods need to adopt an assumption for the pore geometry. This geometrical model of the pores can give the biggest differences in the results obtained based on whether our sample can fit a cylindrical capillary pore model or has another type of geometry that is more difficult to model. However, this limitation is common for most methods of structural characterization of porous materials.

5.1.2.2 Microporous analysis

As we have already mentioned, the Kelvin equation is no longer valid from pore sizes of the order of 40–50 nm (i.e. from macropores). Likewise, in the 2 nm range (i.e. pores in the micropore range), the validity of the hypotheses on which the Kelvin equation is based is no longer guaranteed. Therefore, calculation methods must be developed for the micropore range that do not depend on the Kelvin equation.

The problem is that none of the methods developed for the analysis of micropores is valid in all cases and situations, so a certain flexibility is desirable when dealing with this type of situation.

Among the options that have been proposed for micropore analysis, possibly the simplest is to reformulate the Kelvin equation, including the thickness of the adsorbed layer, but generally with inconsistent results [27].

In view of this, several methods based on a more complete description of the thermodynamics underlying adsorption inside micropores have been developed. Thus, we can comment as main methods:

- The Dubinin–Radushkevich (DR) method [11, 28], starting from the Polanyi potential, postulates a Gaussian function for the adsorption potential inside the pores.
- The Dubinin–Astakhov equation improves the above equation for the case of microporous materials with clearly heterogeneous distributions, as well as for strongly activated carbons [29].

– The Horváth–Kawazoe (HK) method uses a Lennard–Jones-type expression of the potential, especially useful for the case of slit-shaped micropores [30], and can be applied with good results in the low relative pressure region of the isotherm.
– However, there are micropores whose geometry is best modelled by assuming cylindrical pores (this may be the case for many zeolites). In such cases, we have a modification of the HK method, the so-called Saito–Foley method [31].

Both the DR and HK methods and their two improvements are sufficiently accurate and give a good understanding of the adsorption phenomenon in micropores. However, they are methods that require prior knowledge of particular details of the adsorbate–adsorbate interaction in the case of study, as well as a considerable set of mathematical calculations.

In any case, the equations for all these methods can be found explained in a multitude of manuals for commercial GAD equipment. In particular, Quantachrome, in its latest version of the Autosorb iQ manual, reviews the discussed methods and their equations [32].

In this book, we discuss in a little more detail one of the simplest methods for micropore analysis, although, compared to those mentioned above, it does not offer much insight into the phenomenon. This is the so-called Micropore Method (MP).

5.1.2.2.1 Micropore method (MP)

The MP model or MP method by Mikhail, Brunauer and Bodor [11] is based on the De Boer t-plot [33]. The t-plot consists of representing the adsorption isotherm in terms of the thickness of the adsorbed layer, t-layer, instead of using the reduced pressure as the ordinate. For this purpose, De Boer used the Halsey correlation, in order to relate the relative pressure to the statistical thickness of the t-layer [34].

In Fig. 5.5, we can see the t-plot corresponding to two samples (the adsorbed volume has been corrected to standard pressure and temperature conditions), one containing only micropores and the other having only mesopores. The main difference is that, in the case of micropores, the line joining the points gives rise to a non-zero ordinate at the origin. In this way, the study of the t-plot slopes allows us to analyse the presence of micropores.

Based on this idea, Mikhail et al. [11] considered that the t-plot starts to deviate from the initial slope (which should coincide with the total BET surface determined according to the usual procedure) due to the presence of micropores.

Thus, the method consists of determining the successive slopes of the experimental t-plot (see Fig. 5.6). Starting from the first slope (which, as we have indicated, should reasonably coincide with the BET area), all the other slopes are translated into successive values of micropore areas. Thus, considering two points of the t-plot, corresponding to two consecutive values of the thickness t, t_i and t_{i+1}, the volume of micropores existing

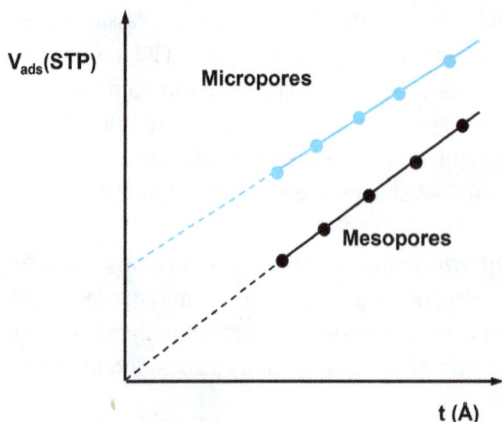

Fig. 5.5: *t*-Plot showing the difference in both samples containing only micropores or only mesopores.

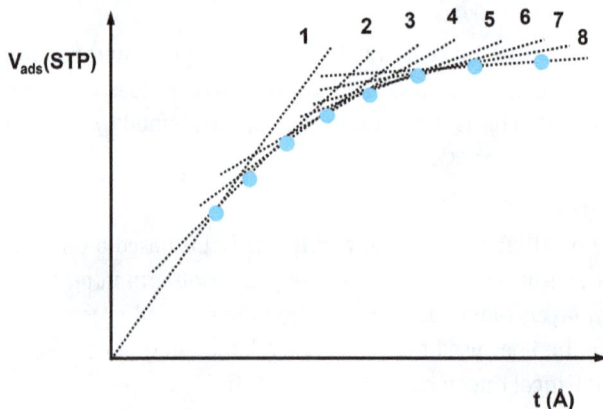

Fig. 5.6: Typical *t*-Plot showing how MP method detects the presence of micropores in a sample from the change in the slope.

between both data will be given, approximately, as a function of the increase in internal surface area and the average between both thicknesses:

$$V_i = (S_i - S_{i+1})(t_i + t_{i+1})/2 \tag{5.14}$$

In this way, we obtain the different volumes of micropores as a function of the average thickness, which we identify with the pore size (remember that gas adsorption in the case of micropores is based on the pore filling mechanism).

As in the case of micropores, it is important that we use a suitable expression to obtain the *t*-values. While De Boer introduced the *t*-plot method using the Halsey correlation, in many cases, a better fit of the experimental data has been observed using the Harkins–Jura correlation. As with the micropore analysis, an important point for reasonable results is the use of a suitable *t*-plot for the given adsorbate. In this case, *t*-values have been evaluated from an empirical relationship given by the Harkins and Jura equation (5.7) [26].

In any case, one way to see which is the most appropriate correlation in a given case is to determine the initial slope of the *t*-plot with the given correlation and see which one best fits the value of the BET area calculated previously [35].

As we can see from the experimental point of view, the calculation of the micropore distribution is quite simple and easy to programme. However, when the isotherm contains a large number of data (to be expected in high-precision equipment capable of increasing the relative pressure very slowly), it is easy to find successive slopes that do not decrease slightly in value and even increase, giving rise to negative volumes. This small problem, which can only be attributed to the experimental margin of error, can be avoided by making an adequate filtering in order to eliminate the data leading to negative volumes. Finally, it should be noted that the micropore distributions obtained by this method are not absolute (at the end, we obtain the contribution in percentage of each of the micropores to the total distribution). If we want to convert these data into absolute results, we must assume a certain geometry for the pores, which is always risky [35].

5.1.2.3 Density functional theory (DFT)

In previous sections, we have briefly discussed various approaches to obtain PSD in samples with mesopores or micropores. All these approaches are macroscopic (although in some cases of micropores with hypotheses on the real interaction potentials), and, as we have said, they involve different hypotheses on the adsorption mechanism or on the geometry of the pores.

However, there are other types of microscopic approaches that realistically describe the phenomenon of adsorption inside pores, based on the laws of statistical thermodynamics and taking into account the interactions (liquid–liquid or liquid–solid) between the different molecules involved in the phenomenon.

Among these methods, we can comment on those that perform molecular interaction simulation such as Monte Carlo (MC) or those based on molecular dynamics. However, here we will analyse the one most commonly used in GAD teams, due to its efficiency in data processing. This is the so-called density functional theory (DFT).

DFT was developed in the 1960s [36, 37], although it was not used effectively until the mid-1980s [38] in the determination of pore sizes and obtaining the PSD [19, 39].

DFT starts from obtaining a functional, $D[\rho(r)]$, that describes the behaviour of the fluid density, $\rho_L(r)$:

$$D[\rho_L(r)] = F[\rho_L(r)] + \int \rho_L(r)[V_{ext} - \mu]dr \qquad (5.15)$$

where $F[\rho(r)]$ is the free energy or Helmholtz potential of the fluid, while V_{ext} refers to the external potential resulting from the solid–fluid interaction, μ is the chemical potential and the subindex L indicates that fluid is a liquid (condensed). Finally, integration must be extended to the whole volume of the adsorbent solid [19].

Evidently, to apply this calculation, we must be able to properly include all implied potentials. In fact, $F[\rho(r)]$ is regarded to be the sum of two different contributions, one attractive and the other repulsive, and the repulsive part of the Helmholtz potential is approximated by the free energy of a fluid composed of hard spheres, though the corresponding potential ($U(r)$) is

$$F[\rho_{L}(r)] = F_{attr}[\rho_{L}(r)] + \int U(r)\,\rho_{L}(r)dr \tag{5.16}$$

Finally, the resulting functional is minimized with respect to the density so as to obtain the density profile in the equilibrium situation:

$$\left(\frac{d\,D[\rho_{L}(r)]}{d\,\rho_{L}(r)}\right)_{equil} = 0 \tag{5.17}$$

On the other hand, the different potentials that will account for the attractive and repulsive terms of fluid–fluid and fluid–solid interactions vary according to the system to be treated. This has given rise to different DFT approaches such as the local DFT, the non-local DFT (NLDFT) or the quenched solid DFT, recently introduced by Quantachrome in its equipment. These various approaches seek to improve the interpretation of the results obtained, adapting more or less successfully to the porous profile of the different samples.

5.1.2.4 Monte Carlo simulation

The other most common microscopic method is an MC simulation. MC-type computational methods are numerical methods of approximating complex mathematical expressions that were first proposed around 1944 in the context of the development of the atomic bomb. Adams [40], in 1975, developed the method for simulations on the grand canonical set (GCMC). In this type of approach, the fluid is simulated as an open system for given conditions of temperature, T, volume, V, and chemical potential, μ. Thus, the equilibrium of a fluid (or even a mixture of fluids) with a solid can be analysed. Operationally, what is done is to vary the pressure and randomly obtain the configuration of adsorbed, desorbed and transferred molecules that lead to the equilibrium situation.

This configuration is associated with a certain probability based on thermodynamic criteria (basically temperature and chemical potential) so that an extensive set of computer simulations can generate the equilibrium profile for the density and, from it, construct the adsorption isotherm, similar to the DFT method.

5.1.3 Analysis of mesopores and micropores by DFT and GCMC

Any of the microscopic methods previously discussed (DFT, usually in its non-local approximation, NLDFT, or MC simulation in the grand canonical ensemble, GCMC) correctly describe the structure adopted locally by the fluid near curved solid walls due to the existence of pores. Thus, adsorption isotherms in model pores are determined from the intermolecular potentials describing the fluid–fluid and solid–fluid interactions.

Finally, to obtain the real PSD in our sample, we must relate the theoretical adsorption isotherms obtained by both microscopic methods with our real data, which is done through the so-called generalized adsorption isotherm:

$$V_{ads,exp} = \int V_{ads,theo}(p_r, r_K) f(r_K) \, dr_K \tag{5.18}$$

where $V_{ads,exp}$ is the experimental isotherm data, r_K is the pore size, $V_{ads,theo}$ is the function that would give us the isotherm in the case of a single pore of size r_K and $f(r_K)$ is the function that gives us the distribution of existing pore sizes in the sample. The integration extending over the entire range of existing pores in the experimental calculation is replaced by the summation extended to all the experimental points of our isotherm.

In any case, the various commercial GAD equipment have the necessary calculation algorithms implemented so that the operator can determine PSD using one or the other approximations.

Another interesting advantage of the microscopic methods discussed in the previous sections is that they can be applied to analyse both mesopores and micropores, since not being based on the Kelvin equation, they do not lose validity in the micropore range.

5.1.4 Commercial devices

GAD has been frequently studied within the framework of the characterization of porous materials. It is therefore normal that there is a wide range of devices available on the market to obtain adsorption isotherms with gases or vapours. The difference between the various existing equipment lies in the method used to determine the amount of gas adsorbed; thus, we can consider three main groups for the equipment designed to apply this technique [9]:
1. Volumetric techniques: In this equipment, the volume and pressure of the gas in equilibrium situation are measured (such determination can be done in static or continuous equilibrium conditions). In this group, we list the following devices:
    –   Belsorp from Bel Co.
    –   Sorpty 1750 from Carlo Erba
    –   Micromeritics ASAP Series

- Nova Series from Quantachrome
- Omnisorp Series by Coulter Ltd.
2. Chromatographic techniques: In these techniques, the equipment uses measurement techniques based on chromatography. Some equipment of this type is:
   - Series 4200 and later from Beta Scientific
   - Rapid Surface Area Analyzer 2300 and Flow II2300, both from Micromeritics
   - Monosorb or Quantasorb, from Quantachrome
3. Gravimetric techniques: In these techniques, the changes in the adsorbed solid mass are determined by using a microbalance. Among them, the following can be mentioned:
   - DVS Analyzer from Surface Measurement Systems
   - Gravimat from Netzsch

5.1.5 Experimental procedure

The procedure to obtain information on the porosity of a sample involves a correct and accurate determination of the adsorption isotherm. Whatever the method of detection of the adsorbed volume, the isotherm is typically obtained in the following steps:
- Degassing of the adsorbent is carried out by the application of a high vacuum to the entire surface to be analysed (usually after prior drying in an oven).
- A convenient step is usually followed to achieve a good surface cleanliness. For this purpose, an inert gas is flowed over the adsorbent at high temperatures.
- Finally, the adsorption isotherm is determined by slowly varying the equilibrium pressure of the gas, while keeping the temperature constant.
- It is important in this last step to ensure that the system is allowed to evolve at each experimental step for the time necessary to actually reach the equilibrium conditions. This is especially critical in pore sizes close to the size of the adsorbate molecule (5 nm, for example, in the case of nitrogen), where the kinetics of gas penetration into the micropores can be very slow.

Finally, we must comment briefly on the adsorbent gas to be used. In principle, any gas that is condensable at the working temperature could be used. However, considerations can be made based on the type of information we wish to obtain from the technique:
a) Determination of structural properties: What we want, as is the general objective of this book, is to obtain structural information about the pore material (basically its PSD as well as its BET area); as a general rule, we will use inert gases that have a condensation temperature, at atmospheric pressure, that is easy to maintain constant.
   - Nitrogen: This is obviously the most widely used due to its low cost to the point that many published works refer directly to the technique as nitrogen adsorption. It is also the standard method for determining the specific sur-

face area of any type of porous material. Its working temperature, in this
case, coincides with the boiling temperature of liquid nitrogen: $T_{tr} = 77$ K.

- Argon: This gas, rather more expensive and not so readily available in many
laboratories, is used in those cases where greater precision is required in the
area between the mesopores and the macropores ($T_{tr} = 87$ K).
- Other gases: In some special cases, and not as a general method, other noble
gases such as He or Kr or also CO_2 have been used.

b) Determination of specific material characteristics based on what we want to know:
- The hydrophobic or hydrophilic character of the analysed sample
- The ratio between the pores to which the gas is accessible and the size of the
gas molecule

In these cases, any condensable gas that fits the characteristics we want to study can
be used: water vapour, alcohols, hydrocarbon compounds and so on.

5.1.6 Advantages and disadvantages

It is well-known that the GAD technique has usually been considered as a routine and
reasonably standardized method for the determination of the surface area of porous
samples, as well as the PSD of these samples. This is especially true for the case of
materials containing mesopores. In 1985, IUPAC issued a manual on the determination
of surface area and porosity, whose conclusions and recommendations have been
broadly accepted by the scientific and industrial community [14]. In 2015, IUPAC published an updated technical report, discussing the advantages and limitations of using
physisorption techniques for studying solid surfaces and pore structures with particular reference to the assessment of surface area and PSD [41]. Even so, the physico-
chemical principles involved in mesopore filling are not fully understood, including
the validity of the Kelvin equation in its lower limit remains unresolved [42].

Moreover, the correct interpretation of the adsorption isotherms and their translation to PSD as a faithful reflection of the reality of the samples is complicated when
we find that they contain both micro- and mesopores, or samples with more complicated pore compositions such as zeolites, aerogels and other newly developed materials [18]. In some samples, such as soft-templated carbons, it is usual to find a different
geometry from mesopores (more or less cylindrically shaped) and micropores (more
likewise to be slit-like shaped), making it difficult to assess a proper interpretation of
adsorption isotherms [43].

From a technical point of view, obtaining accurate adsorption isotherms for their
subsequent interpretation is crucial for good characterization. Thus, although we
mostly use automatic commercial equipment, the results obtained depend on various
measurement parameters that we must know properly to arrive at the correct isotherm in the most accurate way.

On the other hand, the equipment must maintain cryogenic conditions for the entire duration of the experiment. Another important aspect from an operational point of view is that the technique is sensitive to the pore volume contained in the sample, being clearly less accurate for samples with low porosity. This can be a problem for asymmetric polymeric membranes, where the active layer, the one we find most interesting from the point of view of membrane properties, is usually much less porous than the support. Therefore, whenever possible, it is advisable to separate the two layers so that they are analysed separately, and the porosity data of the support does not mask those of the active layer.

5.2 Permporometry (PmP)

The technique called permporometry or permoporometry (PmP) has often been confused with liquid displacement porometry (a different technique that will be discussed in the next chapter), perhaps due to the fact that both are based on determining the permeability of a fluid (gas in the first case, and liquid/gas in the other) through the membrane we want to analyse. But, as we hope it will be obvious for the readers at the end of this book, they are completely different techniques, as we will see throughout this chapter and when we will deal with bubble point-based techniques later on.

Since PmP is based on the capillary condensation of a gas in the membrane pores, it can be considered to be based on Kelvin's equation, although it is really a combination of this condensation plus the permeation of a liquid through the already condensed pores. Firstly, let us proceed with a brief revision of technique fundamentals.

We know that the vapour pressure of a gas inside a capillary tube (consider a pore given the target of our book) depends on the diameter of the pore (as the Kelvin equation tells us). If the relative pressure is sufficiently high, all pores will be filled with the condensed gas in liquid form so that they will be blocked to any other fluid. However, in larger pores, the vapour pressure of the liquid may be higher than the partial pressure of the corresponding vapours so that the liquid in these pores will begin to evaporate. However, in the smaller pores, the partial pressure of the vapour will still be higher than that of the liquid, and these pores will remain filled with liquid. Thus, in the larger pores, which are now free, we can permeate a fluid at low pressure and determine the corresponding permeability (for that situation and that size of pores already open). If we continuously vary the relative pressure so that smaller pores are gradually emptied (by evaporation), the new permeability measurements will be related to the new open pores.

As for the fluid to be permeated, originally Eyraud et al. [44] used condensation of a gas and permeation of a liquid. However, later Cao et al. [45] decided to determine the diffusion permeability of a gas, with better experimental results. Obtaining an equilibrium situation between the liquid phase inside the pores and the gaseous phase outside the pores took an excessively long time [46]. Thus, Katz needed several

days to obtain equilibrium conditions at each experimental point necessary to characterize a Millipore MF membrane [47, 48]. Cuperus et al. [49] modified the method by
using counter diffusion of two gases (nitrogen and air) using methanol as the condensable gas. This resulted in significantly shorter equilibration times (in the order of
30 min for each experimental point). Moreover, since counter diffusion is governed
by a concentration gradient, the diffusive regime is easier to model based on the
Knudsen model of molecular flow:

$$J_i = \frac{n r_i^2 D_i \Delta p_g}{RT \tau \delta}$$
(5.19)

where J_i (mol/m$^2 \cdot$ s) is the gas flux across the cylindrical pore of radius, r_i(m), Δp_g(Pa) is
the partial pressure difference across the membrane, R is the ideal gas constant, T (K) is
the absolute temperature, τ (dimensionless) is the pore tortuosity, δ (m) is the membrane
thickness.

On the other hand, D_i is the Knudsen diffusion coefficient of the gas, which is expressed as follows:

$$D_i = 0.66 \; r_i \sqrt{\frac{8RT}{\pi M}}$$
(5.20)

where M (g/mol) is the molecular weight of the condensable gas.

Figure 5.7 (left) shows a typical PmP experiment, where the flow of the permeating gas is plotted against the relative vapour pressure of the condensing gas. Typically,
this type of curve has three clearly differentiated sections. At high relative pressures
(above 0.8 in the figure), there is condensation of the gas inside the pores with practically no gas permeation. Then there is a sharp increase in gas permeation (for relative
pressures between 0.8 and 0.7), obviously due to pore opening, leading to permeation.
Finally, the last section with a continuous but slow increase in permeation is not attributed to any new open pore so that the distribution of pores present in the sample
is already obtained.

From the above curve, the PSD is obtained using the following expression [45]:

$$f(r) = -\frac{3\tau \delta}{2r^3} \sqrt{\frac{MRT}{8\pi}} \frac{dF}{dr}$$
(5.21)

where $f(r)$ is the size distribution (number of active pores with sizes between r and $r +$
dr, per unit volume, m^{-3}) and F is the diffusion permeability of the non-condensable
gas (mol/m$^2 \cdot$ s $\cdot$ Pa).

As in any other technique based on capillary condensation (thus techniques that
make use of the Kelvin equation in the interpretation of their results), in PmP, it is necessary to take into account the layer of liquid that remains adsorbed on the solid surface of the pore even when most of the liquid has evaporated from the pore. Thus, the
adsorption and/or desorption process will not occur in the entire pore but in a pore in

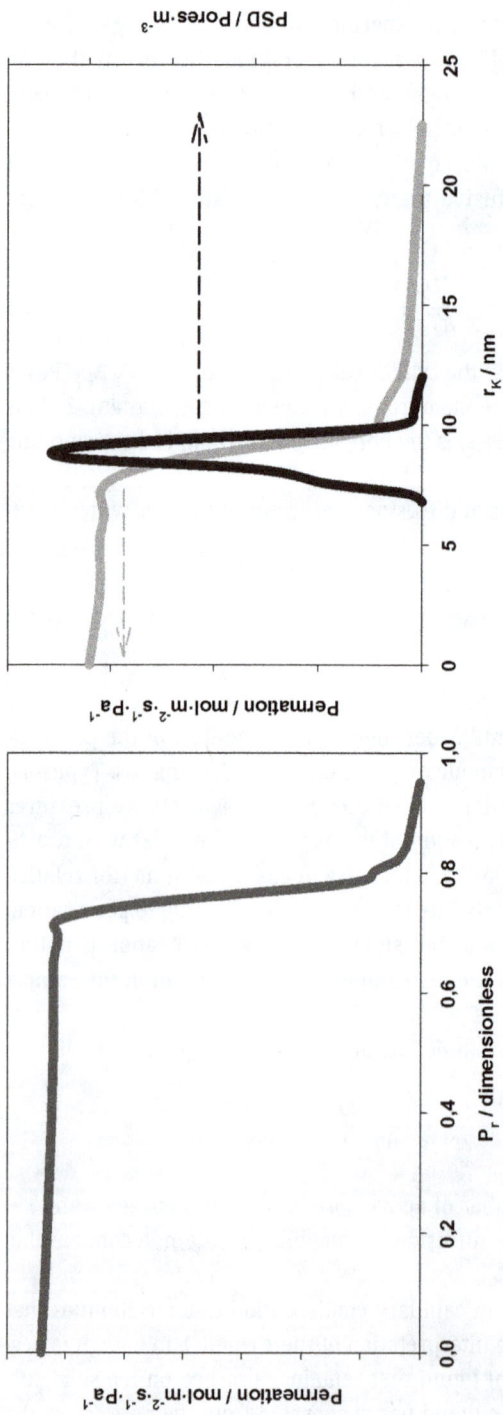

Fig. 5.7: Typical PmP experiment (left) along with the resulting PSD (right), in terms of the Kelvin radius, r_k (adapted from [45]).

which this layer, known as the t-layer, must be discounted, so that if r_K is the radius deduced by applying Kelvin's equation, the real pore radius, r_p, will be given by

$$r_p = r_K + t$$

Obviously, the above expression implies cylindrical pores, a common but not always correct assumption. Thus, for slit-shaped pores, the relation between the actual pore size and the Kelvin radius is different, being in this case:

$$r_p = \frac{r_K}{2} + t$$

Certainly, this implies a certain problem when interpreting PmP results, since the thickness of the t-layer is, in principle, unknown and can vary with the experimental conditions. In practice, what is done is to use some of the empirical correlations that have proven their validity in the GAD technique. In particular, the Halsey correlation [24] is quite convenient as an approximation to the real value [9]:

$$t = 0.354 \left[\frac{5}{\ln\left(\frac{p_0}{p}\right)} \right]^{1/3} \tag{5.22}$$

Another approach to the determination of the thickness of the layer t is given by Cao et al. [45]. Cao considers that the permeability of a pore to non-condensable gas is proportional to the cube of the pore radius. Now, at partial pressures of 0.7 (see Fig. 5.7, left), when the pores are open, the radius to be considered is the Kelvin radius, since there is a deposited t-layer. Whereas if we decrease the relative pressure, approaching zero, the deposited layer is zero and the gas permeability will be related to the actual pore radius. Thus, the thickness of the t-layer can be determined from the determination of the permeabilities at $p_r = 0$ and $p_r = 0.7$:

$$\frac{F_{0.7}}{r_K^3} = \frac{F_0}{(r_K + t)^3} \tag{5.23}$$

Furthermore, the fact that the technique is based on the phenomenon of capillary condensation inside the pores limits its range of validity to those pore sizes for which the Kelvin equation correctly describes this capillary condensation. The Kelvin equation is usually considered to be valid only in the mesopore range. Thus, according to the IUPAC definition, the technique would be limited to pores between 2 and 50 nm. However, the lower limit of applicability of the Kelvin equation is still a matter of debate [42]. Thus, Cuperus et al. [50] consider that the use of the Kelvin equation could be valid up to 1.5 nm, thus extending the range of application of PmP between 1.5 and 100 nm [45]. Tsuru et al. [51] analysed the limits of validity of the Kelvin equation in relation to the use of various condensable vapours (water, methanol, ethanol, isopropanol, carbon tetrachloride and hexane) so that the technique could be extended to the NF range (in this way what the authors call nano-PmP allows the technique to be extended to around 0.6 nm [52]).

The method thus developed and known as permporometry (PmP, although it can also be found in the literature under the name of permoporometry [53, 54]) has been applied with notable success to the characterization of ceramic membranes [55–57].

On the other hand, as mentioned in the introduction, the technique has been confused by some authors with liquid displacement techniques. Thus, for example, Šolcová et al. [58] describe PmP experiments in which they wet a sample with Porofil™ and perform wet and dry curves (which clearly refers to GLDP), although they correctly reference the technique [59, 60]. Even reputed figures such as Fane or Jonsson [61], speak of bi-liquid PmP, which clearly refers to the liquid displacement technique (LLDP), which is quite surprising since Bargeman is co-author of some of the pioneering works on the technique [49, 50].

It is important to note that, as we will see in liquid displacement techniques, a zero or near-zero contact angle is ensured. If this is not the case, the value of the contact angle must be taken into account in the equations governing the technique, in particular in the Kelvin equation, with the attendant difficulty of knowing the value of θ with sufficient precision.

Thus, Cuperus et al. [50] consider that using methanol as the condensing fluid, the contact angle in the Kelvin equation can be considered to be zero.

In many pioneering works, oxygen is used as permeant gas and methanol as condensing fluid [50]. As previously mentioned, many other gases and vapours have been used as condensing gas [51], although the most widely used combination in the literature is probably cyclohexane as condenser and air/oxygen as permeant [62, 63].

5.2.1 Commercial devices

Although the technique has been developed and used with some frequency for a few decades now, there is no commercial equipment that allows PmP experiments to be carried out, so each research group that tackles the use of the technique builds its own, usually based on the work of previous researchers. Thus, from Cuperus or Cao (see Fig. 5.8), various experimental set-ups are published that are quite similar. In some cases, syringes are used to control the flow [51]. In Fig. 5.9, we have a fairly typical set-up for PmP experiments, taken from Higgins et al. [62], using helium and hexane as permeant and condensing gases, respectively.

Figure 5.9 shows a set-up composed of stainless steel tubing and fittings. Purified (high-grade) helium was sent to two flow controllers (MFC), one feeding the hexane bubbler (containing 99% n-hexane) with He, while the other sent pure He to the membrane. Combining the flow of both MFC, hexane activity was adjusted to the desired values. Constant feed pressure in the membrane was settled with a pressure control-

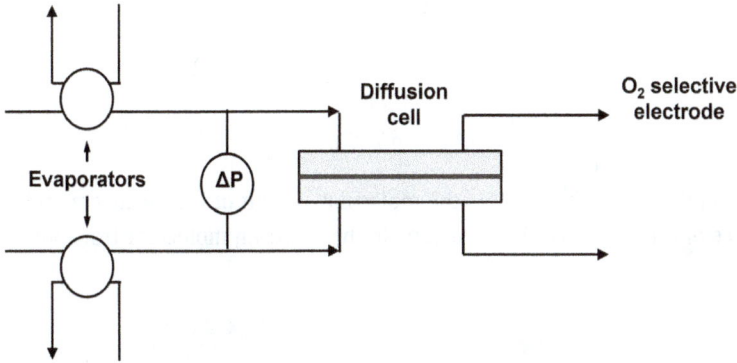

Fig. 5.8: PmP scheme (adapted from [50]).

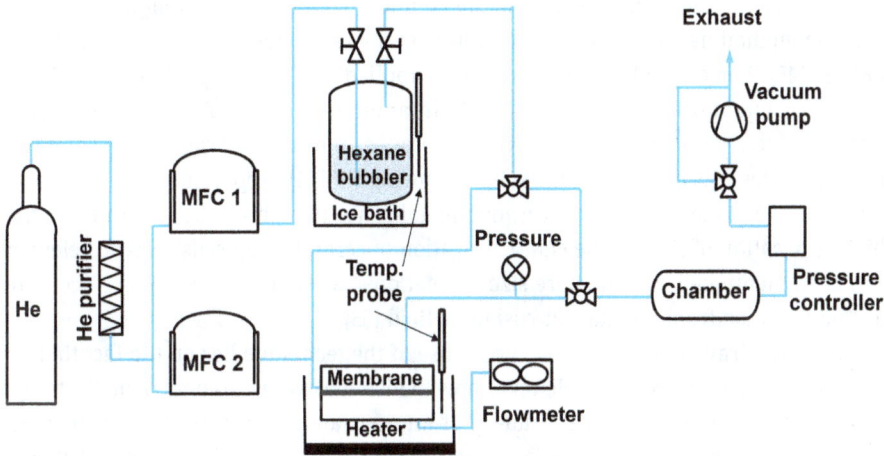

Fig. 5.9: PmP device (adapted from [62]).

ler. Finally, flow rates of He through the membrane were determined by using a soap bubble meter.

The Kelvin equation allows to relate the hexane activity (hexane vapour pressure normalized to saturation pressure at given experimental conditions) with pore size through

$$r_{\mathrm{p}} = \frac{-4\sigma V_{\mathrm{m}}}{R\,T\,\ln a} + t \tag{5.24}$$

where σ is the surface tension of hexane, V_m is its molar volume, R is the ideal gas constant and T is the temperature. Finally, a is the hexane activity and t is the thickness of the adsorbed t-layer [50].

PSD was calculated based on the analysis presented by Cao et al. [45] using the expression:

$$f(r_p) = \frac{-3l}{2\,r_p^3}\sqrt{\frac{MRT}{8\pi}}\frac{dP}{dr_p} \qquad (5.25)$$

where l is the membrane thickness, M is the molecular weight of the permeate and P is the permeance of the inert gas (He) according to the Knudsen molecular transport.

5.2.2 Advantages and disadvantages

As a method based on permeation through the membrane, it is suitable to determine the PSD of the active layer, which is the one governing the flow of the membrane. Thus, ink bottle or closed pores are not taken into account by the technique.

The method has found increasing utility on the characterization of ceramic membranes [45]. For example, Puthai et al. [56] applied PmP to the study of the PSD of SiO_2–ZrO_2 membranes developed for NF. In another example, PmP measurements were used to characterize ceramic NF membranes, and that information was used in modelling the rejection of organic solvents during NF [64]. Apparently, gas condensation works especially well with ceramic membranes (which is also a reason for the high application of GAD in the characterization of ceramic materials). Nevertheless, it has been mentioned that the pore size estimations based on PmP are not so accurate as that obtained from mercury intrusion method [55].

Another drawback to the widespread use of the technique lies on the fact that no commercial devices are available up to now, which makes researchers to dedicate significant time and effort to the build-up of an accurate and precise PmP system. Finally, a proper estimation of the adsorbed t-layer is important in obtaining reliable results.

5.3 Evapoporometry (EP)

Among the indirectly determined porosimetric techniques based on the Kelvin equation, the latest to jump on the bandwagon and surely one of the most promising ones is evapoporometry (EP). This technique originates from an idea by Krantz and Greensberg that they presented at several conferences between 2010 and 2011 [65, 66]. Although the pioneering work on the technique is considered to be the 2013 publication in the *Journal of Membrane Science* [67], in this work, Krantz and his collaborators, after reviewing the various porosimetric techniques including their respective problems and inaccuracies, propose a novel technique based on the use of the Kelvin equation, which describes the vapour pressure for a liquid inside a pore as a function of the curvature.

The technique involves soaking the membrane we are going to characterize with a volatile fluid (a liquid really). If our membrane is saturated with this volatile liquid, there is a natural evaporation of the liquid (due to its volatility) with the evaporation rate being a function of the curvature radius of the surface. In our case, this means that the larger pores will evaporate first, and the smaller pores will evaporate subsequently.

Therefore, the EP technique will determine the mass loss due to the evaporation of the volatile liquid from inside the pores. From the measurement of this evaporated mass, and making use of Kelvin's equation, we will be able to determine the PSD present in the membrane.

In this case, Kelvin's equation tells us that for a pore of diameter d, the interface between the liquid and the evaporated vapour has the following relationship:

$$\ln \frac{P_A}{P_A^0} = -\frac{4\, V\, \gamma\, \cos\theta}{d\, R\, T} \tag{5.26}$$

where P_A and P_A^0 are the vapour pressures of the wetting liquid in the pores that have evaporated and under normal conditions, respectively. On the other hand, V is the molar volume of the liquid, θ is the contact angle of the wetting liquid with the solid contact surfaces, T is the absolute working temperature and R is the well-known ideal gas constant.

Thus, there is a relationship between the vapour pressure of the wetting liquid and the diameter of the pore into which the wetting liquid evaporates. The minus sign in the above equation also indicates that the vapour pressure decreases as the pore becomes smaller.

Taking into account that the vapour pressure is directly proportional to the corresponding mole fractions of the gas phase at the membrane surface during evaporation from the pore (x_A) and of the gas phase at the membrane surface saturated with wetting liquid (x_A^0), we can express the above equation as

$$\ln \frac{x_A}{x_A^0} = -\frac{4\, V\, \gamma\, \cos\theta}{d\, R\, T} \tag{5.27}$$

From which we arrive, by rearranging, at the following expression:

$$d = -\frac{4\, V\, \gamma\, \cos\theta}{R\, T\, \ln\frac{x_A}{x_A^0}} \tag{5.28}$$

On the other hand, these gas phase molar fractions can be determined from the evaporation rates obtained experimentally. To do so, we must consider the mass transfer coefficients due to the diffusion of the wetting fluid between the liquid and solid phases [67], which are related to the mole fractions and the experimental evaporation rates, so that we arrive at the following relation between both magnitudes:

$$\frac{x_A}{x_A^0} = -\frac{1-e^{-\left(\frac{4W_A}{\pi d_c^2 k_x}\right)}}{1-e^{-\left(\frac{4W_A^0}{\pi d_c^2 k_x}\right)}} \qquad (5.29)$$

In most cases, this rigorous deduction can be simplified by taking into account the small value of the exponent, so that by doing a serial development of the exponential one arrives at

$$\frac{x_A}{x_A^0} = -\frac{1-e^{-\left(\frac{4W_A}{\pi d_c^2 k_x}\right)}}{1-e^{-\left(\frac{4W_A^0}{\pi d_c^2 k_x}\right)}} \approx \frac{W_A}{W_A^0} \qquad (5.30)$$

So we can obtain the diameter of the pores present in the PSD directly from the evaporation rates:

$$d = -\frac{4\,V\,\gamma\,\cos\theta}{R\,T\,\ln\frac{W_A}{W_A^0}} \qquad (5.31)$$

In the first published papers on the technique, the above expression was used to determine the PSD from the evaporated mass data. A recent paper by Krantz et al. [68] emphasizes the fact that these early papers do not take into account the contribution of the very thin layer of liquid that has adsorbed on the pore walls and remains there despite the evaporation of the rest of the liquid. This layer, called the t-layer, is very thin, but when we are analysing nanometre-scale pores (such as those found in narrow UF or NF membranes) its contribution is not negligible. In any case, Krantz's article corrects this contribution when determining the PSD.

From a practical point of view, in EP technique what we should determine is to determine the rate of change of the evaporated mass, W_A, by means of a high-precision analytical balance. Therefore, the equipment necessary to carry out an EP experiment consists, together with the analytical microbalance mentioned above, of a suitable cell in which to place the membrane and the wetting liquid (all contained in an insulating chamber placed on an anti-vibration table to avoid inaccuracy in the measurements of the evaporated mass). The schematic of the experimental apparatus is shown in Fig. 5.10, as well as the detail of the measuring cell. The lid of the cell has an orifice of controllable diameter so that the evaporation rate can be varied. The measuring cell is isolated from lateral liquid losses by means of vacuum grease. The microbalance is connected to a computer with suitable software to take measurements of the mass variation on the balance at pre-set intervals. Finally, the chamber containing the cell and the microbalance is thermostatically controlled in order to maintain a constant temperature throughout the experiment. It is also convenient to use granular activated carbon to reduce the vapour concentration of the volatile wetting liquids outside the test cell.

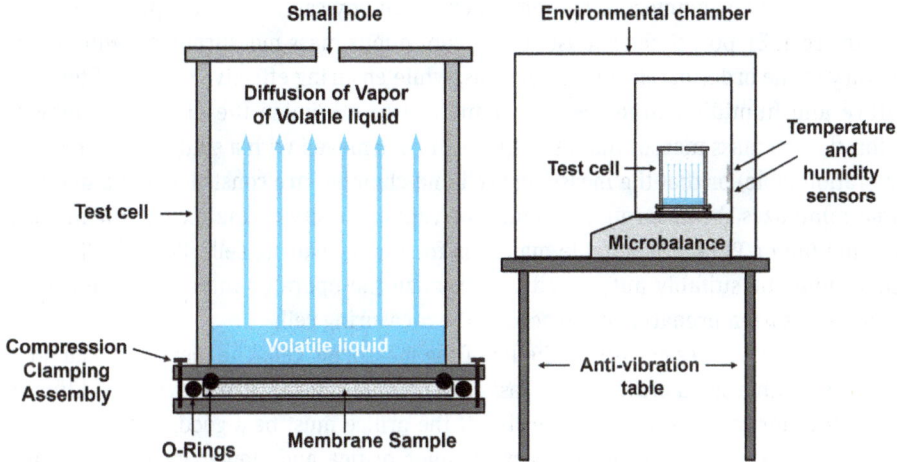

Fig. 5.10: EP chamber (left) and complete device (right) (adapted from [69]).

The location of the sample at the bottom of the cell, and covered by the wetting liquid, allows the cell to be saturated in the gas phase of the volatile wetting liquid. This means that the surface of the membrane is saturated in the gas phase with respect to the liquid in those pores that are emptying, while it is supersaturated with respect to all pores smaller than this. Thus, evaporation, as indicated by the Kelvin equation, progresses from the larger to the smaller pores [67].

Figure 5.11 presents the results of a typical EP experiment, where the net mass collected by the balance is represented in Fig. 5.11a, from which the evaporation rates are determined (Fig. 5.11b). In the latter representation, three distinct zones are clearly visible, which are called, respectively, (a) transient zone, (b) free liquid evaporation zone and (c) liquid evaporation zone inside the pores. The transitional period corresponds to the time necessary for the system to equilibrate, while in the second section the evaporation of the liquid above the membrane takes place, so that the loss of mass is linear with time. Finally, the last section is due to the evaporation of the liquid present in the pores and is the one that will allow us to determine the PSD.

As shown in Fig. 5.11, a complete experiment requires considerable time (around 14 h in the case considered), including the stabilization of the system, which could result in excessive or uncontrolled evaporation distorting results.

In any case, the implementation of the technique and its correct execution requires a considerable degree of experience, which makes it difficult to use as a routine characterization technique. Thus, the first published works on PD were based on prototypes built in a relatively artisanal way in research laboratories dealing with this type of measurements.

The most recent work by Krantz et al. [68] proposes the prototype equipment for EP analysis, which could be converted into the commercial equipment. However, at present, no such commercial equipment exists. In any case, a suitable prototype for a commercial EP porosimeter must ensure continuous mass measurements with an accuracy in the order of tens of micrograms, while ensuring effective control of temperature and humidity. Other requirements are to minimize the effects of ambient vibration on mass measurements (which can be achieved with a suitably located anti-vibration table) or that the measuring cell and chamber are constructed of a material that minimizes the absorption of liquid or vapour, falsifying the measurements (e.g. aluminium or PTFE are suitable materials for the measuring cell). Finally, all equipment must be suitably automated to minimize the operator intervention once the sample has been prepared and placed in the measuring cell.

An important design issue is the lid of the measuring cell. The measuring cell lid is provided with a small orifice whose task is to provide resistance to convective–diffusive mass transfer in the gas phase. The size of the orifice must be a good compromise between higher measurement accuracy (smaller orifice and slower evaporation) and shorter test duration (larger orifice and faster evaporation) [68].

As a wetting liquid, one should opt for a liquid that has a relatively high vapour pressure (quite volatile) and also wets the membranes well, ensuring zero contact angle with most of the membrane-forming materials. The first choice was isopropyl alcohol [67], and it is still the most widely used liquid, although other liquids such as water, 1-butanol, acetone, ethanol or ethylene glycol have also been tested, with quite diverse results, as was to be expected from their very different characteristics [71]. In any case, the comprehensive review by Tanis-Kanbur et al. [72] presents a summary of the work published on EP up to that time, including the different liquids used as volatile wetting liquids.

5.3.1 Experimental procedure

Typically, the following steps are followed to perform an EP experiment [69]:
– First, the sample to be analysed is conveniently wetted with a volatile liquid (in many cases, distilled water is a good option). In any case, a complete wetting is a necessary condition to obtain reliable results that include all the pores present in the sample.
– Once the sample is placed in the cell of the EP apparatus on an analytical balance of suitable accuracy, evaporation begins. Data are usually collected from the balance reading every 10 s.
– The experiment continues as all the wetting liquid (first from the measuring cell itself and then from the interior of the pores) gradually evaporates. The process, as men-

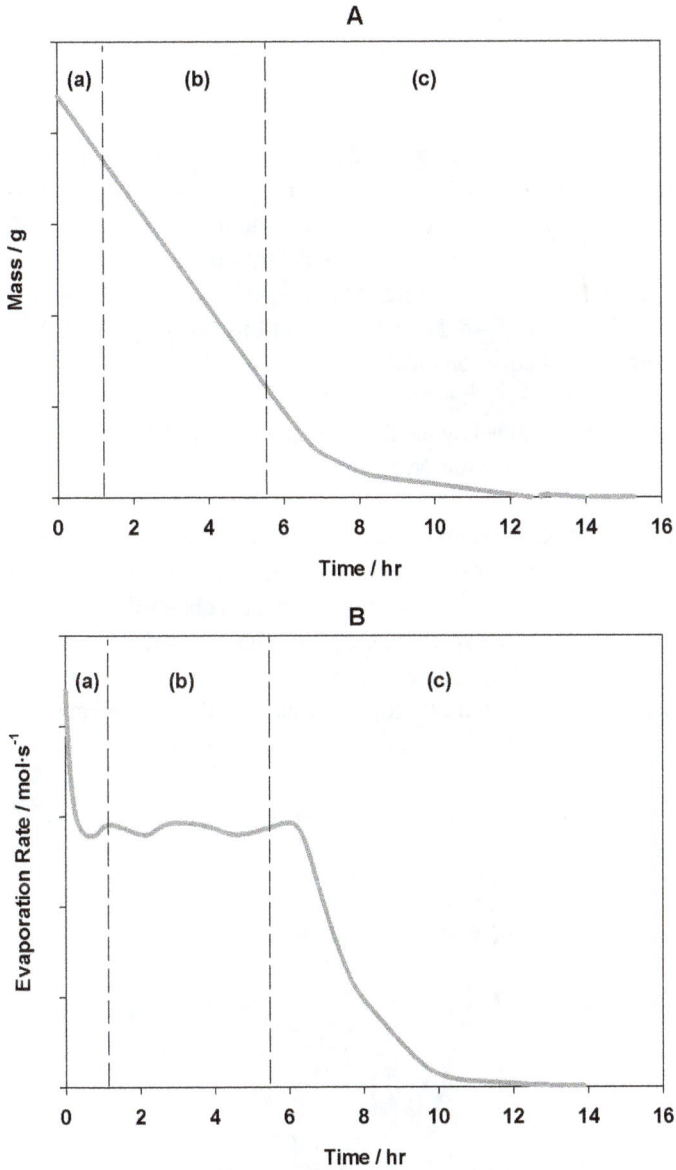

Fig. 5.11: Typical EP experiment: (a) evaporated mass versus time and (b) evaporation rate versus time. Three characteristics zones are shown: (a) transient zone, (b) free liquid evaporation zone and (c) liquid evaporation zone inside the pores (adapted from [70]).

tioned above, can take considerable time, so it must be ensured that the ambient conditions in the laboratory do not change abruptly to influence the evaporation rates.

– From the data in Fig. 5.11a, the evaporation rate (W_A) is determined as the slope of a reasonable number of consecutive data.

– The evaporation rate of the free-standing liquid layer (W_A^0) is taken as the value corresponding to a period in which W_A can be considered constant. This period is the so-called free liquid evaporation zone in Fig. 5.11b. The standard deviation (δ) of the data used in the determination of W_A^0 is also determined.

– The experiment is considered to have entered the pore emptying phase when the evaporation rate meets $W_A < W_A^0 - \delta$. From the data of this last stage, the PSD is obtained by applying Kelvin's equation (5.31).

Figure (5.12) shows a typical PSD obtained by the EP technique. Next to it, the pore distribution results obtained by LLDP for the same Nylon membrane are presented. It is interesting to note the good agreement between both techniques in terms of the mean value of the distribution [70], although the distributions obtained by EP are generally quite wider. Although this does not a priori mean worse results, it can be significant when assessing the retention properties of the membrane, based on the information obtained on the PSD.

In Chapters 7 and 8, we will discuss various possibilities for estimating the molecular weight cut-off of a membrane based on knowledge of PSD, an estimate that, although only approximate, can be really useful for the prior selection of a membrane suitable for our industrial application.

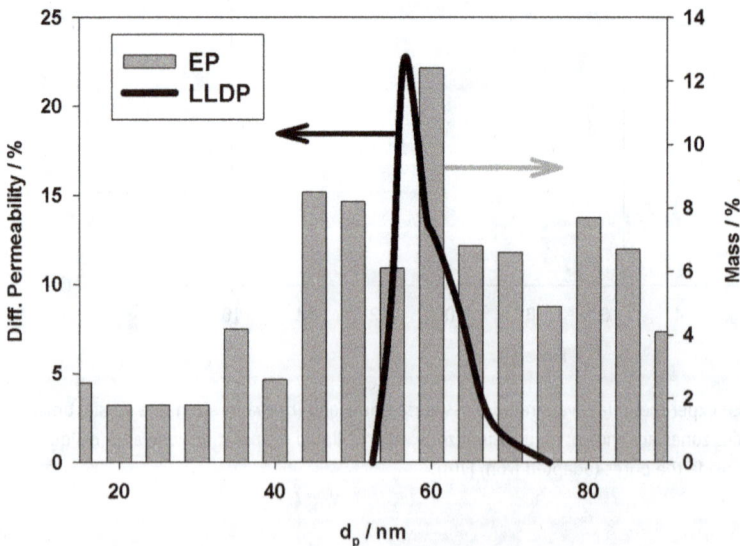

Fig. 5.12: Resulting PSD from the data of a typical EP experiment, along with corresponding results obtained for the same membrane by using LLDP (adapted from [70]).

5.3.2 Advantages and disadvantages

The EP technique, although not yet commercial, has beneficial aspects that make it interesting as a membrane characterization technique. Firstly, its range extends to MF and UF membranes, and it is not useful for analysing NF membranes, where the validity of the Kelvin equation is compromised.

On the other hand, EP does not need to assume geometrical models of the pore, which is an advantage over most indirect methods (such as those based on the bubble point). However, when dealing with asymmetric membranes, this statement is no longer true. In such a case, it is necessary to correct the results by assuming a geometry for the pores.

The equipment to carry out EP experiments is simple and relatively cheap, although it has the disadvantage that there is currently no company that commercializes such equipment. Therefore, the researcher (or company) interested in this technique will have to design and build their own equipment, with the consequent time needed for proper testing of its correct operation.

The researchers who have developed the technique also consider it to be simple to execute and accurate in its results. The fact that it analyses a relatively large sample area makes its results more representative than microscopic methods in which the area studied is always small. On the other hand, it does not allow working with membrane modules, so the analysis of the samples must always be done after extracting them from the module. Merriman et al. [73] tested the validity of the technique applied to hollow fibres by comparing their results with those obtained with a Nuclepore membrane (0.1 µm).

Another advantage relies on the fact that the EP technique is claimed to be able to characterize not only the continuous pores but also the dead-ended ones.

One of the main problems of the EP technique is that its experiments, as already mentioned (Fig. 5.11), require a rather long time (typically 10–14 h), which could distort the results obtained.

Other limitations of the technique are:
- It starts to become less accurate as the pore size increases, being inadequate above 150 nm.
- To ensure uniform saturation conditions, it requires relatively large pore densities.
- It also needs high porosities in order to have sufficient measurable mass of gas evaporating.

References

[1]   Cranston R.W., Inkley F.A. Determination of pore structures from nitrogen adsorption isotherms. In: D.D. Eley, H. Pines, P.B. Weisz (Eds.), Advances in Catalysis, 9. Academic Press, New York, USA, (1957) 143–154. ISBN: 9780080565132.
[2]   Ohya H., Imura Y., Moriyama T., Kitaoka M. A study on pore size distribution of modified ultrathin membranes. J Appl Polym Sci, 18 (1974) 1855. https://doi.org/10.1002/app.1974.070180621.
[3]   Ohya H., Konuma H., Negishi Y. Post treatment effects on pore size distribution of Loeb–Sourirajan-type modified cellulose acetate ultrathin membranes. J Appl Polym Sci, 21 (1977) 2515. https://doi.org/10.1002/app.1977.070210919.
[4]   Zeman L.J., Tkacik G. Pore volume distribution in ultrafiltration. In: D.R. Lloyd (Eds.), Materials Science of Synthetic Membranes. American Chemical Society, Washington, USA, (1985).
[5]   Smolders C.A., Vugteveen E. New characterization methods for asymmetric ultrafiltration membranes. In: D.R. Lloyd (Eds.), Materials Science of Synthetic Membranes. American Chemical Society, Washington, USA, (1985).
[6]   Scheele C.W. Chemical Observations and Experiments on Air and Fire with a Prefatory Introduction by Torbern Bergman to Which are Added Notes, by Richard Kirwan with a Letter to Him from Joseph Priestley. J.R. Forster, J. Johnson Trans. by, London, UK, (1780).
[7]   Kaiser H. Ueber die Verdichtung von Gasen an Oberflächen in ihrer Abhängigkeit von Druck und Temperatur. Wied. Ann, 14(451) (1881) https://doi.org10.1002/andp.18812480404.
[8]   Brunauer S., Deming L.S., Deming W.S., Teller E. On a Theory of the van der Waals Adsorption of Gases. J Am Chem Soc, 62 (1940) 1723–1732. https://doi.org/10.1021/ja01864a025.
[9]   Calvo J.I., Bottino A., Prádanos P., Palacio P., Hernández A. Porosity. In: E.M.V. Hoek, V.V. Tarabara (Eds.), Encyclopedia of Membrane Science and Technology. Wiley Intersci. Pub, New York, USA, (2013), 1062–1086.
[10] IUPAC Manual of Symbols and Terminology, Appendix 2, Pt. I, Colloid and Surface Chemistry. Pure Appl Chem, 31 (1972) 578. http://dx.doi.org/10.1351/pac197231040577.
[11] Mikhail R.Sh., Brunauer S., Bodor E.E. Investigation of a complete pore structure analysis. I. Analysis of micropores. J Colloid Interface Sci, 26 (1968) 45–53. https://doi.org/10.1016/0021-9797(68)90270-1.
[12] Dubinin M.M. Fundamentals of the theory of adsorption in micropores of carbon adsorbents: Characteristics of their adsorption properties and micropores structure. Carbon, 278 (1989) 457–467. https://doi.org/10.1016/0008-6223(89)90078-X.
[13] Gregg S.J., Sing K.S.W. Adsorption, Surface Area and Porosity, 2nd ed., Academic Press, London, UK, (1982). https://doi.org/10.1002/bbpc.19820861019.
[14] Sing K.S.W., Everett D.H., Haul R.A.W., Moscou L., Pieroti R.A., Rouquerol J. Siemieniewska T. Reporting physisorption data for gas/solid systems with special reference to the determination of surface area and porosity (Recommendations 1984). Pure Appl Chem 57 (1985). http://dx.doi.org/10.1351/pac198557040603.
[15] Rouquerol J., Avnir D., Fairbridge C.W., Everett D.H., Haynes J.M., Pernicone N., Ramsay J.D.F., Sing K.S.W., Unger K.K. Recommendations for the characterization of porous solids (Technical Report). Pure & Appl Chem, 66 (1994) 1739–1758. https://doi.org/10.1351/pac199466081739.
[16] Langmuir I. The adsorption of gases on planes surfaces of glass, mica and platinum. J Am Chem Soc, 40 (1918) 1361–1368. https://doi.org/10.1021/ja02242a004.
[17] Burgess C.G.V., Everett D.H., Nuttall S. Adsorption hysteresis in porous materials. Pure & Appl Chem, 61 (1989) 1845–1852. http://dx.doi.org/10.1351/pac198961111845.
[18] Fraile Sainz J.C. Utilización de técnicas de adsorción de gases para la caracterización textural de materiales micro y mesoporosos. PhD Thesis, Universidad Autónoma de Barcelona, Barcelona (Spain) 2022.

[19] López R.H. Caracterización de Medios Porosos y Procesos Percolativos y de Transporte. PhD Thesis, Universidad Nacional de San Luis, San Luis (Argentina), 2004.

[20] IUPAC. Porosity and pore size distribution of materials. Method of evaluation by gas adsorption. British Standard 7591, 1992.

[21] Brunauer S., Emmett P.H., Teller E. Adsorption of gases in multimolecular layers. J Am Chem Soc, 60 (1938) 309–319. https://doi.org/10.1021/ja01269a023.

[22] Barrett E.P., Joyner L.G., Halenda P.P. The determination of pore volume and area distributions in porous substances. I. Computations from nitrogen isotherms. J Am Chem Soc, 73 (1951) 373–380. https://doi.org/10.1021/ja01145a126.

[23] Dollimore D., Heal G.R. An improved method for the calculation of pore size distribution from adsorption data. J Appl Chem, 14 (1964) 109–114. https://doi.org/10.1002/jctb.5010140302.

[24] Halsey G.D. Physical adsorption on non-uniform surfaces. J Chem Phys, 16 (1948) 931–937. https://doi.org/10.1063/1.1746689.

[25] De Boer J.H., Lippens B.C., Linsen B.G., Broekhoff J.C.P., van den H.A., Osinga Th.V. J. The t-curve of multimolecular N2-adsorption. Colloid Interface Sci, 21 (1966) 405–414. https://doi.org/10.1016/0095-8522(66)90006-7.

[26] Harkins W.D., Jura G.J. The decrease of free surface energy as a basis for the development of equations for adsorption isotherms; and the existence of two condensed phases in films on solids. J Chem Phys, 12 (1944) 112–113. http://dx.doi.org/10.1021/ja01236a046.

[27] Kaneko K. Determination of pore size and pore size distribution. 1. Adsorbents and Catalysts. J Membrane Sci, 96 (1994) 59. https://doi.org/10.1016/0376-7388(94)00126-X.

[28] Dubinin M.M., Radushkevich L.V. The equation of the characteristic curve of activated charcoal. Dokl Akad Nauk SSSR, 55(331) (1947).

[29] Stoeckli H.F. Microporous Carbons and their characterization: The present State of the Art. Carbon, 28(1) (1990) 1–6. https://doi.org/10.1016/0008-6223(90)90086-E.

[30] Horváth G., Kawazoe K. Method for the calculation of effective pore size distribution in molecular sieve carbon. J Chem Eng Japan, 16(6) (1983) 470–475. https://doi.org/10.1252/jcej.16.470.

[31] Saito A., Foley H.C. Curvature and parametric sensitivity in models for adsorption in micropores. AIChE Journal, 37 (1991) 429–436. https://doi.org/10.1002/aic.690370312.

[32] Autosorb iQ: Gas Sorption System Operating Manual. Quantachrome Instruments, Corporate Drive Boynton Beach FL 33426 USA, April 2020.

[33] Lippens B.C., de Boer J.H. Studies on pore systems in catalysts V. The T-method J Catal, 4 (1965) 319–323. https://doi.org/10.1016/0021-9517(65)90307-6.

[34] De Boer J.H., Linsen B.G., Plas T., van der, Zondervan G.J. Studies on pore systems in catalysts: VII. Description of the pore dimensions of carbon blacks by the t method. J Catalysis, 4 (1965) 649–653. https://doi.org/10.1016/0021-9517(65)90264-2.

[35] Calvo J.I., Prádanos P., Hernández A., Bowen W.R., Hilal N., Lovitt R.W., Williams P.M. Bulk and surface characterization of composite UF membranes: Atomic force microscopy, gas adsorption–desorption and liquid displacement techniques. J Membrane Sci, 128 (1997) 7–21. https://doi.org/10.1016/S0376-7388(96)00304-3.

[36] Casco M.E., Cheng Y.Q., Daemen L.L., Fairén-Jiménez D., Ramos-Fernández E.V., Ramírez-Cuesta A.J., Silvestre-Albero J. Gate-opening effect ion ZIF-8: The first experimental proof using inelastic neutron scattering. Chem. Commun, 52 (2016) 3639–3642. https://doi.org/10.1039/C5CC10222G.

[37] López-Domínguez P., López-Periago A.M., Fernández-Porras F.J., Fraile J., Tobias G., Domingo C. Supercritical CO2 for the synthesis of nanometric ZIF-8 and loading with hyperbranched aminopolymers. Applications in CO2 capture. J CO2 Utilization, 18 (2017) 147–155. http://dx.doi.org/10.1016%2Fj.jcou.2017.01.019.

[38] Portolés-Gil N., López-Periago A.M., Borrás A., Fraile J., Solano E., Vallcorba O., Giner-Planas J., Ayllón J.A., Domingo C. Tuning the structure and flexibility of coordination polymers via solvent control of

tritopic triazine conformation during crystallization. Cryst Growth Des, 20 (2020) 3304–3315. https://doi.org/10.1021/acs.cgd.0c00088.

[39] Seaton N.A., Walton J.P.R.B., Quirke N. A new analysis method for the determination of the pore size distribution of porous carbons from nitrogen adsorption measurements. Carbon, 27 (1989) 853–861. https://doi.org/10.1016/0008-6223(89)90035-3.

[40] Adams D.J. Grand canonical ensemble Monte Carlo for a Lennard–Jones fluid. Molecular Physics, 29 (1975) 307–311. https://doi.org/10.1080/00268977500100221.

[41] Thommes M., Kaneko K., Neimark A.V., Olivier J.P., Rodriguez-Reinoso F., Rouquerol J., Sing K.S.W. Physisorption of gases, with special reference to the evaluation of surface area and pore size distribution (IUPAC Technical Report). Pure Appl Chem, 87(9–10) (2015) 1051–1069. https://doi.org/10.1515/pac-2014-1117.

[42] Rouquerol F., Rouquerol J., Sing K. Adsorption by Powders and Porous Solids. Academic Press, London, UK, (1999). ISBN: 9780080970356.

[43] Choma J., Jagiello J., Jaroniec M. Assessing the contribution of micropores and mesopores from nitrogen adsorption on nanoporous carbons: Application to pore size analysis. Carbon, 183 (2021) 150–157. https://doi.org/10.1016/j.carbon.2021.07.020.

[44] Eyraud C., Betemps M., Quinson J.F., Chatelet F., Brun M., Rasneur B. Détermination de la répartition des rayon de pores d'un ultrafilter. Bull Soc Chim France, 9–10 (1984) I237–I244.

[45] Cao G.Z., Meijerink J., Brinkman H.W., Burggraaf A.J. Permporometry study on the size distribution of active pores in porous ceramic membranes. J Membr Sci, 83 (1993) 221–235. https://doi.org/10.1016/0376-7388(93)85269-3.

[46] Mey-Marom A., Katz M. Measurement of the active pore size distribution of microporous membranes A new approach. J Membrane Sci, 27 (1986) 119. https://doi.org/10.1016/S0376-7388(00)82049-9.

[47] Katz M.G. Measurement of pore size distribution in microporous filters and membranes. In Harnessing Theory for Practical Applications, World Filtration Congress III. The Filtration Society, Uplands Press, Croydon, UK, (1982), Vol. II, 508.

[48] Katz M., Baruch G. New insights into the structure of microporous membranes obtained using a new pore size evaluation method. Desalination, 58 (1986) 199. https://doi.org/10.1016/0011-9164(86)87004-7.

[49] Cuperus F.P., Bargeman D., Smolders C.A. Characterization of anisotropic UF-membranes: Top layer thickness and pore structure. J Membr Sci, 61 (1991) 73–83. https://doi.org/10.1016/0376-7388(91)80007-S.

[50] Cuperus F.P., Bargeman D., Smolders C.A. Permporometry, The determination of the size distribution of active pores in UF membranes. J Membr Sci, 71 (1992) 57–67. https://doi.org/10.1016/0376-7388(92)85006-5.

[51] Tsuru T., Hino T., Yoshioka T., Asaeda M. Permporometry characterization of microporous ceramic membranes. J Membrane Sci, 186 (2001) 257–265. https://doi.org/10.1016/S0376-7388(00)00692-X.

[52] Tung K.-.L., Chang K.-.S., Wu T.-.T., Lin N.-.J., Lee K.-.R., Lai J.-.Y. Recent advances in the characterization of membrane morphology. Current Opinion in Chemical Engineering, 4 (2014) 121–127. http://dx.doi.org/10.1016/j.coche.2014.03.002.

[53] Vaidya A.M., Haselberger N.J. A Novel Technique for the Characterization of Asymmetric Membranes by Permoporometry. Sep Sci Technol, 29 (1994) 2523–2531. https://doi.org/10.1080/01496399408002206.

[54] Hazri A., Vaidya A.M. A new computational technique for the analysis of diffusión permoporometric data for asymmetric membranes. J Membrane Sci, 101 (1995) 61–66. https://doi.org/10.1016/0376-7388(94)00276-5.

[55] Rahman M.A., Mutalib M.A., Li K., Othman M.H.D. Pore Size measurements and distribution for ceramic membranes. In: N. Hilal, A.F. Ismail, T. Matsuura, D. Oatley-Radcliffe (Eds.), Membrane Characterization. Elsevier, The Netherlands, (2017). ISBN: 978-0-444-63776-5.

[56] Puthai W., Kanezashi M., Nagasawa H., Wakamura K., Ohnishi H., Tsuru T. Effect of firing temperature on the water permeability of SiO2eZrO2 membranes for nanofiltration. J Membrane Sci, 497 (2016) 348–356. https://doi.org/10.1016/J.MEMSCI.2015.09.040.

[57] Ren X., Kanezashi M., Nagasawa H., Tsuru T. Plasma-assisted multi-layered coating towards improved gas permeation properties for organosilica membranes. RSC Adv, 5 (2015) 59837–59844. https://doi.org/10.1039/C5RA08052E.

[58] Šolcová O., Hejtmánek V., Šnajdaufová H., Schneider P. Liquid Expulsion Permporometry – A Tool for Obtaining the Distribution of Flow-Through Pores. Part Part Syst Charact, 23 (2006) 1–8. https://doi.org/10.1002/ppsc.200601014.

[59] Sah A., Castricum H.L., Bliek A., Blank D.H.A., ten Elshof J.E. Hydrophobic modification of γ-alumina membranes with organochlorosilanes. J Membrane Sci, 243 (2004) 125–132. https://doi.org/10.1016/j.memsci.2004.05.031.

[60] Chowdhury S.R., Schmuhl R., Keizer K., Ten Elshof J.E., Blank D.H.A. Pore size and surface chemistry effects on the transport of hydrophobic and hydrophilic solvents through mesoporous γ-alumina and silica MCM-48. J Membrane Sci, 225 (2003) 177–186. https://doi.org/10.1016/j.memsci.2003.07.018.

[61] Kim K.J., Fane A.G., Ben A.R., Liu M.G., Jonsson G., Tessaro I.C., Broek A.P., Bargeman D. A comparative study of techniques used for porous membrane characterization: Pore characterization. J Membrane Sci, 81 (1994) 35–46. https://doi.org/10.1016/0376-7388(93)E0044-E cup.

[62] Higgins S., Kennard R., Hill N., DiCarlo J., DeSisto W.J. Preparation and characterization of non-ionic block co-polymer templated mesoporous silica membranes. J Membrane Sci, 279 (2006) 669–674. https://doi.org/10.1016/j.memsci.2006.01.014.

[63] Qiu H., Xu N., Kong L., Zhang Y., Kong X., Wang M., Tang X., Meng D., Zhang Y. Fast synthesis of thin silicalite-1 zeolite membranes at low temperature. J Membrane Sci, 611 (2020) 118361. https://doi.org/10.1016/j.memsci.2020.118361.

[64] Blumenschein S., Böcking A., Kätzel U., Postel S., Wessling M. Rejection modeling of ceramic membranes in organic solvent nanofiltration. J Membrane Sci, 510 (2016) 191–200. https://doi.org/10.1016/j.memsci.2016.02.042.

[65] Kujundzic E., Yeo A., Hosseini S.S., Krantz W.B., Greenberg A. Evapoporometry, a novel technique for characterizing membrane pore-size distribution. Proc. 20th Annual Meeting of the North American Membrane Society and 11th International Conference on Inorganic Membranes 2010, NAMS/ICIM 2010, 436–437 2010. ISBN: 978-161738851-4.

[66] Krantz W.B., Greenberg A.R., Kujundzic E., Yeo A., Hosseini S.S. Evapoporometry – A novel method for determining the pore-size distribution of MF and UF membranes. Separations Division – Core Programming Topic at the 2011 AIChE Annual Meeting, 1, 354–355, 2011.

[67] Krantz W.B., Greenberg A.R., Kujundzic E., Yeo A., Hosseini S.S. Evapoporometry: A novel technique for determining the pore-size distribution of membranes. J Membrane Sci, 438 (2013) 153–166. https://doi.org/10.1016/j.memsci.2013.03.045.

[68] Krantz W.B., Lua M.DCh., Absalon J.L., Narayanswamy B. Prototype commercial evapoporometer instrument. J Membrane Sci, 655 (2022) 120573. https://doi.org/10.1016/j.memsci.2022.120573.

[69] Tanis-Kanbur M.B., Zamani F., Krantz W.B., Hu X., Chew J.W. Adaptation of evapoporometry (EP) to characterize the continuous pores and interpore connectivity in polymeric membranes. J Membrane Sci, 575 (2019) 17–27. https://doi.org/10.1016/j.memsci.2018.12.050.

[70] Tanis-Kanbur M.B., Peinador R.I., Hu X., Calvo J.I., Chew J.W. Membrane characterization via evapoporometry (EP) and liquid–liquid displacement porosimetry (LLDP) techniques. J Membrane Sci, 586 (2019) 248–258. https://doi.org/10.1016/j.memsci.2019.05.077.

[71] Zamani F., Jayaraman P., Akhondi E., Krantz W.B., Fane A.G., Chew J.W. Extending the uppermost pore diameter measurable via evapoporometry. J Membrane Sci, 524 (2017) 637–643. https://doi.org/10.1016/j.memsci.2016.11.082.

[72] Tanis-Kanbur M.B., Peinador R.I., Calvo J.I., Hernández A., Chew J.W. Porosimetric membrane characterization techniques: A review. J Membrane Sci, 619 (2021) 118750. https://doi.org/10.1016/j.memsci.2020.118750.

[73] Merriman L., Moix A., Beitle R., Hestekin J. Carbon dioxide gas delivery to thin-film aqueous systems via hollow fiber membranes. Chem Eng J, 253 (2014) 165–173. https://doi.org/10.1016/j.cej.2014.04.075.

Chapter 6
Gibbs–Thomson equation-based techniques

6.1 Thermoporometry (ThP)

6.1.1 Introduction

When a liquid is introduced into a porous material, the curvature of the liquid interface changes. The phase equilibrium thermodynamics shows that the curvature of the solid–liquid interface can be related to the phase change temperature. Thus, if a porous material is embedded in a liquid, when an attempt is made to freeze the liquid, it can be observed that, inside the pores, the freezing point is lower than in the nonporous areas. This difference is governed by the Gibb–Thomson equation, and as a consequence of this fact, if we can manage to determine these variations of the melting (or freezing) temperature, we will be able to determine the pore size distribution (PSD) of the sample.

There are several ways to do this: one possibility is to determine the heat fluxes associated with these phase transitions, using a differential scanning calorimeter (DSC), which gives rise to the technique usually known as thermoporosimetry or thermoporometry (ThP). We can also determine the amount of liquid remaining during the phase change, either by means of nuclear magnetic resonance (thus we will have NMR cryoporometry, or NMRP) or by measuring the amplitude of neutron scattering in the crystalline or liquid phases embedded in the sample (in this latter case, we will speak of Neutron diffraction cryoporometry). In this section, we discuss the ThP technique, which is older and has been more widely used in the characterization of membranes, leaving the relatively new NMRP for the next section.

Although the study of the dependence of the freezing point on the capillarity of a material was already addressed in early works such as Tamman in 1920 [1] or Kubelka in 1932 [2], it was not until the 1950s that the phenomenon was properly understood and modelled [3–5]. From this point on, Maurice Brun's doctoral thesis, carried out at the Claude Bernard University in Lyon [6], can be considered the starting point for the establishment of ThP as an independent porosimetric technique.

In fact, the first papers published on the technique came all from Brun, Quinson and Eyraud's own group [7, 8]. Subsequently, the technique experienced strong interest, which resulted in an increasing number of publications in the 1980s [9–11]. Not too long ago a good review on the topic, in two parts, was published by Riikonen et al. [12, 13].

https://doi.org/10.1515/9783110792195-006

6.1.2 Gibbs–Thompson equation and pore radius

The principle of ThP relies on the change in the freezing and melting temperatures of a liquid when it is confined inside a pore or porous material. The equation that governs this process is the Gibbs–Thompson equation. This equation has been obtained in the section devoted to some thermodynamic fundamentals as follows:

$$T = T_0 - \frac{2T_0\gamma_{s\ell}\cos\theta_{s\ell}}{\rho_\ell\Delta H_f}\frac{1}{r_p} \tag{6.1}$$

where T is the temperature of the actual phase transition of the confined liquid, T_0 is the transition temperature of the non-confined liquid, r_p is the pore radius, ρ_ℓ is the density of the liquid, $\gamma_{s\ell}$ and $\theta_{s\ell}$ are the surface tension and the contact angle in the solid (pore wall)–liquid interface, and ΔH_f is the enthalpy of the phase change per unit mass of dry material.

Equation (6.1) can be written, equivalently, by relating ΔT to the temperature increase, $\Delta T = T - T_0$, to obtain

$$\Delta T = -\frac{2T_0\gamma_{s\ell}\cos\theta_{s\ell}}{\rho_\ell\Delta H_f}\frac{1}{r_p} \tag{6.2}$$

It is necessary to accept that there could exist a certain thickness of permanent liquid in contact with the pore walls, δ, that would leave an effective freezable radius that effectively decreases the radius $(r_p - \delta)$, and consequently:

$$\Delta T = -\frac{2T_0\gamma_{s\ell}\cos\theta_{s\ell}}{\rho_\ell\Delta H_f\left(r_p - \delta\right)} \tag{6.3}$$

This ΔT versus r_p relationship would allow to extract information on pore radii from the temperature depression in the melting/freezing process of a given liquid within the pores of a porous material. The most common liquid used to freeze/melt to study the capillarity effects on the temperature of the phase change has been water, but several organic liquids such as chlorobenzene, cyclohexane and 1,4-dioxane can also be used. Some other organics that have been tested are *cis*-decalin, *trans*-decalin, benzene, chlorobenzene, naphthalene, and heptane [14]. Pore radii as large as 1 μm may be quantitatively measured with appropriate experimental conditions.

Given that, usually, ΔT for a porous material would be measured by using DSC, let us start by explaining briefly the basic principles of DSC.

6.1.3 Differential scanning calorimetry

DSC is a technique that measures the difference between the heat flow into a substance and a reference. This is done as a function of temperature while the substance and reference are subjected to a controlled temperature protocol. In fact, there are no heat flow metre, so indirect techniques must be used for this purpose. DSC uses the temperature difference developed between the sample, and a reference to allow the calculation of the heat flow.

Two types of DSC [15] instruments are heat flux and power compensation, as shown in Fig. 6.1.

Originally, heat flux DSC developed from differential thermal analysis (DTA). The basic design of a DTA includes a single furnace containing both the sample and the reference holders while the furnace is heated according to a pre-established linear programme. The signals from the DTA sensors, usually thermocouples, are then fed to an amplifier. Given that the sample and the reference are heated from the outside, the DTA response is now susceptible to undesired effects of heat transport through the sample. In DTA, some of the factors giving spurious heat transport could be the amount, packing or thermal conductivity of the sample. These problems are reduced for DSC when the sample is separated from direct contact with the sensor and encapsulated in a pan constructed of a high thermal conductivity material (metals such as platinum and aluminium are used frequently).

Power compensation DSCs, unlike the heat flux DSC, have two separate identical holders (sample and reference holders), each with its own heater and sensor. These two calorimeters are placed in a common block of constant temperature. The sample holder contains the sample in a pan, and the reference container contains an empty sample pan as a reference. Platinum resistance thermometers measure the temperature. These thermometers are built into the base of the calorimeter holder. There are two individual heaters.

In both cases, an inert purge gas of constant and predetermined rate flows through the cells.

In a power compensation DSC, the difference in power applied to the sample and the reference furnaces is directly registered. In the case of heat flux DSCs, when the arrangement is ideally symmetrical (samples of the same kind), equally high heat flow rates flow into the sample and the reference sample. The differential temperature signal ΔT (originally a difference between electric potentials) is then zero. If this steady-state equilibrium is disturbed by a sample transition, a differential signal is generated which is proportional to the difference between the heat flow rates to the sample and to the reference sample.

In both cases, the differential power flow per unit mass $\dot{W}$ is registered and represented as a function of temperature showing a peak whenever there is a phase change in the sample, which is not happening in the reference,

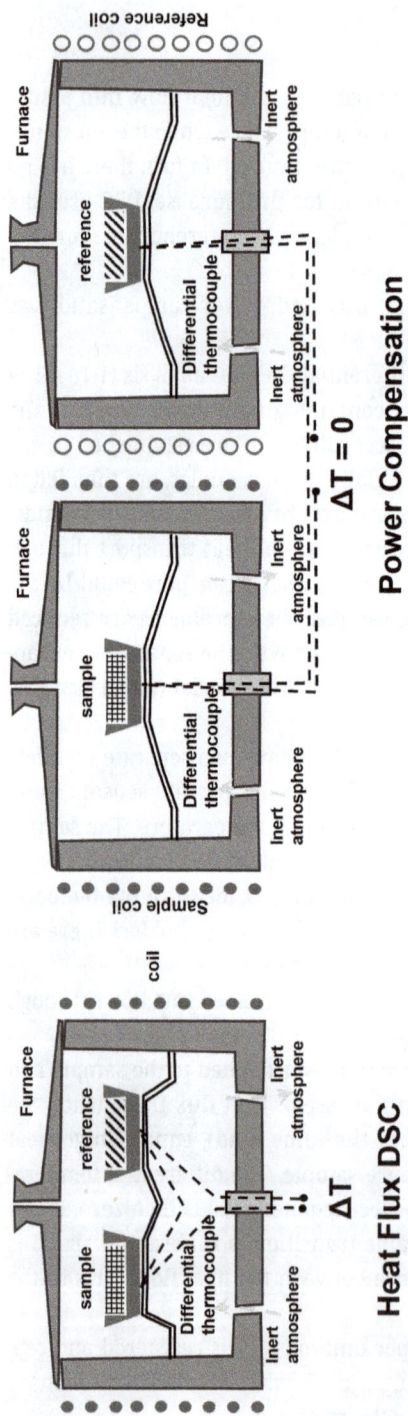

Fig. 6.1: The two standard procedures to perform DSC.

6.1.4 DSC thermoporometry

The general approach to the DSC measurements required in ThP can be described as follows. The samples are usually cooled far below the freezing temperature until detecting a large exothermic response revealing that the sample is completely frozen. ThP measurements on frozen samples are performed afterwards by heating through the melt transitions of both the pore and the excess liquid, not penetrating the pores, and noting the different temperatures corresponding to these two steps. Other procedures involve heating/cooling cycles. This technique requires slowly heating through the pore melting point until melting of the excess (external) phase just commences, then cooling back at the same slow rate through the freezing of the liquid inside the pores or dispersed phase. Further, very slow heating/cooling cycles can be performed. This procedure has the advantage of decreasing undercooling that could blur the melting point, because the still frozen external phase serves to nucleate the crystallization of the confined liquid once the equilibrium transition temperature corresponding to the pore size is reached. This technique minimizes the differences in the freezing and melting DSC responses.

Although it is important to be aware that melting/freezing hysteresis is unavoidable and always appears experimentally for any experimental conditions, hysteresis has been frequently attributed to pore shape and geometry. This is most commonly interpreted in terms of pore shape and geometry; for example, Brun et al. [7] predicted that cylindrical pores would give a temperature depression for melting that should be half that obtained when freezing, while both the temperature depressions would be equal for pure spherical pores. Enüstün et al. [16] stated that melting should be controlled by the radius of the pore, while freezing should be determined by the radius of the opening to the pore cavity. This idea is based on the occurrence of freezing due to nucleation triggered by the advancing solid phase. Denoyel and Pellenq [17] proposed that the phenomenon is due to meta-stabilities induced by differences in surface tensions between the liquid and solid phases and between the liquid phase and the pore walls that would be different in nature for the advancing solid phase (freezing) and the advancing liquid phase (melting). In summary, it seems clear that the melting temperature is greater than the equilibrium freezing temperature. Thus, it would be necessary to test both melting and freezing ThP.

The procedure is illustrated in Fig. 6.2, where the different possible measures of ΔT are shown, for the melting process, depending on whether we fix our attention on the onset, or the peaks, or from the onset to the peak.

Unfortunately, the linear dependence predicted by eq. (6.3) without any independent term is not totally fulfilled but it is necessary to allow for an independent term, possibly due to the dependence on the temperature of the terms within the slope in a ΔT versus r_p plot. Therefore, eq. (6.3) should be modified to

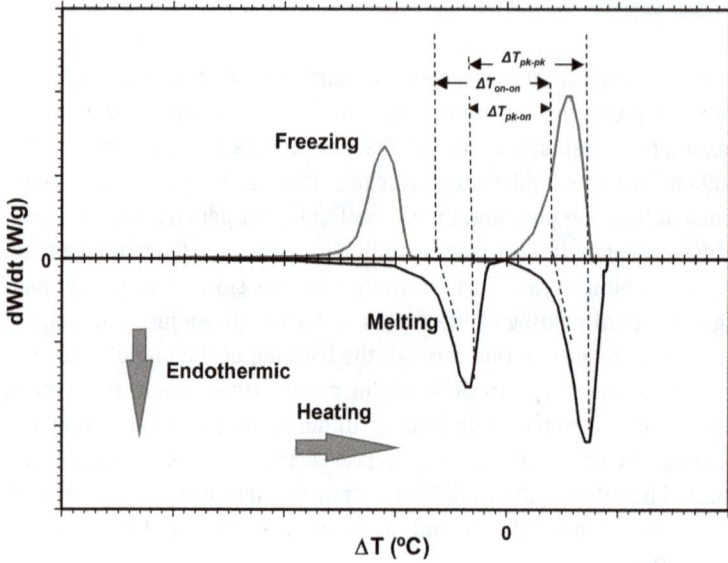

Fig. 6.2: Qualitative melting–freezing DSC response for a porous material.

$$\Delta T = A\left(\frac{1}{r_p - \delta}\right) + B \tag{6.4}$$

and

$$r_p = A\left(\frac{1}{\Delta T - B}\right) + \delta \tag{6.5}$$

Of course, these constants depend on the liquid filling the pores and also on whether the process is melting or freezing. Brun et al. [7] arrived, for water, to

$$(r_p)_m = -32.33\,\frac{1}{\Delta T_m} + 0.68 \tag{6.6}$$

$$(r_p)_f = -64.67\,\frac{1}{\Delta T_f} + 0.57 \tag{6.7}$$

with r_p given in nm and ΔT in K. Ishikiriyama et al. [18,19] proposed that

$$(r_p)_m = -33.30\,\frac{1}{\Delta T_m} + 0.32 + \delta_m \tag{6.8}$$

$$(r_p)_f = -56.36\,\frac{1}{\Delta T_f} + 0.90 + \delta_f \tag{6.9}$$

where $0.5 < \delta_m < 2.2$ and $0.6 < \delta_f < 2.8$.

While Landry [20] obtained

$$(r_p)_m = -19.082 \left(\frac{1}{(\Delta T)_m + 0.1207} \right) + 1.12 \tag{6.10}$$

$$(r_p)_m = -38.558 \left(\frac{1}{(\Delta T)_m - 0.1719} \right) + 0.04 \tag{6.11}$$

It is worth taking into account that these correlations have been obtained by different methods [20]. The correlations of Brun and Ishikiriyama were obtained by more or less theoretical correlations for the parameters in eq. (6.3), the last one also comparing the size distributions between ThP and nitrogen adsorption measurements. Landry conducted a fitting procedure for calibrated materials with known pore sizes assumed to be perfectly homogeneous. In Fig. 6.3, eqs. (6.6)–(6.11) are plotted. Note that, especially for melting, the temperature depression is more sensible to pore size for narrow pores (high $1/r_p$).

The equivalent plots for chlorobenzene, cyclohexane and water are compared in Fig. 6.4. Note that the pore radius ranges to be detected could advise to use one or the other filling liquid. For example, for large pores, cyclohexane shows that a higher slope for low $1/r_p$ should be preferable. Cyclohexane should be similarly sensitive for pores around 5 nm than water for pores in the 2 nm range, because they show similar slopes within these ranges in Fig. 6.4.

6.1.5 Pore volumes and thermoporometry

If we want to get information on the porous volume per unit mass of the dry sample, we can start by neglecting δ and differentiating eq. (6.3), or equivalently from eq. (6.2) to obtain

$$\frac{dV_p}{dr_p} = \frac{dV_p}{dT} \left(\frac{dT}{dr_p} \right) = \frac{2T_0 \gamma_{s\ell} \cos \theta_{s\ell}}{\rho_\ell \Delta H_f} \frac{1}{r_p^2} \left(\frac{dV_p}{dT} \right) \tag{6.12}$$

$$\frac{dV_p}{dr_p} = \frac{K_{GT}}{r_p^2} \left(\frac{dV_p}{dT} \right) \tag{6.13}$$

where

$$K_{GT} = \frac{2T_0 \gamma_{s\ell}}{\rho_\ell \Delta H_f} \cos \theta_{s\ell} \tag{6.14}$$

Moreover, the derivative dV_p/dr_p can be evaluated by

$$\frac{dV_p}{dr_p} = \frac{dV_p}{dt} \left(\frac{dt}{dT} \right) \frac{dT}{dr_p} \tag{6.15}$$

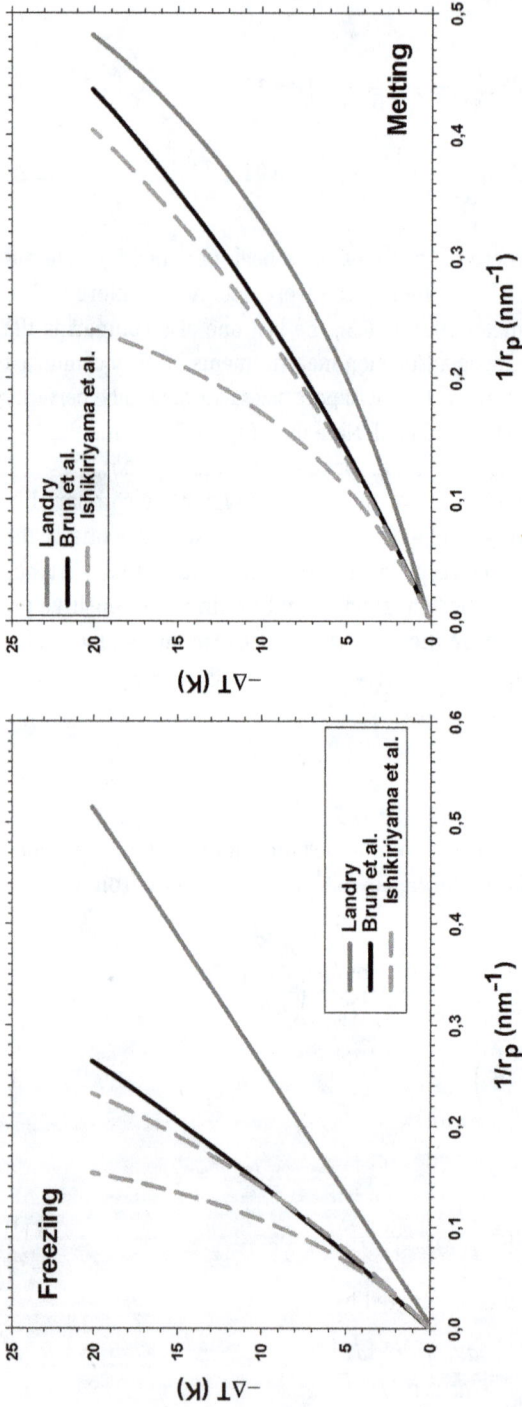

Fig. 6.3: Phase change temperature reduction for water (during freezing, left, and melting, right) as a function of $1/r_p$ according to eqs. (6.6)–(6.11). The curves representing the work of Ishikiriyama et al. correspond to the minimum and maximum values of δ that they proposed.

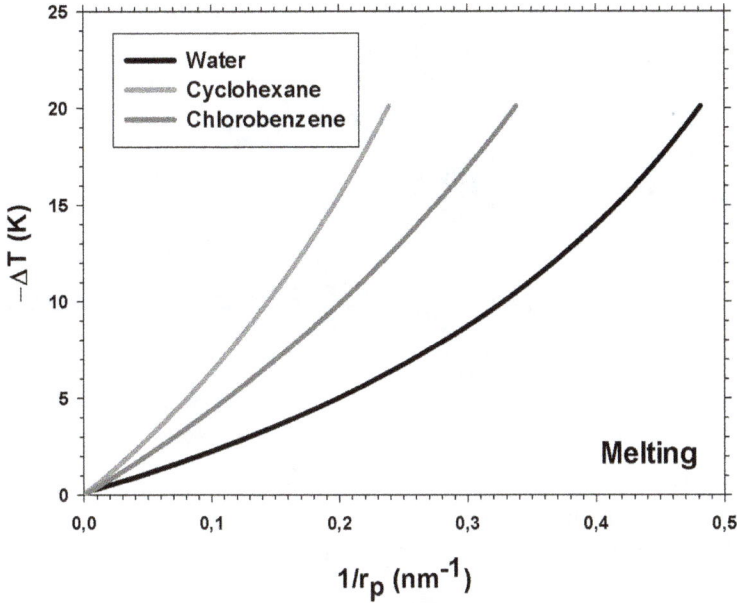

Fig. 6.4: Change in the melting temperature as a function of $1/r_p$ for some organics when compared with water, according to the data given by Landry [20].

But

$$\frac{dV_p}{dt} = \left(\frac{1}{m\Delta H_f \rho}\right) \frac{dW}{dt} \tag{6.16}$$

where W is the energy required to heat or cool the sample and $dW/dt = \dot{W}$ is the power transferred. Finally, m is the mass and ρ_ℓ is the density of the liquid. Then

$$\frac{dV_p}{dr_p} = \left(\frac{1}{m\Delta H_f \rho_\ell}\right) \frac{\dot{W}}{\dot{T}} \frac{dT}{dr_p} \tag{6.17}$$

With $dt/dT = 1/\dot{T}$, $\dot{T}$ is the temperature speed.

According to eqs. (6.1), (6.3) and (6.14):

$$\frac{dT}{dr_p} = \frac{K_{GT}}{r_p^2} \tag{6.18}$$

Thus,

$$\frac{dV_p}{dr_p} = \left(\frac{1}{m\Delta H_f \rho}\right) \frac{\dot{W}}{\dot{T}} \frac{K_{GT}}{r_p^2} \tag{6.19}$$

This method has been used to characterize porous materials, specifically ultrafiltration membranes, giving good results for PSD in the range of 2–30 nm [21–23].

Once the PSD given by eq. (6.19) has been evaluated, we can calculate the total porous volume and surface as well as the average pore radius by

$$V_p = \int_0^\infty \left(\frac{dV_p}{dr_p} \right) dr_p \tag{6.20}$$

$$S_p = \int_0^\infty \frac{2}{r_p} \left(\frac{dV_p}{dr_p} \right) dr_p \tag{6.21}$$

$$\langle r_p \rangle = 2 \left(\frac{V_p}{S_p} \right) \tag{6.22}$$

Equations (6.21) and (6.22) assume that the pores are cylindrical.

Some sources of error in the technique are:

1. The Gibbs–Thomson equation is approximate because it comes from the approximation: $\ln \frac{T}{T^0} = -\frac{\Delta T}{T_0}$.
2. $K_{GT} = \frac{2T_0 \gamma_{s\ell}}{\rho_\ell \Delta H_f} \cos \theta_{s\ell}$ could be not constant. Rather, this parameter would depend on the pore radius and thus on temperature in a difficult-to-know manner. The number 2 should correspond to hemispheric interfaces that would appear only for cylindrical pores.
3. It is difficult to have perfectly homogeneous porous materials in a wide enough range to perform the fitting of eqs. (6.4) and (6.5).

These drawbacks need to be taken into account when using this technique. Probably due to these drawbacks and to the laborious process involved in the technique, it is worth noting that this technique is not being used extensively.

6.1.6 Applications

Figure 6.5 shows the pore range that can be studied by ThP, whereas Fig. 6.6 shows the number of publications concerning ThP and membranes within their topic in the WOS web.

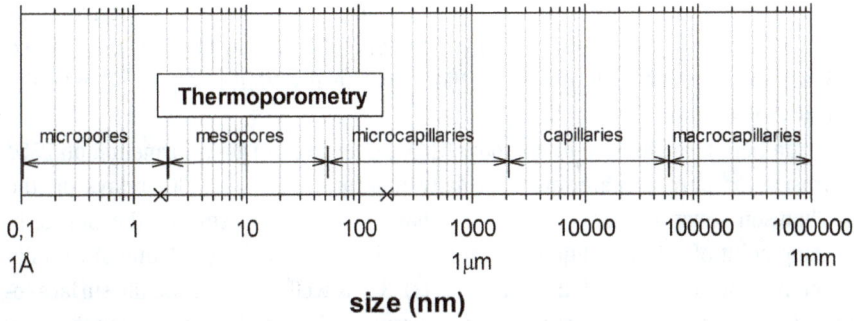

Fig. 6.5: Range of pores detectable by ThP.

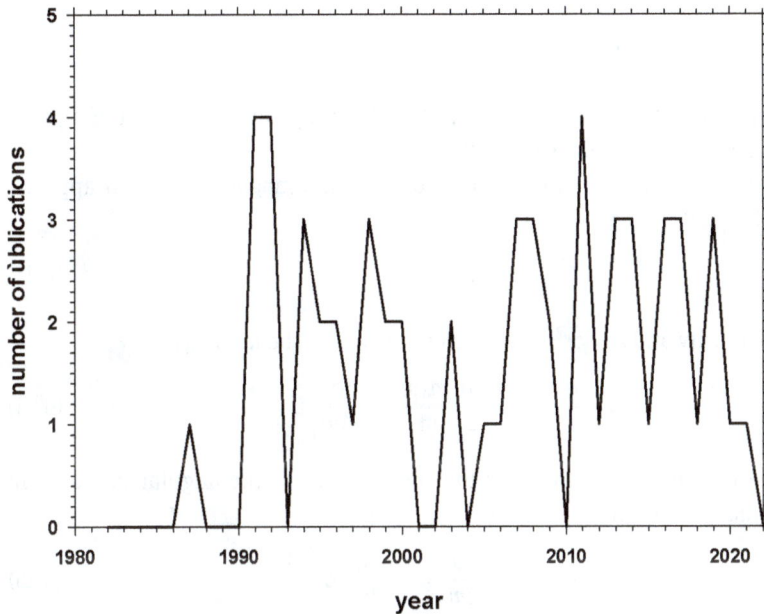

Fig. 6.6: Number of publications in the WOS web, from Clarivate, on topics such as thermoporometry and membranes.

6.2 NMR porometry

In 1938, I.I. Rabi of Columbia University accurately measured the nuclear magnetic moments by means of magnetic resonance absorption of molecular beams [24]. In 1944, I.I. Rabi was bestowed with the Nobel Prize in Physics for this work. Shortly afterwards, in 1946, Bloch [25] and Purcell et al. [26] demonstrated NMR of solids and shared the Nobel Prize in Physics in 1952. This was the starting mark for the development of NMR. Actually, neither the further developments of NMR as a spectroscopic technique [27, 28] nor the medical applications of NMR, which have reached an extraordinary development [29] and made NMR, specifically NMR imaging, an essential diagnostic tool, are our objectives.

Our aim here is to describe briefly how NMR can be used to get information on PSDs and porosity for porous materials, especially membranes. NMRP [30] relays on the Gibbs–Thomson equation that correlates the characteristic pore sizes and the change in the freezing point of a certain liquid, or melting point of its solid crystal, due to confinement within a porous matrix. NMR techniques [31, 32] as well as ThP probe the surface-to-volume (S/V) ratio of the pore structure and assumptions have to be made about the pore geometry to evaluate pore radii or diameters of the pores in a given porous material [33].

6.2.1 Nuclear magnetism and resonance

Let us start by analysing the interaction of the orbital dipolar momentum of a particle, with a charge q, with an external magnetic field $\vec{B}$.

For a charge describing a circular orbit of radius, a magnetic moment appears, which is (see Fig. 6.7)

$$\vec{\mu}_q = IA\hat{z} \tag{6.23}$$

Here I is the intensity and $A = \pi r^2$ is the area of the orbit. The intensity is

$$I = \frac{q}{\tau} = \frac{q}{\frac{2\pi r}{v}} = \frac{q}{2A}\frac{m_q vr}{m_q} = \frac{q}{2Am_q}L \tag{6.24}$$

where τ is the period in the orbit, v is the speed of q, L is the angular momentum and m_q is the mass of the charge q. Then

$$\vec{\mu}_q = IA\hat{z} = \frac{q}{2m_q}L\hat{z} = \frac{q}{2m_q}\vec{L} \tag{6.25}$$

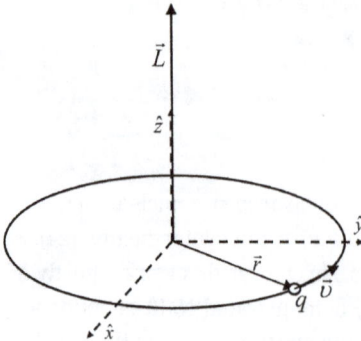

Fig. 6.7: A charge q, moving in a circular orbit.

According to mechanics of rotation, the moment of forces and angular momentum are correlated by

$$\vec{M} = \frac{d\vec{L}}{dt} \tag{6.26}$$

The moment acting on the dipole $\vec{\mu}_q$

$$\vec{\mu}_q = \frac{q}{2m_q}\vec{L} = \gamma_q\vec{L} \tag{6.27}$$

would be

$$\vec{M} = \vec{\mu}_q \times \vec{B} \tag{6.28}$$

Thus,

$$L \sin\theta \frac{d\phi}{dt} = \mu_q B \sin\theta \tag{6.29}$$

and, in consequence

$$\frac{d\phi}{dt} = \gamma_q B \tag{6.30}$$

This angular precession speed is the Larmor's angular frequency:

$$\omega_L = \gamma_q B \tag{6.31}$$

For an electron orbiting (see Fig. 6.8), we obtain:

$$\vec{\mu}_e = -\frac{|q_e|}{2m_e}\vec{L} = \gamma_e\vec{L} \tag{6.32}$$

Thus γ_e is negative. In atoms and molecules, the global orbital angular momenta usually cancel and give a zero resultant orbital magnetic moment. In fact, electrons also have an intrinsic angular momentum or spin. Spins do not cancel, and a molecule may have a total electron spin angular momentum. It was shown experimentally by Stern and Gerlach, and theoretically by Dirac, that associated with spin momentum $\vec{S}$ is a magnetic moment of magnitude $\vec{\mu}_{s'}$ which is

$$\vec{\mu}_S = -g_S\mu_B\frac{\vec{S}}{\hbar} \tag{6.33}$$

where $\mu_B = |e|\hbar/2m_e$ is the Bohr magneton:

$$\mu_B = 9.27 \times 10^{-24} J/T = 5.79 \times 10^{-5} eV/T \tag{6.34}$$

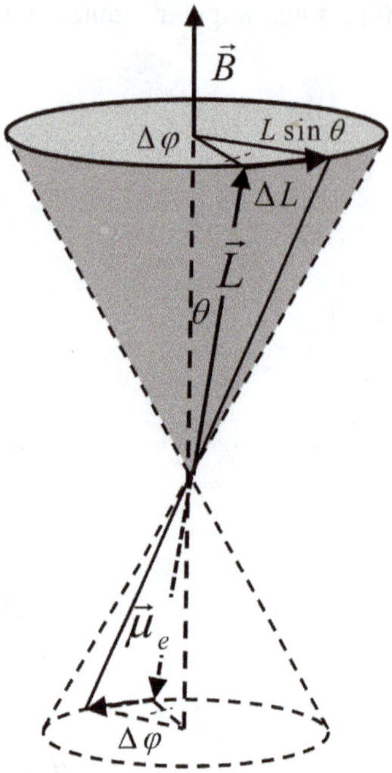

Fig. 6.8: NMR relaxation processes in magnetization.

g_s is a factor for free electrons, that is, $g_s = 2$ according to Dirac's equation. For an electron $s = S/\hbar = -1/2$, with s being the quantum number of spin, therefore, $\mu_S = \mu_B$.

A similar situation exists for nuclei. The spins of all nucleons lead to a global spin angular momentum $\hbar I$. I is an integral or half-integral number that can be called spin quantum number for the nucleus. The small magnetic moment μ_I associated with $\hbar I$ is given by

$$\mu_I = \gamma_I \hbar I \tag{6.35}$$

$$\gamma_I = \frac{g_I |q_e|}{2m_p} \tag{6.36}$$

where m_p is the mass of the proton. In early experiments, g_I (called Lande's splitting factor for the nucleus) was a phenomenological dimensionless constant found to be between about 0.5 and 5. The coefficient γ_I, closely related to g_I, according to eq. (6.36), is called nuclear magnetogyric ratio. Since I, like s, is of the order of unity while m_p is some thousand times m_e, the nuclear magnetic moments are typically of the order of 1,000 times smaller than those of electrons.

When $\vec{B}$ is applied to a magnetic nucleus, the nucleus will precess around $\vec{B}$. The precession frequency is the Larmor's frequency as obtained for the orbital resulting dipole. Moreover, when there is an external magnetic field $\vec{B}$, the nuclear spin angular momentum can appear in several quantum states along different orientations such that its projection on the magnetic field vector is $m_s \hbar$ where m_s runs from $-I$ to $+I$ in steps of unity. Thus, for $I = 3/2$, m_s takes the values $-3/2$, $-1/2$, $+1/2$, $+3/2$. The number m_s

may be used as a label to distinguish the various quantum states and their energies. The potential energy of a magnetic moment $\vec{\mu}_I$ in a magnetic external field $\vec{B}$ is

$$E_I = -\vec{\mu}_I \cdot \vec{B} \tag{6.37}$$

$$E_{m_s} = -\gamma_I m_s \hbar B \tag{6.38}$$

In our example with m_s taking the values $-3/2$, $-1/2$, $+1/2$ and $+3/2$, energy goes from $-\gamma_I \hbar B\, 3/2$ (for $m_s = 3/2$) to $\gamma_I \hbar B\, 3/2$ (for $m_s = -3/2$) in equal steps:

$$\Delta E = \gamma_I \hbar B \tag{6.39}$$

Thus, a transition from level to level would happen when

$$\Delta E = 2\pi \hbar v = \hbar \omega \tag{6.40}$$

$$\omega = \gamma_I B \Rightarrow v = 2\pi \gamma_I B \tag{6.41}$$

Here ω is the angular frequency (v is the linear frequency measured in Hz) of an electromagnetic radiation capable of being absorbed while making the energy of the spin state of the nucleus to jump one step. Note that this angular frequency is the Larmor's frequency ω_L for the spin precession (see eq. (6.31)):

$$\omega_L = \gamma_I B \tag{6.42}$$

Then the nuclear magnetogyric ratio $\gamma_I = \omega_L/B$ expressed as $\gamma_I/2\pi = v_L/B$ would be measured in Hz/T (hertz/tesla). Some gyromagnetic ratios are given in Tab. 6.1 as taken from the literature [34, 35].

Tab. 6.1: Some gyromagnetic constants.

	Isotopic abundance (%)	Spin	$\gamma_I = \omega_L/B$ (Hz/T)
Hydrogen ^{1}H	99.98	1/2	42.58
Carbon ^{13}C	1.108	1/2	10.81
Nitrogen ^{14}N	99.63	1	3.08
Oxygen ^{17}O	0.37	5/2	5.77
Fluorine ^{19}F	100	1/2	40.08
Phosphorus ^{31}P	100	1/2	17.25
Sulfur ^{33}S	4.7	3/2	3.27

For an electron, v_L lies in the microwave range from 10^9 to 10^{11} Hz, while for nuclei it lies in the radiofrequency (RF) range from 10^6 to 10^9 Hz. In both cases, the frequency depends on the externally applied magnetic field. Anyway, to get NMR, we need to bombard the nuclei with RF radiation.

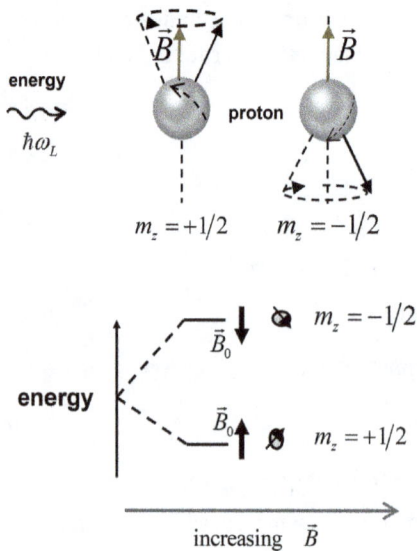

Fig. 6.9: Proton spin energy levels. Note that according to Eq. (6.39) the energy gap increases with B.

6.2.2 Magnetization

Because of the phenomena described above, a neat global magnetization appears that can be evaluated as follows [36]. Considering a collection of particles of spin ½, each particle can have two possible spin states (as shown in Fig. 6.9): $m_s = +½$ and $m_s = -½$ corresponding, respectively, to spin parallel and antiparallel to the external magnetic field. The total number of particles N is

$$N = N_{\uparrow\uparrow} + N_{\uparrow\downarrow} \tag{6.43}$$

Or

$$\left.\begin{aligned} N_{\uparrow\downarrow} &= N \Big/ \left(\frac{N_{\uparrow\uparrow}}{N_{\uparrow\downarrow}} + 1\right) \\ N_{\uparrow\uparrow} &= N \Big/ \left(\frac{N_{\uparrow\downarrow}}{N_{\uparrow\uparrow}} + 1\right) \end{aligned}\right\} \tag{6.44}$$

But

$$\frac{N_{\uparrow\uparrow}}{N_{\uparrow\downarrow}} = e^{\Delta E / k_B T} \tag{6.45}$$

where k_B is the Boltzmann constant and T is the temperature. Thus

$$\left.\begin{array}{l} N_{\uparrow\downarrow} = \dfrac{N}{e^{\gamma_I \hbar B/k_B T} + 1} \\[4mm] N_{\uparrow\uparrow} = \dfrac{N}{e^{-\gamma_I \hbar B/k_B T} + 1} \end{array}\right\}$$

(6.46)

But the corresponding magnetization per unit volume will be

$$\mathcal{M} = \mu\left(N_{\uparrow\uparrow} - N_{\uparrow\downarrow}\right) = N\mu\left(\frac{e^{\gamma_I \hbar B/k_B T} - 1}{e^{\gamma_I \hbar B/k_B T} + 1}\right) = N\mu \tanh\left(\frac{\gamma_I \hbar B}{2k_B T}\right)$$

(6.47)

But by doing a Taylor expansion of

$$\tanh\left(\frac{\gamma_I \hbar B}{2k_B T}\right) = \frac{\gamma_I \hbar B}{2k_B T} - \frac{1}{3}\left(\frac{\gamma_I \hbar B}{2k_B T}\right)^3 + \cdots$$

(6.48)

$$\mathcal{M} \approx N\mu \frac{\gamma_I \hbar B}{2k_B T}$$

(6.49)

In accordance with eq. (6.35), $\mu_l = \gamma_I I$ and for $I = 1/2$:

$$\mathcal{M} \approx N\frac{(\gamma_I \hbar)^2 B}{4k_B T}$$

(6.50)

This is the Curie's law, which states that magnetization for small applied magnetic fields is proportional to the field and inversely proportional to temperature.

6.2.3 Relaxation times: depolarization, tipping and FID

When protons get aligned in parallel to the static magnetic field, $\vec{B}$, they are polarized. But, of course, this polarization is a process that is not instantaneous but rather develops with a time constant known as the longitudinal relaxation time, T_1. This time evolution of magnetization can be identified by a receiver coil that measures magnetization in the longitudinal direction.

In order to isolate and measure transverse relaxation times we can use tipping and FID procedures. The technique can be described as follows. Firstly, we tip the magnetization from the longitudinal direction to a transverse plane. This tipping is accomplished by applying an oscillatory magnetic field $\vec{B}_\perp$ transversal to the primarily applied external field $\vec{B}$, which is a static magnetic field. But for the tipping to be efficient, the frequency of $\vec{B}_\perp$ must be equal to the Larmor frequency relative to $\vec{B}$.

If a nucleon is in the low-energy state, it may absorb energy provided by $\vec{B}_\perp$ and jump to a higher energy state. The application of $\vec{B}_\perp$ also causes the nucleons to precess in phase with one another due to resonance.

The angle through which the global magnetization vector is tipped is given by

$$\theta = \omega_L t = \gamma_I B_\perp t = \frac{\Delta E}{\hbar} t \tag{6.51}$$

with t being the time over which the oscillating magnetic acts. Pulses of the $B_\perp$ field help to analyse relaxation times.

When a $\pi/2$ pulse is applied after the population of protons is fully polarized, the protons precess in phase perpendicularly to the $\vec{B}$ field. Immediately after switching off $\vec{B}_\perp$ the protons' spin starts to dephase and the overall transversal magnetization will decay in what is called FID (free induction decay). FID is caused by the spin–spin coupling and diffusion (and also by the inhomogeneity of $\vec{B}$) causing different Larmor frequencies. This relaxation time is τ_2, which can be detected by a receiver coil that measures magnetization in the transverse direction.

From eqs. (6.25) and (6.32) along with eqs. (6.26) and (6.28), but now for the nuclear spin, we obtain

$$\frac{d\vec{\mu}_I}{dt} = \gamma_I \vec{\mu}_I \times \vec{B} \tag{6.52}$$

And in terms of the total magnetization corresponding to the global action of all the nuclear spin dipoles

$$\frac{d\vec{\mathcal{M}}}{dt} = \gamma_I \vec{\mathcal{M}} \times \vec{B} \tag{6.53}$$

Now we can write

$$\frac{d\vec{\mathcal{M}}}{dt} = \gamma_I \left(\mathcal{M}_x \hat{x} + \mathcal{M}_y \hat{y} + \mathcal{M}_z \hat{z} \right) \times B\hat{z} \tag{6.54}$$

$$\left. \begin{array}{l} \dot{\mathcal{M}}_x = \gamma_I \mathcal{M}_y B \\ \dot{\mathcal{M}}_y = -\gamma_I \mathcal{M}_x B \\ \dot{\mathcal{M}}_z = 0 \end{array} \right\} \tag{6.55}$$

Then, in order to include relaxation times, the Bloch equations for magnetization must be written as

$$\left. \begin{array}{l} \dot{\mathcal{M}}_x(t) = \gamma_I \mathcal{M}_y(t)B - \frac{\mathcal{M}_y(t)}{T_2} \\ \dot{\mathcal{M}}_y(t) = -\gamma_I \mathcal{M}_x(t)B - \frac{\mathcal{M}_x(t)}{T_2} \\ \dot{\mathcal{M}}_z(t) = \frac{\mathcal{M}_z^0 - \mathcal{M}_z(t)}{T_1} \end{array} \right\} \tag{6.56}$$

or

$$
\left.\begin{array}{l}
\dot{\mathcal{M}}_x(t) = \omega_L \mathcal{M}_y(t) - \frac{\mathcal{M}_y(t)}{T_2} \\[4pt]
\dot{\mathcal{M}}_y(t) = -\omega_L \mathcal{M}_x(t) - \frac{\mathcal{M}_x(t)}{T_2} \\[4pt]
\dot{\mathcal{M}}_z(t) = \frac{\mathcal{M}_z^0 - \mathcal{M}_z(t)}{T_1}
\end{array}\right\}
\tag{6.57}
$$

This should be solved with the following initial conditions:

$$
\left.\begin{array}{l}
\mathcal{M}_x^0 = \mathcal{M}_{xy}^0 \sin \varphi \\[4pt]
\mathcal{M}_y^0 = \mathcal{M}_{xy}^0 \cos \varphi \\[4pt]
\mathcal{M}_z^0 = 0
\end{array}\right\}
\tag{6.58}
$$

Thus, the solutions are

$$
\left.\begin{array}{l}
\mathcal{M}_x(t) = \mathcal{M}_{xy}^0 \cos \varphi\, e^{-t/T_2} \sin(\omega_L t) \\[4pt]
\mathcal{M}_y(t) = \mathcal{M}_{xy}^0 \sin \varphi\, e^{-t/T_2} \cos(\omega_L t) \\[4pt]
\mathcal{M}_z(t) = \mathcal{M}_z^f \left(1 - e^{t/T_1}\right)
\end{array}\right\}
\tag{6.59}
$$

with

$$
\left.\begin{array}{l}
\mathcal{M}_x^0 = \mathcal{M}_{xy}^0 \sin \varphi \\[4pt]
\mathcal{M}_y^0 = \mathcal{M}_{xy}^0 \cos \varphi
\end{array}\right\}
\tag{6.60}
$$

and

$$
\mathcal{M}_{xy} = \sqrt{\left(\mathcal{M}_x\right)^2 + \left(\mathcal{M}_y\right)^2}
\tag{6.61}
$$

The heat transfer to the surroundings through the spin–lattice interaction corresponds to τ_1, which is the longitudinal relaxation time, while the spin–spin coupling corresponds to τ_2, which is the transversal relaxation time. According to eq. (6.59), after time τ_2, the transversal magnetization decreases to $1/e \approx 36.79\%$, the initial one, while after τ_1, the longitudinal magnetization increases to $1 - 1/e \approx 63.21\%$, the final one. The decay processes are shown in Fig. 6.10.

The oscillation in Fig. 6.10 (bottom) corresponds to an ω_L angular frequency. If there is a set of different spin oscillations or different decaying times, we could obtain a spectrum in terms of frequency by performing a Fourier transform. Ernst and Anderson [37], for the first time, applied the Fourier transform to improve the primitive NMR analysis comprising a slow variation in frequencies of the exciting signal by the application of short pulses and an ulterior Fourier analysis to change the time decay to a frequency spectrum. This awarded R.R. Ernst the Nobel Prize in Chemistry in 1991.

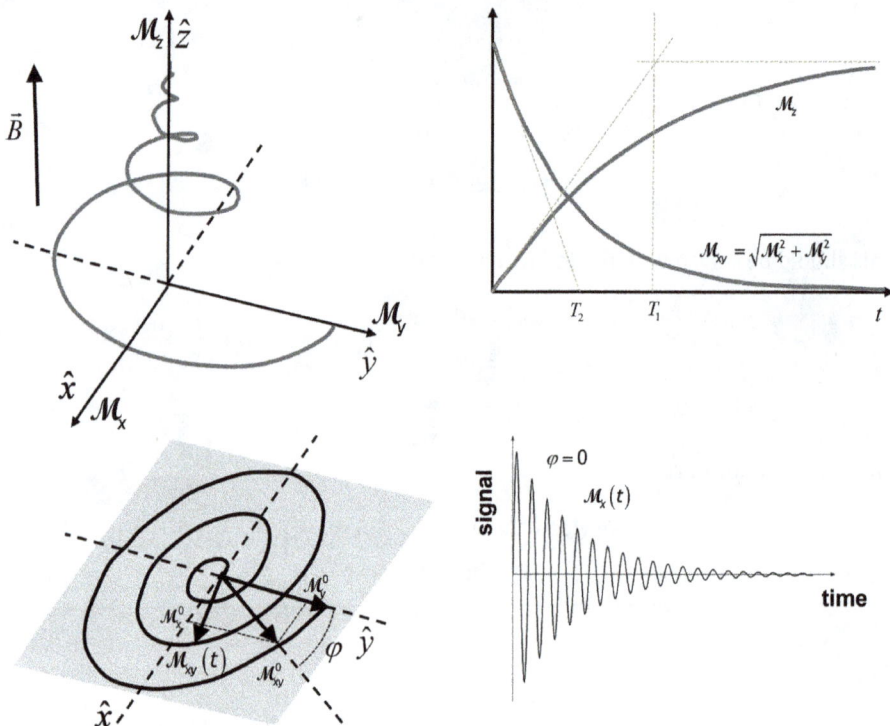

Fig. 6.10: Relaxation NMR processes in magnetization.

The process explained here corresponds to FID [25, 38]. The relaxation time in FID is the reciprocal of π times the linewidth at half height:

$$T_2 = \frac{1}{\pi \omega_{1/2}} \tag{6.62}$$

Because a pure decay with a single τ_2 gives a Lorentzian peak, this relaxation time is actually affected by many factors such as mainly inhomogeneities of the magnetic field.

6.2.4 Spin echoes

Apart from FID, other techniques that can be used to measure the signal intensity in NMR experiments are the individual spin echoes [39] and spin echo trains [40, 41]. In effect, because the FID relaxation time is very short, the so-called CPMG (after Carr, Purcell, Meiboom and Gill) series of pulses can be used. When total dephasing is completed after a $\pi/2$ pulse, spins are re-phased by using a π-pulse that inverts the dephasing order of spins to put them again in phase until relaxing and dephasing again. Then another π-pulse is applied to re-phase the spins so successively. In the process, called spin echo, the successive re-phasing procedures compensate the reversible dephasing caused by inhomogeneity but not those irreversible ones linked to spin–spin interaction. Thus, successive echoes show decaying amplitudes of spin along the perpendicular of the external field direction. If the application of the 90° pulse is followed after time τ by the first 180°pulse, then it would be necessary to wait for the same time τ until the appearance of the spin echo. That is, re-phasing time equals the dephasing time, and the spin echo peak occurs at 2τ. This time scale of 2τ corresponds to the abscissa of successive points along the decay linked to spin–spin coupling, thus allowing a more accurate, and mostly independent of inhomogeneities, calculation of τ_2.

6.2.5 NMR relaxometry

A simple modelling of relaxation times describes them as an average of surface and bulk contributions according to

$$\frac{1}{T_{1,2}} = \frac{1}{T_{1,2}^S} + \frac{1}{T_{1,2}^b} \tag{6.63}$$

Here $T_{1,2}^S$ corresponds to the contribution of the surface, while $T_{1,2}^b$ describes the contribution of the bulk. Experimentally, it is seen that $T_{1,2}^S \ll T_{1,2}^b$; thus, $1/T_{1,2} \approx 1/T_{1,2}^S$. Surface relaxation times are approximately proportional to the surface-to-volume ratio with a proportionality constant called relaxivity $\rho_{1,2}$ according to

$$\frac{1}{T_{1,2}} \simeq \frac{1}{T_{1,2}^S} = \rho_{1,2} \frac{S}{V} \tag{6.64}$$

The surface-to-volume ratio clearly depends on the geometry of material spin-responsive particles. Table 6.2 lists some possibilities. Note that all V/S ratios are inversely proportional to a characteristic length a; thus, it can be assumed that $S/V = K/a$.

In accordance with eq. (6.55) and according to the easiest measurement of T_2, and because $S/V = K/a$, we get

$$\frac{1}{T_2} = \frac{K\rho_2}{a} = \frac{K'}{a} \tag{6.65}$$

Thus, a calibration process should be performed to get $K\rho_2$ for each material. Moreover, not a single $1/T_2$ is obtained but a distribution of relaxing times. This leads to the distribution of T_2 and correspondingly of a. Of course, it would be theoretically possible to use a perfectly known material with particles of a quite regular geometry to get K. This would allow an experimental determination of ρ_2 and, if relaxivity would not depend on the particle's geometry, this would eventually lead to acquire information

Tab. 6.2: Volume-to-surface ratios for some shapes.

		Surface	Volume	S/V
Tetrahedron	Side = a	$a^2\sqrt{3}$	$a^3\sqrt{2}\big/12$	$6\sqrt{6}/a \cong 14.70/a$
Cube	Side = a	$6a^2$	a^3	$6/a$
Octahedron	Side = a	$2\sqrt{3}a^2$	$\left(\sqrt{2}\big/3\right)a^3$	$3\sqrt{6}/a \cong 7.35/a$
Icosahedron	Side = a	$5\sqrt{3}a^2$	$\left[5\left(3+\sqrt{5}\right)\big/12\right]a^3$	$12\sqrt{3}\big/\left(3+\sqrt{5}\right)a \cong 3.97/a$
Dodecahedron	Side = a	$3\sqrt{25+10\sqrt{5}}a^2$	$\left[\left(15+7\sqrt{5}\right)\big/4\right]a^3$	$12\sqrt{25+10\sqrt{5}}\big/\left(15+7\sqrt{5}\right)a$ $\cong 2.69/a$
Sphere	Diameter = a	πa^2	$(\pi/3)a^3$	$0.33/a$
Open cylinder	Diameter = a Length = l	$\pi a \ell$	$(\pi/4)a^2\ell$	$0.25/a$

on K/a, for an unknown sample of the same material. Then, assuming that the shape of the particles is constant, we could get the distribution of a and the most probable one, that is, the size distribution curve.

The NMR porosimetry relaxometric technique is extensively used as a geophysical analysis methodology both in the oil and gas industry [42] and building technologies [43]. NMR relaxometry can evaluate pore sizes from 10 nm to 100 μm [44].

6.2.6 NMR cryoporometry

If we deal with a certain porous material, we can impregnate it with water or some highly protonated organic liquids to fill the pores and then freeze and melt this water. In solids τ_2 is short (of the order of some microseconds), while it is much longer (in the seconds range) for liquids. This is shown in Fig. 6.11 [45, 46]. This difference in τ_2 allows an easy separation of the signal coming from the solid and liquid components, whose relative amounts change with temperature [33].

The Gibbs–Thompson equation gives the fusion temperature as follows:

$$T = T_0 - \frac{2(\gamma_{s\ell}\cos\theta_{s\ell})}{\rho_\ell \Delta s_f}\left(\frac{v_\ell}{r_p}\right) \tag{6.66}$$

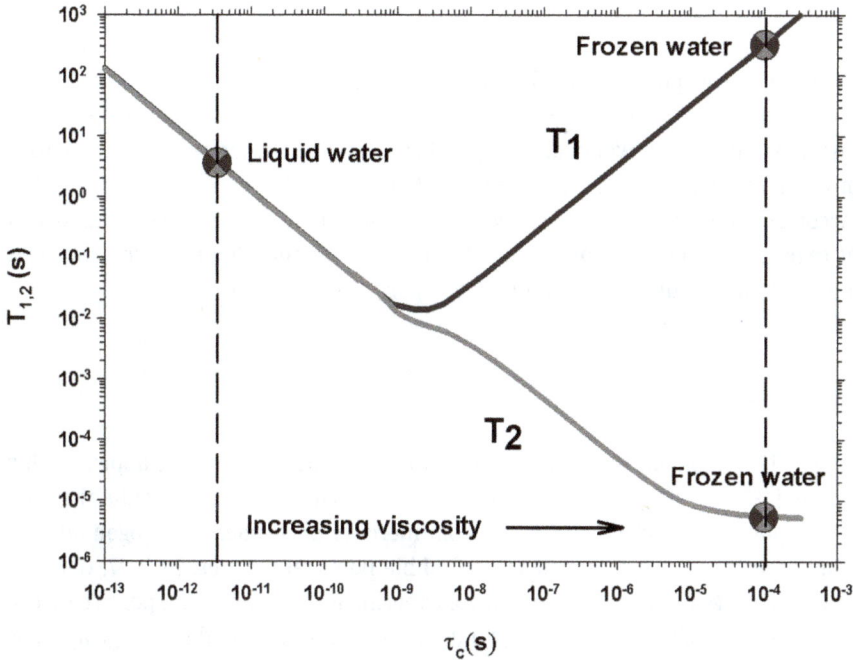

Fig. 6.11: NMR relaxation times as a function of correlation times.

Here the pore radius r_p would be the characteristic size a. Then:

$$\frac{dV}{da} = \frac{dV}{dT}\left(\frac{dT}{da}\right) = \frac{2(\gamma_{s\ell}\cos\theta_{s\ell})v_\ell}{a^2\Delta S_f}\frac{dV}{dT} \tag{6.67}$$

$$\frac{dV}{da} = \left(\frac{K_{GT}}{a^2}\right)\frac{dV}{dT} \tag{6.68}$$

where the Gibbs–Thompson constant has been defined as

$$K_{GT} = \frac{2(\gamma_{s\ell}\cos\theta_{s\ell})v_\ell}{\Delta S_f} \tag{6.69}$$

Here, the 2-factor can change based on the shape of the pores.

Because the maximum height of the spin echo at time 2τ can be assumed to be proportional to the volume of liquid in the sample if τ is small compared to the transverse relaxation rate. By choosing an appropriate value of τ, it is often possible to observe the liquid signal without any residual magnetization from the frozen solid, even when the solid is a plastic crystal. The integral intensity of the spectral peak obtained after a Fourier transform of the FID corresponding to the liquid fraction of the sample is plotted as a function of temperature to generate the so-called $I(T_2)$ curve [47].

Because $I(T_2)$ is proportional to the volume of the liquid phase, using eq. (6.68) would allow to get the pore volume distribution as a function of a and, after integration, the cumulative pore volume distribution. If the shape of the pores and other parameters in eq. (6.69) are unknown, we can use a well-known porous sample to calibrate and use as a reference. It would also be possible to use the technique to follow changes in porosity and focus on relative porosity alterations.

According to Rottreau et al. [48], NMRP can use water as a freezing liquid, which seems to be appropriate for pores below 10 nm, and t-butanol should detect pore sizes from 10 to 60 nm, while menthol could detect pores from 60 to 100 nm.

6.2.7 Applications

As early as 1981, a patent on the use of NMR to determine pore volume appeared due to Lauffer [49]. In 1989, Glaves and Smith [50] compared the NMR results they obtained for porous materials with those from mercury porosimetry, nitrogen adsorption/condensation and flow permeability/bubble point techniques. In 1987, Gallegos et al. proposed NMR relaxation spin–lattice experiments to analyse the pore structure of membranes [51, 52]. Schmidt et al. [47] used NMR complemented by image analysis

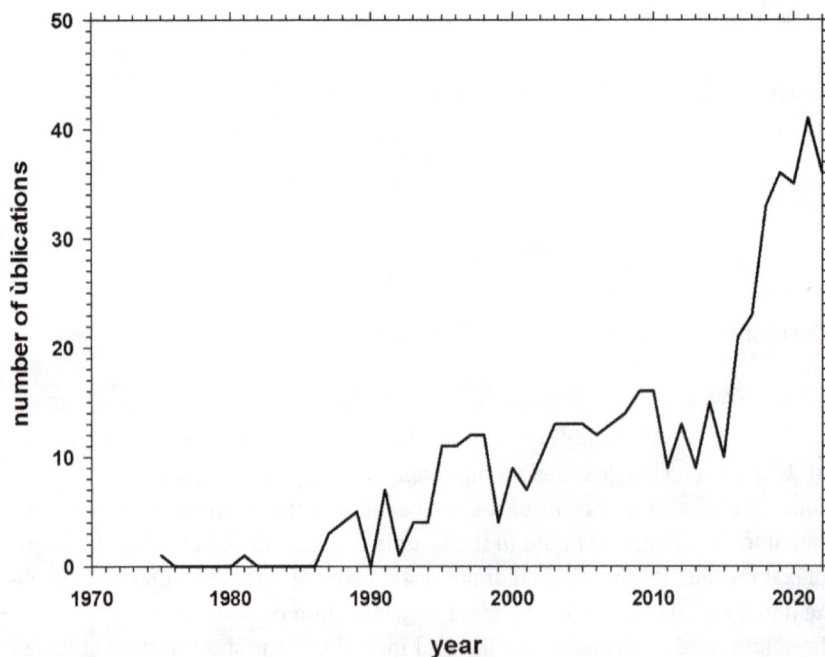

Fig. 6.12: Number of publications in the WOS web, from Clarivate, with NMR and pore in their titles.

and N_2 adsorption to study mesoporous materials, whereas the cryoporometric point of view appeared in the mid-1990s [30, 53, 54].

Figure 6.12 shows the number of publications including NMR and pore in their titles in the WOS web from Clarivate.

6.2.8 Advantages and disadvantages

NMRP is a technique for non-destructively determining PSDs in porous media through the observation of the depressed melting point of a confined liquid. It is suitable for measuring pore diameters in the range of 2 nm to 1 µm, depending on the absorbate. Whilst NMRP is a perturbative measurement, the results are independent of spin interactions at the pore surface and can offer direct measurements of pore volume as a function of pore diameter. PSDs obtained with NMRP have been shown to compare favourably with those from other methods such as gas adsorption, DSC thermoporosimetry and SAXS. The applications of NMRP include studies of silica gels, bones, cements, rocks and many other porous materials. It is also possible to adapt the basic experiment to provide structural resolution in spatially dependent PSDs, or behavioural information about the confined liquid [33].

References

[1]   Tamman G. Über eine Methode zur Bestimmung der Abhiingigkeit des Schmelzpunktes einer Kristallamelle von ihrer Dicke. Z. Anorg. Allg. Chem, 110 (1920) 166–168.

[2]   Kubelka P. Über den Schmellpunkt in sehr Engen Kapillaren. Zeitschrift Fur Elektrochemie Und Angewandte Physikalische Chemie, 38(8a) (1932) 611–614.

[3]   Defay R., Prigonine I. Tension Superficielle Et Absorption, L. Desoer (Eds.), Belgium, (1951). ISBN mkt0006861384.

[4]   Everett D.H. Thermodynamics of frost damage to porous solids. Trans Faraday Sot, 57 (1961) 1541–1549. https://doi.org/10.1039/TF9615701541.

[5]   Williams P.J. Properties and Behaviour of Freezing Soils, Vol. 72, Norvegian Geotechnical Institute Publ., Oslo, (1967).

[6]   Brun M. *Contribution à l'etude du changement d'état liquide–solide dans les milieux poreux. Méthode calorimétrique de détermination d'une répartition de rayons de pores*, Thesis, Claude Bernard University, Lyon, France, (1973).

[7]   Brun M., Lallemand A., Quinson J.F., Eyraud C. A new method for the simultaneous determination of the size and shape of pores: The thermoporometry. Thermochim Acta, 21(1) (1977) 59–88. https://doi.org/10.1016/0040-6031(77)85122-8.

[8]   Jallut C., Lenoir L., Bardot C., Eyraud C. Thermoporometry: Modelling and simulation of a mesoporous solid. J Membrane Sci, 68(3) (1992) 271–282. https://doi.org/10.1016/0376-7388(92)85028-H.

[9]   Brun M., Quinson J.F., Blanc R., Negre M., Spitz R., Bartholin M. Caractérisation texturale de résines en milieu réactionnel. Makromol Chem, 182 (1981) 873–882. https://doi.org/10.1002/macp.1981.021820319.

[10] Quinson J.F., Astier M., Brun M. Determination of surface areas by thermoporometry. Appl Cat, 30 (1987) 123–130. https://doi.org/10.1016/S0166-9834(00)81016-7.

[11] Eyraud C., Jallut C. Caractérisation texturale de membranes mésoporeuses minérales et organominérales: Importance du choix du modèle et des méthodes associées. In: Proc. First Int. Conf. On Inorg. Membranes, Montpellier, July 3–6,1989, L. Cot (Eds.), MTR Languedoc–Roussillon, 193.

[12] Riikonen J., Salonen J., Lehto V.-.P. (a) Utilising thermoporometry to obtain new insights into nanostructured materials. Review part 1. J Therm Anal Calorim, 105 (2011) 811–821. https://doi.org/ 10.1007/s10973-010-1167-0.

[13] Riikonen J., Salonen J., Lehto V.-.P. (b) Utilising thermoporometry to obtain new insights into nanostructured materials. Review part 2. J Therm Anal Calorim, 105 (2011) 823–830. https://doi.org/ 10.1007/s10973-011-1337-8.

[14] Jackson C.L., McKenna G.B. The melting behavior of organic materials confined in porous solids. J Chem Phys, 93 (1990) 9002–9011. https://doi.org/10.1063/1.459240.

[15] Hahne G.W.H., Hemminger W., Flammersheim H.J. Differential Scanning Calorimetry. An Introduction for Practitioners, Springer-Verlag, Berlin–Heidelberg, Germany, (1996). ISBN 978-3-662-03304-3.

[16] Enüstün B.V., Sentürk H.S., Yurdakul O. Capillary freezing and melting. J Colloid Interface Sci, 65 (1978) 509–516. https://doi.org/10.1016/0021-9797(78)90103-0.

[17] Denoyel R., Pellenq R.J. Simple Phenomenological Models for Phase Transitions in a Confined Geometry. 1: Melting and Solidification in a Cylindrical Pore. Langmuir, 18 (2002) 2710–2716. https://doi.org/10.1021/la015607n.

[18] Ishikiriyama K., Todoki M., Motomura K. Pore Size Distribution (PSD) Measurements of Silica Gels by Means of Differential Scanning Calorimetry: I. Optimization for Determination of PSD. J Colloid Interface Sci, 171 (1995) 92–102. https://doi.org/10.1006/jcis.1995.1154.

[19] Ishikiriyama K., Todoki M. Pore Size Distribution Measurements of Silica Gels by Means of Differential Scanning Calorimetry: II. Thermoporosimetry J Colloid Interface Sci, 171 (1995) 103–111. https://doi.org/10.1006/jcis.1995.1155.

[20] Landry M.R. Thermoporometry by differential scanning calorimetry: Experimental considerations and applications. Thermochim Acta, 433 (2005) 27–50. https://doi.org/10.1016/j.tca.2005.02.015.

[21] Eyraud C., Bontemps M., Quinson J.F., Chatelut F., Brun M., Rasneur B. Determination of the pore–size distribution of an ultrafilter by gas–liquid permporometry measurement comparison between flow porometry and condensate equilibrium porometry. Bulletin De La Societe Chimique De France Partie I–Physicochimie Des Systemes Liquides Electrochimie Catalyse Genie Chimique, 9–10 (1984) I237–I244.

[22] Cuperus F.P., Bargeman D., Smolders C.A. Critical points in the analysis of membrane pore structures by thermoporometry. J Membrane Sci, 66 (1992) 45–53. https://doi.org/10.1016/0376-7388 (92)80090-7.

[23] Broek A.P., Teunis H.A., Bergeman D., Sprengers E.D., Smolders C.A. Characterization of hollow fiber hemodialysis membranes–pore–size distribution and performance. J Membrane Sci, 73 (1992) 143–152. https://doi.org/10.1016/0376-7388(92)80124-3.

[24] Rabi I.I., Zacharias J.R., Millman S., Kusch P. A new method of measuring nuclear magnetic moment. Phys Rev, 53 (1938) 318. https://doi.org/10.1103/PhysRev.53.318.

[25] Bloch F. Nuclear induction. Phys Rev, 70 (1946) 460–474. https://doi.org/10.1103/PhysRev.70.460.

[26] Purcell E.M., Torrey H.C., Pound R.V. Resonance absorption by nuclear magnetic moments in a solid. Phys Rev, 69(1–2) (1946) 37–38. https://doi.org/10.1103/PhysRev.69.37.

[27] Mitchell T.N., Costisella B. NMR – From Spectra to Structures, Springer, Berlin, Germany, (2007). ISBN: 978-3-540-72195-6.

[28] Keeler J. Understanding NMR Spectroscopy, Wiley, New York, USA, (2010). ISBN: 978-0470746080.

[29] Jordan D. State of the art in magnetic resonance imaging. Phys Today, 73(2) (2020) 34–40. https://doi.org/10.1063/PT.3.4408.

[30] Strange J.H., Rahman M., Smith E.G. Characterization of porous solids by NMR. Phys Rev Lett, 71(21) (1993) 3589–3591. https://doi.org/10.1103/PhysRevLett.71.3589.

[31] Watson A.T., Chang C.T.P. Characterizing porous media with NMR methods. Prog Nucl Mag Res Sp, 31 (1997) 343–386. https://doi.org/10.1016/S0079-6565(97)00053-8.

[32] Barrie P.J. Characterization of porous media using NMR methods. Annual Reports on NMR Spectroscopy, 41 (2000) 265–316. https://doi.org/10.1016/S0066-4103(00)41011-2.

[33] Mitchell J., Webber B., Strange J.H. Nuclear magnetic resonance cryoporometry. Phys Rep, 461(1) (2008) 1–36. https://doi.org/doi:10.1016/j.physrep.2008.02.001.

[34] Cheng Y.-.C.N., Haacke E.M. Magnetic Moment of a Spin, Its Equation of Motion, and Precession. Current Protocols in Magnetic Resonance Imaging, (2007) B1.1.1–B1.1.10. https://doi.org/10.1002/0471142719.mib0101s14.

[35] Beall P.T., Amtey S.R., Kasturi S.R. NMR Data Handbook for Biomedical Applications, Pergamon Press, New York, (1984). ISBN–13: 978-0080307756.

[36] Cowan B. Nuclear Magnetic Resonance and Relaxation, Cambridge University, Press, Cambridge, U.K, (1997). ISBN: 978-0470746080.

[37] Ernst R.R., Anderson W.A. Application of Fourier transform spectroscopy to magnetic resonance. Rev Sci Instrum, 37 (1966) 93–102. https://doi.org/10.1063/1.1719961.

[38] Hahn E.L. Free nuclear induction. Phys Today, 6(11) (1953) 4–9. https://doi.org/10.1063/1.3061075.

[39] Hahn E.L. Spin echoes. Phys Rev, 80(4) (1950) 580–594. https://doi.org/10.1103/PhysRev.80.580.

[40] Carr H., Purcell E. Effects of diffusion on free precession in NMR experiments. Phys Rev, 94 (1954) 630–638. https://doi.org/10.1103/PhysRev.94.630.

[41] Meiboom S., Gill D. Modified spin–echo method for measuring nuclear relaxation times. Rev Sci Instrum, 29 (1958) 668–691. https://doi.org/10.1063/1.1716296.

[42] Elsayed M., Isah A., Hiba M., Hassan A., Al Garadi K., Mahmoud M., El Husseiny A., Radwan A.E. A review on the applications of nuclear magnetic resonance (NMR) in the oil and gas industry: Laboratory and field scale measurements. J Pet Explor Prod Technol, 12 (2022) 2747–2784. https://doi.org/10.1007/s13202-022-01476-3.

[43] Nagel S.M., Strangfeld C., Kruschwitz S. Application of 1H proton NMR relaxometry to building materials – A review. J Magn Reson Open, 6–7 (2021) 100012. https://doi.org/10.1016/j.jmro.2021.100012.

[44] Jaeger F., Bowe S., Van as H., Schaumann G.E. Evaluation of 1H NMR relaxometry for the assessment of pore–size distribution in soil samples. Eur J Soil Sci, 60 (2009) 1052–1064. https://doi.org/10.1111/j.1365-2389.2009.01192.x.

[45] Besghini D., Mauri M., Simonutti R. Domain NMR in Polymer Science: From the Laboratory to the Industry. Appl Sci, 9 (2019) 1801. https://doi.org/10.3390/app9091801.

[46] Goldman M. Quantum Description of High–Resolution NMR in Liquids, Oxford University Press, Oxford, UK, (1988). ISBN: 9780198556527.

[47] Schmidt R., Stöcker M., Hansen E., Akporiaye D., Ellestad O.E. MCM–41: A model system for adsorption studies on mesoporous materials. Microporous Mater, 3 (1995) 443–448. https://doi.org/10.1016/0927-6513(94)00055-Z.

[48] Rottreau T.J., Parlett C.M.A., Lee A.F.R., Evans R. Extending the range of liquids available for NMR cryoporometry studies of porous materials. Microporous and Mesoporous Mater, 274 (2019) 198–202. https://doi.org/10.1016/j.micromeso.2018.07.035.

[49] Lauffer D.E. Determining pore–size and fluid distribution in porous media – from spin echo amplitudes using nuclear magnetic resonance spectrometer, US4291271–A, 1981–75678D.

[50] Glaves C.L., Smith D.M. Membrane pore structure analysis via NMR spin lattice relaxation experiments. J Membrane Sci, 46(2–3) (1989) 167–184. https://doi.org/10.1016/S0376-7388(00)80333-6.

[51] Gallegos D.P., Munn K., Smith D.M., Stermer D.L. A NMR Technique for the Analysis of Pore Structure: Application to Materials with Well–Defined Pore Structure. J Colloid Interface Sci, 119(1) (1987) 127–140. https://doi.org/10.1016/0021-9797(87)90251-7.

[52] Gallegos D.P., Smith D.M., Brinker C.F. An NMR technique for the analysis of pore structure: Application to mesopores and micropores. J Colloid Interface Sci, 124(1) (1988) 186–198. https://doi.org/10.1016/0021-9797(88)90339-6.

[53] Strange J.H., Webber J.B.W., Schmidt S.D. Pore size distribution mapping. Magn Reson Imaging, 14 (1996) 803. https://doi.org/10.1016/S0730-725X(96)00167-1.

[54] Hansen E.W., Schmidt R., Stöcker M., Akporiaye D. Water–saturated mesoporous MCM–41systems characterized by 1H NMR spin–lattice relaxation times. J Phys Chem, 99 (1995) 4148. https://doi.org/10.1021/j100012a040. https://chemistry.mit.edu/wp-content/uploads/2018/08/NMR-Frequency-Table.pdf. http://triton.iqfr.csic.es/guide/eNMR/chem/NMRnuclei.html.

Chapter 7
Young–Laplace equation-based (fluid penetration) techniques

The Young–Laplace equation relates, as demonstrated in Chapter 4, the pressure required to empty (or fill) a capillary of a fluid to the radius of the capillary. It seems obvious that this equation could be used, departing from experimental information about the pore filling or emptying to determine the size of those pores (assumed cylindrical capillaries) present in a membrane sample.

In fact, based on this equation [1], a set of porosimetric methods of undoubted success and interest have been developed, methods that we will conveniently develop throughout this chapter.

Interestingly, the Young–Laplace equation has been given different names when used in porosimetry (Washburn's equation or Cantor's equation, as alternative or particular names for certain cases, can also be found in the bibliography), although they are always particular cases of the previously derived equation.

7.1 Bubble point

The first idea related to the use of the Young–Laplace equation to determine the pore size of a membrane was proposed by Bechhold in 1908 [2]. In a series of articles published between 1907 and 1908, Bechhold presented a method of making membrane filters from collodion [3]. It should be noted that collodion membranes are practically the first synthetic filters manufactured and marketed for analytical and laboratory applications [4]. He proposed that these filters, which he called ultrafilters (the name will be the origin of the term "ultrafiltration" (UF)), can be characterized by applying air pressure to the filter previously soaked in water. The idea proposed by Bechhold would be the origin of what is now called the bubble point method. The name is obvious since the observer determined the time (and the air pressure required) for the first bubbles to appear on the surface of the ultrafilter.

The method proposed by Bechhold only served to determine the maximum pore size present in a membrane. This is because once the first bubbles appear on the surface, further increases in the applied air pressure result in the appearance of more bubbles but there is no way to correlate them with the pore size opened at that time. Nevertheless, Peinemann and Pereira Nunes [5] still name the characterization methods that have emerged from Bechhold's original idea, as the bubble point methods (in particular, this term is often used to refer to the gas–liquid displacement method, which will be the subject of the next section of this book).

https://doi.org/10.1515/9783110792195-007

Taking into account the inverse relationship between pore size and the pressure applied to introduce air into the pores (previously occupied by a wetting liquid), it is evident that the first bubbles will be detected by applying a minimum pressure corresponding to the maximum pore size. To see this, let us consider the Young–Laplace equation:

$$P = \frac{2\gamma \cos \theta}{r_p} \tag{7.1}$$

where P is the applied pressure, γ is the interfacial tension (at the membrane–air–water interface), θ is the contact angle of the water with the membrane material and r_p is the radius of a given pore. If we consider that the liquid completely wets the membrane, we can assume that $\theta = 0$, so the above equation is simplified (in some texts, the simplified equation, valid for zero contact angle, is called Cantor's equation), as follows:

$$P = \frac{2\gamma}{r_p} = \frac{4\gamma}{d_p} \tag{7.2}$$

Although the size of a pore is perfectly explained by giving its equivalent radius or, alternatively, diameter, the truth is that, in the literature on membranes, it is very common that when we talk about pore size, we expressly refer to the diameter of the pores. The reason is that the diameter of a pore gives us the size of the largest molecule that can pass through it, information that is essential for the different applications of membranes as sieves for molecules.

Let us now consider the case of a membrane made up of pores of assorted sizes, which we have wetted with the wetting liquid (water in original Bechhold's experiments). If we now apply air pressure to the membrane (from the bottom, suitably sealed) and very slowly increase the air pressure, we will see that initially this pressure is not sufficient to pass through any of the pores, so we will not detect any bubbles on its surface. When the first bubble appears on the surface, we can say that it has occurred at the minimum applied pressure, which will be related to the open pore by the previous expression. We can then assume that this pore will be the maximum of the distribution present on the membrane:

$$r_{max} = \frac{2\gamma \cos \theta}{P_{min}} \tag{7.3}$$

Although the bubble point test (or method) only tells us the value of the largest pore size present in our membrane, this information can be really useful. Firstly, if the membrane is sufficiently homoporous (i.e. if it has a narrow pore size distribution (PSD)), the maximum pore value will not be too far away from the average pore size and then the bubble point test will give us a good indication of that average pore size of our filter.

But more frequently, the bubble point method has been used as an integrity test of the filters obtained. The integrity test is a check of the performance of a filter before it is used in the application for which it was designed. This type of test is especially important for filters that are to be used with fluids that could be contaminated with bacteria or viruses. In industries such as pharmaceuticals, the presence of pathogens in process water can be a serious risk to the final products obtained [6]. However, it is a matter of ensuring that the filter we are going to use to sterilize our fluid is free of such pathogens. Although we have acquired the appropriate filter for this purpose, it can always happen that it arrives defective or that it is damaged by excess pressure or by the chemical attack of substances present in the solution or used in the cleaning process. This is why a prior check of the filter's integrity is very convenient. Basically, three types of tests are used to check the integrity of a filter:

–   Diffusion test
–   Pressure hold test
–   Bubble point test

Apart from those mentioned above, other types of tests such as microbial challenge tests can be performed to check the functional integrity of the filter. A very comprehensive and exhaustive review of integrity methods applied to virus retention membranes is found in the work of Guo et al. [7].

The bubble point test consists of soaking the filter to be tested with an appropriate wetting liquid (it is necessary that the liquid does not damage the membrane and that it can be easily removed after the test so that the filter can be later used without interference or contamination during further use) and then, once sealed and connected to a clean air source, increasing the air pressure until the first air flow through the filter is determined. This pressure, according to the above equation, determines the maximum pore size present in our membrane. If the bubble point value is too low, it means that the maximum pore size is much larger than expected and suitable for our filter and that the filter is possibly damaged or defective (we can say that it is likely the filter has some kind of pinhole).

It is very common for membrane manufacturers to carry out this type of test on a regular basis as a quality control method for their products. For this reason, many membrane manufacturers (Millipore, Pall, Sartorius, etc.) have developed and commercialized automated equipment so that the user can conduct these integrity tests in situ. Some of the porometer manufacturers (PMI, IFTS, etc.) also design their equipment to perform the bubble point test as part of the porosimetric study of a given membrane. Some of the equipment are adapted to analyse different membrane configurations (flat, hollow fibre, tubes, etc.) and with different types of cartridges or modules (generally small membrane area modules), by adaptations of the basic devices. These modifications are important in order to be able to test the different cartridges or filters as they are marketed, and then use them in the desired applications. In terms of commercial membrane integrity testing equipment, without claiming to be exhaustive, we could name the following: Sartocheck4 (Sartorius), Flowstar XC (Palltronic), BpTester (PMI) or IntegrityTest II (Millipore) [8].

The biggest problem with bubble point testing has to do with the pressure required to obtain the first bubbles in membranes with very small pores. Recalling the original article by Bechhold and knowing that the interfacial tension of the air–water interface is approximately 73 mN/m, which means that to test a sample whose maximum pore (diameter) is 0.1 μm, we need an applied pressure of 29 bar (surely not too high), but for pores of the order of 0.05 μm, we need almost 60 bar, a pressure high enough to damage the filter or the cassette or module that contains it. Thus, the bubble point test is only regularly used for microfiltration (MF) membranes, or even wide UF membranes, presenting pores around 50 nm, while UF membranes with narrower pores must be tested for integrity by other methods. Obviously, when using the bubble point test to analyse such small pores, water cannot be used as a wetting liquid.

The bubble point technique has been so well-recognized and accepted that it has given rise to several characterization standards: ASTM F316-03(2019) – Standard Test Methods for Pore Size Characteristics of Membrane Filters by Bubble Point and Mean Flow Pore Test; BS 3321 – Method for measurement of the equivalent pore size of fabrics (bubble pressure test); EN 24003 – Permeable sintered metal materials – Determination of bubble test pore size (ISO 4003: 1977); DIN 58355-Part 2 – Membrane filters – Part 2: Testing for bubble point; ASTM E128 – Standard Test Method for Maximum Pore Diameter and Permeability of Rigid Porous Filters for Laboratory Use.

Referring in particular to the ASTM F316 Standard, the basis and reference for most of the other standards, it indicates that the method is designed for the determination of pore size properties for membranes with sizes between 0.1 and 15.0 μm. The standard actually proposes two methods:

- Test method A presents a test method for measuring the maximum limiting pore diameter of non-fibrous membranes. The limiting diameter is the diameter of a circle having the same area as the smallest section of a given pore (see Fig. 7.1 for a schematic illustration of the measuring sample holder).
- Test method B measures the relative abundance of a specified pore size in a membrane, defined in terms of the limiting diameter.

The standard document stands that the method may be used to:
- determine the maximum pore size of a filter;
- compare the maximum pore sizes of several filters; and
- determine the effect of various processes such as filtration, coating or autoclaving on the maximum pore size of a membrane.

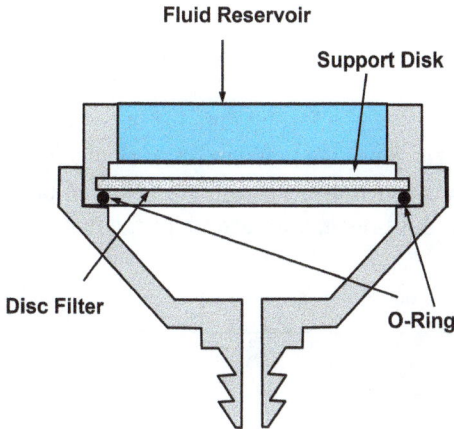

Fig. 7.1: Sample holder suitable for applying ASTM F316 Standard on flat membrane discs.

7.2 Gas–liquid displacement porosimetry (GLDP)

The bubble point technique, as proposed by Bechhold, is only useful for determining the maximum pore size present in a filter, as mentioned in the previous section. However, it seems natural to extend this study to all the pores present in the membrane, simply by increasing the applied pressure in small steps and detecting the air flowing through the membrane. Evidently, simply observing the bubbles coming out of the membrane does not serve this purpose, but we will have to design some kind of system that allows us to measure, for each increase in pressure applied, the flow of air that passes through the membrane. Thus, the technique that allows us to determine the distribution of pore sizes present in a membrane by combining bubble point and solvent permeability measurement is called gas–liquid displacement porosimetry (GLDP). The technique was proposed by Erbe in 1933 [9] and basically consists of applying air pressure to a membrane that has been previously impregnated completely with a wetting liquid. If the pressure is applied slowly with gradual increments, once the bubble point is reached, the following increments will result in an increase in the air flow through the membrane. By measuring this air flow, we can determine the contribution of the open pores, at each stage, to the air flow.

The resulting technique is called GLDP although in many works it is called capillary flow porometry following the terminology used, among others, by PMI [10]. However, both terms refer to the same characterization technique.

To obtain PSD, we simply apply the Young–Laplace equation to our measurements of the applied pressure versus The obtained airflow.

Graphically, the necessary process can be understood from Fig. 7.2:

-   Initially, the membrane is completely soaked in the wetting fluid and when we
    increase the pressure from a low or zero value, it will not be enough to reach the
    bubble point of the membrane (Fig. 7.2A).
-   As the pressure is gradually increased (always allowing time for the system to
    equilibrate), there will be a point (P_{min}) at which the pressure is sufficient to push
    the liquid out of the largest pores in the membrane (r_{max}). We will then have
    reached the bubble point (Fig. 7.2B). At this pressure, we determine the first mea-
    surable value of the air flow (F_0).

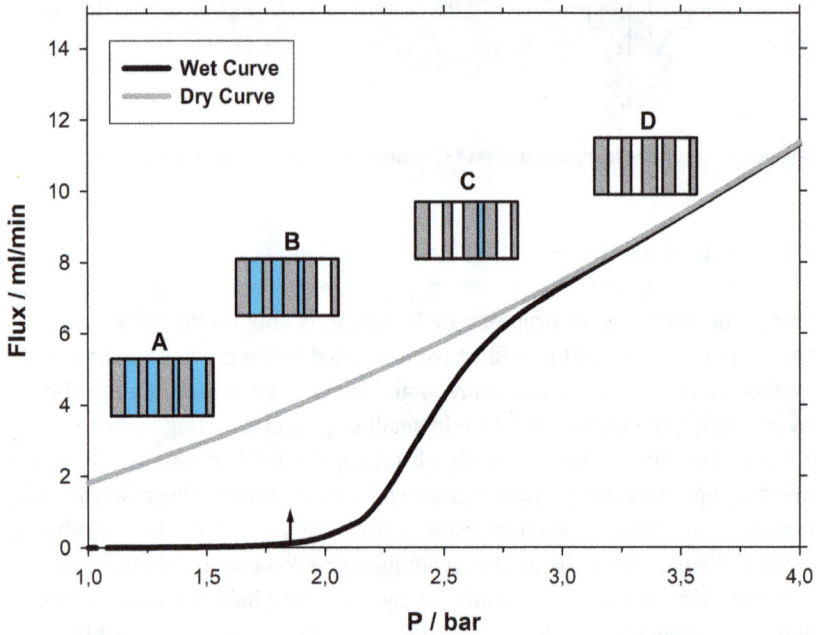

Fig. 7.2: Typical GLDP experiment.

-   If we continue to increase the pressure, new pores will open, and the flux will
    increase at each stage (Fig. 7.2C).

This can be analysed as follows:

-   Let there be two points on the pressure flow curve, corresponding to two succes-
    sive pressures (P_i and P_{i+1}), at which the flow rates (F_i and F_{i+1}) have been mea-
    sured. Consequently, at the ith stage, which connects the data: (i, $i+1$), the flow
    has increased by an amount: $\Delta F_i = F_{i+1} - F_i$. This increase is partly due not only to
    the increase in pressure (since the higher the pressure, the higher the flow) but
    also to the fact that at this stage new pores have been opened to the flow that
    were not previously open. However, we are only interested in the increase in
    flow due to the opening of new pores. Therefore, we will consider the air perme-

ability at each stage, so that we can define the increase in permeability at the ith stage as

$$\Delta L_i = \frac{\Delta F}{\Delta P} = \frac{F_{i+1} - F_i}{P_{i+1} - P_i} \tag{7.4}$$

We can say that this increase in permeability is only due to the opening of new pores. This is not strictly true since the permeability of a porous material to air or any other gas is not constant, due to the compressibility of the gas, although in the first instance this change in permeability can be considered negligible.

- Finally, this increase in permeability must be related to the size of the open pores in the interval. To do this, we will consider, without too much error (especially if the pressure measurements are sufficiently slow and the increments are sufficiently low), that an average pressure acts in the interval considered:

$$P_{i,\text{med}} = \frac{P_{i+1} + P_i}{2} \tag{7.5}$$

At this pressure, there will be open pores whose size must be given by the Young–Laplace equation:

$$d_{i,\text{med}} = \frac{4\gamma}{P_{i,\text{med}}} \tag{7.6}$$

- In this way we can construct a curve formed by pairs of values (ΔL_i, $d_{i,\text{med}}$) that we can consider the PSD in terms of flux for our membrane.
- Obviously, if we continue to increase the pressure, there will come a time when all the pores of the membrane will have been emptied of the wetting liquid and the flux will increase linearly due to the simple increase in pressure (Fig. 7.2D).

In most cases, commercial GLDP equipment, once the curve represented in the figure (typically called wet curve) has been completed, performs a new sweep from the minimum pressure applied to the maximum pressure, measuring the flow again at the same pressure values used in the wet curve. These new values of air flow allow the construction of the so-called dry curve. Its main task is to ensure that the measurement range used is adequate to contain all the information on the pores present (see explanatory figure).

The choice of measuring range is an important first step in obtaining reliable results. Most commercially available GLDP porometers require a relatively short time to perform a complete measurement. However, it is necessary to make a preliminary test with a sample of the membrane to be analysed, so that we can adjust the measuring range. If the chosen pressure range starts at a too high pressure, we will not find the bubble point (Fig. 7.3A) and the resulting distribution will miss the larger pores. Conversely, a pressure value that is too low, being selected as the maximum pressure

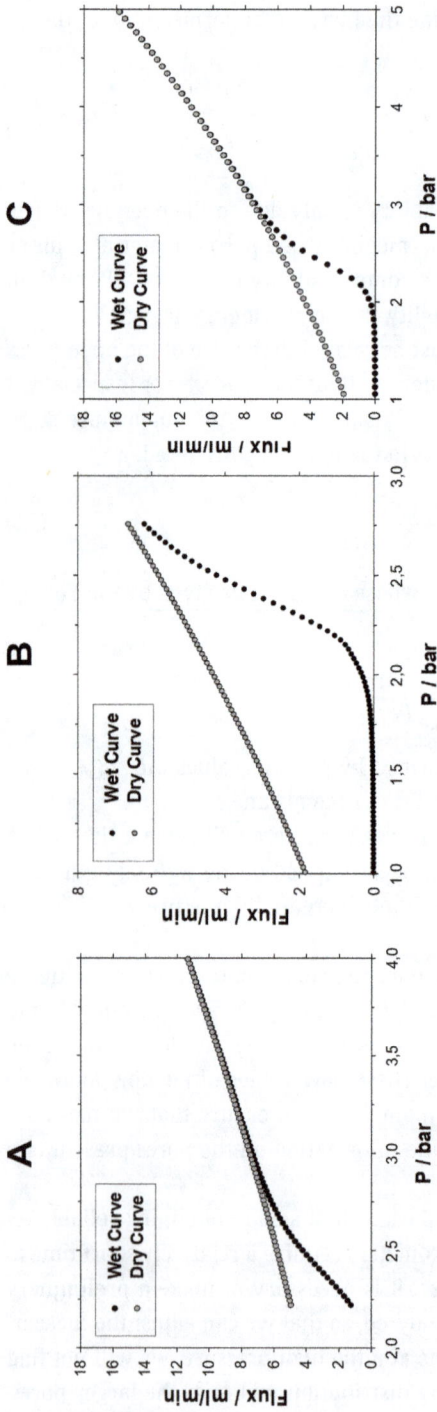

Fig. 7.3: Comparison between different GLDP runs.

limit, will cut off the smallest pores from the distribution (Fig. 7.3B). Once we are sure that the whole curve is present in the chosen pressure range, we can try to adjust it to obtain as much data as possible (Fig. 7.3C).

The flux distribution thus obtained is a reasonable representation of the contribution of each of the pore sizes present in the sample to the total permeability of the sample. The advantage of such a distribution is that to obtain it, it is not necessary to make any assumptions about the transport model through the pores, assumptions that are necessarily based on geometrical considerations whose validity we cannot always be sure of. Finally, from this PSD, we can determine its characteristic values, such as the mode, median and standard deviation.

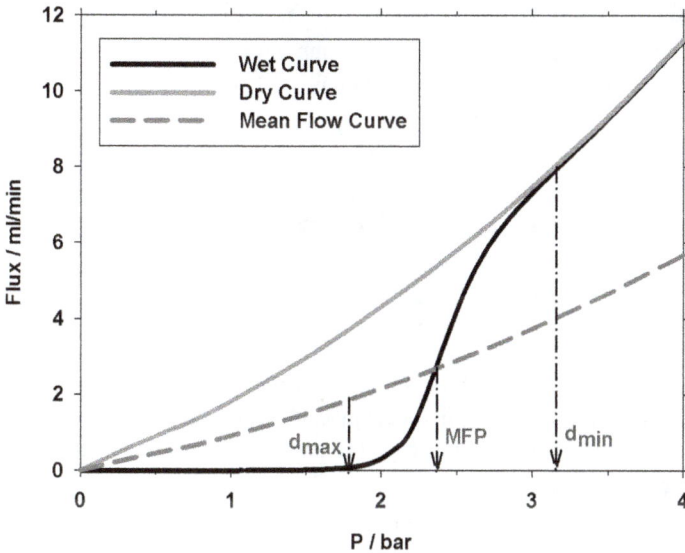

Fig. 7.4: Calculation of MFP from a GLDP experiment.

Another possibility to obtain information about the pore size is to determine the mean flow curve. Basically, this curve, in Fig. 7.4, consists of taking, for each pressure data with which the dry flow curve was constructed, the corresponding flow value and dividing it by half. The curve thus constructed (which we can call half or mean flow curve) will cut the wet flow curve at a point whose abscissa value (pressure value) determines (by means of the Young–Laplace equation) the so-called mean flow pore (MFP). Its meaning is that pore sizes smaller than MFP will contribute 50% of the total membrane flux while larger pores will account for the other 50%. Thus, the MFP value should coincide with the median of our distribution obtained by the above procedure.

Although this so-called direct distribution is quite reliable and easy to interpret, it is still common to be interested in the pore number distribution, that is, the one that tells us how many pores of each size exist in the sample. This should be needed to obtain, for example, the density of porosity of the sample being analysed. To arrive at this information, it is necessary, as we have said, to consider a transport model in which we must specify the geometry of our pores. It is usual to consider this geometry as cylindrical pores, perfectly normal to the membrane surface (capillary pore model). The possible validity of this assumption is one of the most frequent criticisms made to GLDP, and also to all related porosimetries.

The main problem in determining the absolute pore number distribution lies in the choice of the transport model to be applied to the air flow through the membrane.

Basically, there are two types of gas flows that can be applied inside a pore:

– The convective flow, governed by the Hagen–Poiseuille model
– Molecular flow, which follows the so-called Knudsen model

The difference between the two models lies in the assumption of the mean free path of the gas molecules inside the pore. If this mean free path, λ, is much smaller than the capillary diameter, we can assume that a fluid velocity profile develops inside the pore, in which we can be sure that the transport velocity at the capillary walls is zero. In that case the volume flow through the pores of size d_p will be given by the Hagen–Poiseuille equation which, for a cylindrical capillary pore, is expressed as follows:

$$J_v(d_p) = \frac{\pi \, N(d_p) \, d_p^4 \Delta p}{128 \eta \psi l} \tag{7.7}$$

where $J_v(d_p)$ is the air flow through all pores of size d_p, $N(d_p)$ is the number of pores of size d_p in the membrane, Δp is the applied pressure, η is the gas viscosity and ψ is the factor to account for the possible tortuosity of the pores (where $\psi = 1$ for perfectly cylindrical pores normal to the surface).

In case the mean free path is not much smaller than the pore diameter or both values are of the same order of magnitude, the Hagen–Poiseuille model must be discarded. The value of λ is given by the following expression:

$$\lambda = \frac{kT}{\pi \sqrt{2} \, p' d_m^2} \tag{7.8}$$

with k being Boltzmann's constant, T the temperature of the gas, p' its pressure and d_m the molecular diameter of the gas.

Thus, when the size of the capillary, or the pressure of the gas inside, is so small that it greatly exceeds the capillary diameter, the flow must be treated by the kinetic theory of gases and will be mostly driven by the set of collisions of the gas molecules with the capillary tube wall. The resulting model, called the Knudsen flow or free molecular diffusion, predicts that the gas flow through the capillary will be given by the following expression:

$$J_v\left(d_{\mathrm{p}}\right) = \frac{2\pi\, N\left(d_{\mathrm{p}}\right)}{3}\left(\frac{RT}{8\pi M_{\mathrm{w}}}\right)^{1/2}\frac{d_{\mathrm{p}}^3 \Delta p}{p\psi l} \tag{7.9}$$

where M_{w} is the molecular weight of the gas (28.97 g/mol for dry air) and p is the pressure at the capillary outlet.

Between these two limits ($\lambda \ll d_{\mathrm{p}}$ and $\lambda > d_{\mathrm{p}}$), neither model is completely valid and a transition model between the two should be used.

In a paper by Hernández et al. [9], this transition was carefully analysed, and it was shown that for the usual conditions of temperature and pressure used in the usual air porometers (the work made use of the pioneer Coulter Porometer II), the transition zone is located on pore sizes of the order of 1 μm. Hence, in practice, all MF membranes are affected by a certain error if only convective transport is considered in obtaining the PSD curve in terms of pore number.

In any case, using the conversion algorithms implemented in the software of the existing commercial equipment or improving these algorithms with suitable transition models, the result is to obtain a pore number distribution as a function of pore size, which also allows us to estimate magnitudes such as porosity (at least surface porosity, i.e. in the active layer of the membrane) or the pore density of the membrane. It is evident from the above that the PSD in number will always be affected by the validity of the assumptions made when applying the corresponding transport model. In particular, the assumption of a capillary pore model for the membrane structure may be quite reasonable for track-etched or anodized symmetric membranes, but not so for more or less tortuous polymeric membranes.

It is therefore advisable (as will be discussed in the rest of the porometries) to present the direct data, avoiding geometrical assumptions about the pores as much as possible. Thus, the MFP is a perfectly valid value, at least as far as we can assume the Young–Laplace equation to be valid.

A very important role in the GLDP technique, as was the case in the bubble point test or later we will analyse in the liquid–liquid displacement porosimetry (LLDP) technique, is played by the wetting liquid. In fact, the range of applicability of the technique will depend on the characteristics of the liquid. The appropriate characteristics to demand from potential wetting liquid candidates are:

– Low surface tension, the lower interfacial tension and the smaller pores can be analysed using the same applied pressure.
– Small contact angle: obviously, the contact angle depends on the membrane material. Thus, for hydrophobic membranes, it would be advisable to use an equally hydrophobic liquid (of low dielectric constant), while hydrophilic liquids will be preferable for wetting membranes of a hydrophilic nature. Finally, the most suitable liquids will be those with amphoteric characteristics, so that they are equally suitable for wetting hydrophilic and hydrophobic membranes. These include the most commonly used liquids such as alcohols or halogenated compounds.
– Low vapour pressure avoids evaporation problems.

- Low viscosity allows a dynamic character of the measurement.
- Has high chemical compatibility.

With these restrictions, we can find, among the most commonly used liquids, the following:
- Isopropyl alcohol: one of the first choices for GLDP technique, it presents characteristics that make it recommendable:
  - $\gamma = 20.86$ mN/m
  - Vapour pressure = 5.8 kPa
  - Amphoteric character
  - Moderate viscosity
  - High chemical compatibility
- Porofil® (halogenated): it has been used as a wetting liquid, with excellent performance, since the 1980s. Its most notable characteristics are:
  - $\gamma = 16$ mN/m
  - Vapour pressure = 0.4 kPa
  - $\theta \approx 0$ for most materials.

With basically the same composition, it is marketed by several porometer manufacturers with trade names as varied as Porefil™, Porewick™ and slight variants thereof (Galpore™ (15.9 mN/m), Silwick™ (20.1 mN/m), Galwick™ (15.9 mN/m), etc.) as can be found in the PMI, Quantachrome or Porometer webpages. Any of these liquids can be used to conveniently analyse membranes in the MF range (with pore sizes of at least 0.05 μm) using air pressures of the order of 13 bar, which is quite reasonable.

The same international standards governing the use of the bubble point test apply to the technique, as well as some more specific ones related only to this technique or to specific materials (ASTM D6767 – Standard Test Method for Pore Size Characteristics of Geotextiles by Capillary Flow Test).

7.2.1 Commercial devices

As mentioned previously, the GLDP technique has reached a standardized status for pore sizing characterization, being useful in many types of porous materials, not only membranes but also other conventional filters, including paper, geotextiles or carbon beds. This is the reason why several companies afforded the task of building and putting in the market automated porosimeters based on the GLDP technique, which covers the gap existing in these devices. In the following lines, we will enumerate the well-known commercial devices, starting from the pioneer Porometer II (there should be a Porometer I, but to the best of our knowledge, we have not found such initial device):

- Porometer II (Coulter–Beckman): one of the pioneering equipment, it appeared at the end of the 1980s of the previous century and presented the main characteristics of later equipment. It is now discontinued.
- Porometer 4 (Porvair): a continuation of the Coulter technology, it has similar characteristics in a more modern design with updated software.
- Porometer 3G (Quantachrome, also marketed by Anton Paar): its models can measure pores down to 0.07, 0.03 or 0.02 µm, depending on the different versions, reaching a maximum pressure of 500 psi (34.5 bar).
- IPore (PMI): the different models measure minimum pores between 0.3 and 0.013 µm, working with maximum pressures between 20 and 500 psi (1.38–34.5 bar), according to the manufacturer's specifications.
- Porolux (Porometer): the series of models covers maximum pressures from 1.5 to 35 bar. This allows pores between 0.42 and 0.013 µm to be analysed.

7.2.2 Criticisms of the GLDP technique

– Among the most frequently made criticisms of the GLDP technique (which can also be extended to the LLDP technique discussed later) is the fact that the technique is not able to differentiate the internal geometry of the pores. It is true that the Young–Laplace equation, which is strictly derived for cylindrical capillaries, when applied to pores of variable geometry along their path, will only give information on the narrowest diameter that the fluid encounters along its path [11].

It is evident that the technique has limitations in determining the true geometrical characteristics of those pores that are interconnected, a situation that is quite frequent in polymeric or ceramic filters, as an early work by Gijsbertsen-Abrahamse et al. [12] showed.

On the other hand, the purpose of membranes must be taken into account when discussing such critical issues. Given that their role is selective separation of several components, and that in many cases, they perform this function through a sieving effect, it is clear that when deciding which molecules pass or do not pass through a given membrane pore, we are interested in the narrowest section of the pore. This is the one that will decide the size of the molecule to be retained and, consequently, separated.

Certainly, for other uses of membranes (not merely as sieves) such as membrane contactors, where the internal structure is more important than the pores of the active layer, researchers could use different membrane characterization techniques that provide a better picture of the inner of the pores:

– Another important criticism against liquid displacement porometries, raised against both GLDP and LLDP, concerns the assumption of a certain geometry to interpret and convert the measured data into PSD. Indeed, when using any transport model (Knudsen, Poiseuille or a combination of both), it is generally assumed that the pores are

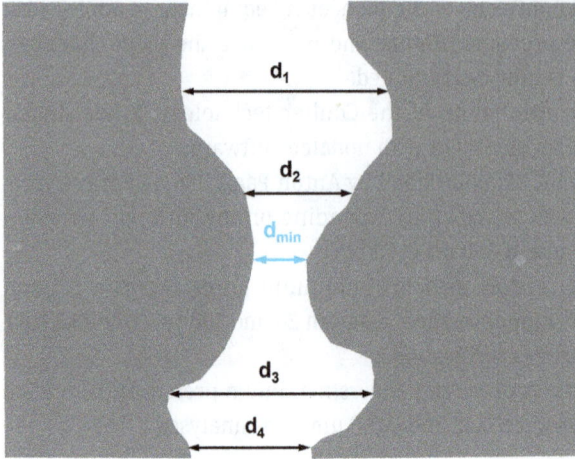

Fig. 7.5: A simplified cross-sectional view of a non-cylindrical-shaped pore structure.

perfect cylindrical capillaries that cross the membrane normally from one side to the other. However, the cylindrical geometry is only true for very few test membranes, while more complicated geometries exist for most membranes. This problem has sometimes been circumvented by using different types of structural parameters, on a more or less empirical basis, to account for the differences between real and cylindrical geometry. These issues are similar to the usual criticisms attributed to all techniques based on the Washburn equation (which defines the differential flow rate of the capillary fluid in a cylindrical tube). To get a clearer idea of the relevance of such an assumption, a few comments are warranted. In particular, the pores are often so irregularly shaped that they cannot be considered as tubes at all, but rather as tortuous pathways that remain open between the complex structures formed during the membrane fabrication process. These paths, along with cross sections that clearly differ from circular ones, exhibit significant diameter variability, making it nearly impossible to mimic any portion of any ideal tube. Since Laplace's equation gives the highest pressure required for the fluid to penetrate the interior of the porous structure, this corresponds to the pressure required to pass the narrowest section of the pore. Even in the case of non-tube-shaped pores, the cross sections of some part of the pore, especially around the neck, can be considered more or less constant (see Fig. 7.5). The Washburn equation refers to such cross sections and thus provides information on the minimum cross section inside the pore, which governs the flow and separation potential of that pore.

– Another most frequent criticisms of the technique have more to do with the validity of the Young–Laplace equation itself. Thus, in the commonly used expression of this equation, the meniscus formed between the two interfaces is considered to be a spherical surface, a consequence of assuming a circular section for the pore. Although we have already mentioned that this assumption is not generally true, it is clear that, whatever the shape of the cross section, the surface of the meniscus formed will be given by a curvature characterized by its two principal radii of curvature, R_1 and R_2.

These principal radii of curvature are given by the dimensions (maximum and minimum) of the cross section of the pore. Thus, for an elliptical pore, the principal radii would coincide with the two semi-axes of the ellipse, while for a rectangular pore, they would be its length and width.

However, the complete Young–Laplace equation uses both radii of curvature and only when we go from a more or less irregular surface to a spherical surface, these two values are equalized. On the other hand, we can roughly evaluate the error made in this simplification. To do this, let us consider that on a spherical surface we will take a single radius of curvature, R_p, which will obviously take a value situated between the two extreme values R_1 and R_2. Let us assume that R_1 is the largest principal radius. Thus we can say that

$$R_1 \sim R_p + \Delta R$$
$$R_2 \sim R_p - \Delta R$$

(7.10)

Thus, the Young–Laplace equation correctly expressed should contain the following term:

$$\frac{1}{R_1} + \frac{1}{R_2} \approx \frac{1}{R_p + \Delta R} + \frac{1}{R_p - \Delta R} = \frac{1}{R_p + \Delta R}\left(\frac{1}{1 + \dfrac{\Delta R}{R_p}} + \frac{1}{1 - \dfrac{\Delta R}{R_p}} \right)$$

(7.11)

where it seems reasonable to assume that the quantity $\Delta R/R_p$ can be very small. Taking this into account, we can make a power series development of the above expression which leads us to

$$\frac{1}{1 + \dfrac{\Delta R}{R_p}} = 1 - \frac{\Delta R}{R_p} + \frac{1}{2}\left(\frac{\Delta R}{R_p}\right)^2 + \cdots$$

$$\frac{1}{1 - \dfrac{\Delta R}{R_p}} = 1 + \frac{\Delta R}{R_p} + \frac{1}{2}\left(\frac{\Delta R}{R_p}\right)^2 + \cdots$$

(7.12)

Here, taking into account the value of $\Delta R/R_p$, we can neglect the terms higher than the quadratic power. So finally, we have

$$\frac{1}{R_1} + \frac{1}{R_2} \approx \frac{2}{R_p}\left[1 + \left(\frac{\Delta R}{R_p}\right)^2\right]$$

(7.13)

If we consider the second term of the bracket to be negligible, we arrive at the usual expression of the Young–Laplace equation for capillaries of circular cross section. This assumption, based on considering $\Delta R/R_p \ll 1$, is very reasonable in most of the cases we may encounter.

Thus, Kaneko [13] classifies the pores that can be found in a material according to their geometric shape into the following categories (see Fig. 7.6): cylinder, slit, conical shape and inkwell. Of all these geometries, the one that departs most clearly from the simplification is the slit-shaped pore (B), for which the radii of curvature take the extreme values: $R_1 = R_p$; $R_2 = \infty$. In such a case, the Young–Laplace equation would take the following expression:

$$\Delta p = \gamma/R_p \tag{7.14}$$

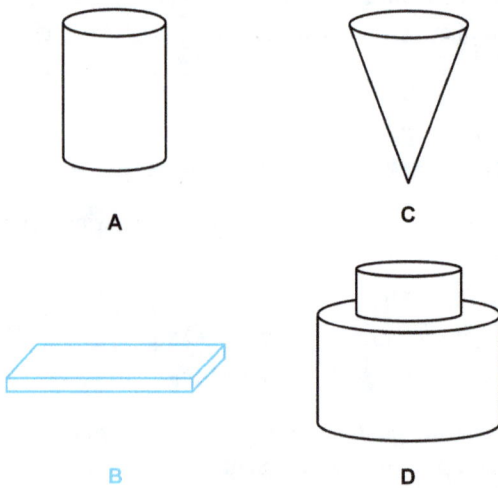

Fig. 7.6: Pore shape classification according to Kaneko [13].

But in all other cases, the error made is reasonable since, what we obtain by using the simplified Laplace equation is a pore size that reasonably averages the extreme real dimensions of the real pores. This can easily be assumed to be an acceptable and convenient standard, leaving aside the possibility that research may allow us to refine this approximation in cases where it is deemed really necessary.

7.3 Liquid extrusion–intrusion porometry (LEIP)

The GLDP technique has become, as mentioned above, a standard in the characterization of porous materials, for sizes between several tens of nanometres and hundreds of micrometres. However, Bechhold himself [14] already observed that the results obtained depend on the rate of pressure increase in the successive experimental steps. In fact, he derived the following expression to account for this dependence:

$$d_p = \frac{4\gamma}{\Delta p} \left[1 + \frac{2l}{\gamma} \left(\frac{d(\Delta p)}{dt} \bar{\eta} \right)^{1/2} \right] \tag{7.15}$$

Obviously, one possibility to minimize this drift of the obtained value is to extend the duration of the pressure steps by keeping the system at constant pressure for a sufficient time. However, this procedure runs up against the possibility of the wetting liquid evaporating from the pores of the membrane and thus falsifying the results. Thus, most commercial porometers typically perform the analysis within a few minutes.

In 1994, Volfkovich [15, 16] proposed a new method for determining the PSD, also based on Young's equation but slightly modified from the GLDP so that a longer equilibrium time could be used for each measurement, thus minimizing the associated errors.

The technique, which Volfkovich initially called the standard porosimetry method and later standard contact porosimetry [17], has been referred to by many authors as liquid extrusion porosimetry (LEP) or more generally liquid extrusion–intrusion porometry (LEIP) (this term includes the possibility of determining extrusion and further intrusion curves).

To understand the LEIP technique a little better, we must pause for a moment to consider the concept of capillary pressure:

We can define capillary pressure as the difference in pressure existing at the interface between two immiscible fluids when they come into contact through a porous medium, so that one of the fluids wets the surface of the medium and the other fluid is considered non-wetting.

Obviously, this distribution of roles will depend on the wettability of the porous medium to one or the other of the fluids considered. Thus, in the case of water rising in a glass capillary tube, water is the wetting phase and air is the non-wetting phase, whereas if we replace water with mercury, mercury will now play the role of the non-wetting fluid and air will be the wetting fluid. We can see the different behaviour in Fig. 7.7, where we can observe the different concavity or convexity that a wetting fluid (water) or a non-wetting fluid (mercury) gives rise to.

Returning to the capillary pressure, it can be defined in general terms as follows:

$$P_c = P_{\text{Non-wetting}} - P_{\text{Wetting}} \tag{7.16}$$

As discussed in the corresponding chapter, capillary pressure can be determined from the Young–Laplace equation, which relates this pressure to the surface tension (γ) between the two fluids and to the mean curvature of the interface formed between the phases (H). Thus,

$$P_c = 2\gamma H \tag{7.17}$$

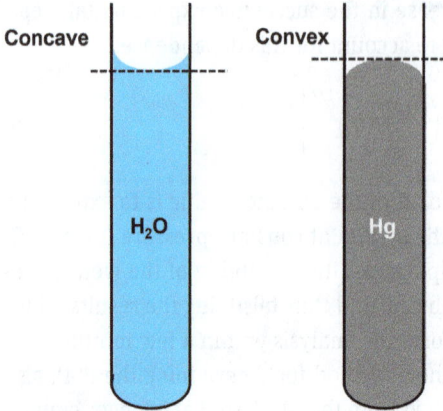

Fig. 7.7: Different meniscus curvatures depend on the capillary wettability by different liquids. Note: The concept of fluid is broader than simply considering liquids in contact. This will allow us to include a liquid in contact with a gas (air) as well as the contact between two liquids, increasing the possibilities of analysis.

This last term (H) can easily be added as a function of the principal radii of curvature of the surface. Thus,

$$H = \frac{1}{2}\left[\frac{1}{R_2} + \frac{1}{R_2}\right] \tag{7.18}$$

where R_1 and R_2 are the principal radii of curvature.

If now, as we have been doing in this chapter, we consider cylindrical tubes (capillary pore model), the expression of the mean curvature of the interface is simplified, based on geometrical considerations, as a function of the capillary radius (r_c) and the contact angle (θ) that forms the solid–liquid interface. Thus, the equation simplifies to

$$P_c = -\frac{2\gamma\cos\theta}{r_c} \tag{7.19}$$

Here the minus sign simply indicates that we must apply pressure to introduce the liquid into the capillary, forcing the interface resistance.

This special form of the Young–Laplace equation is commonly known as the Washburn equation and, as we will see later, is the basis for interpreting the results of mercury porosimetry (HgP). In fact, mercury intrusion is one of the methods commonly considered to measure the capillary pressure of a filter.

When determining the capillary pressure curves of a given porous sample, there are two different experimental approaches, although they lead to comparable results. Thus, we can determine the capillary pressure by controlling the applied pressure or by controlling the volume of fluid injected [18].

7.3.1 Pressure-controlled LEIP

Gostick et al. [19], following Volfkovich's idea, proposed a method by which the capillary pressure in the sample is controlled by the gas pressure on it.

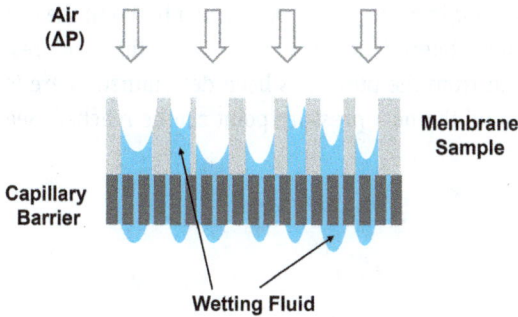

Fig. 7.8: LEIP principle (adapted from [26]).

Basically, the technique consists of the same procedure as used in the bubble point-derived methods (GLDP or LLDP) (See Fig. 7.8). The sample to be analysed is completely soaked in a wetting liquid to ensure that all pores of the sample are filled. Then, the sample is placed in a measuring cell soaked in the wetting liquid and a porous material (also soaked in the wetting liquid) is placed underneath it to act as a capillary barrier. This material has to be such that its largest pore is clearly smaller than the smallest pore size of the sample to be analysed, as well as being a hydrophilic material which facilitates wetting. This can complicate the measurement somewhat if there is no prior information on the pore distribution of the sample to be analysed.

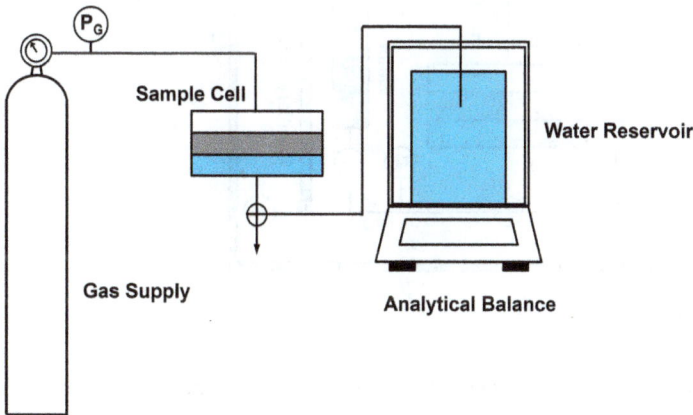

Fig. 7.9: Example of experimental set-up to perform LEIP (pressure-controlled) analysis.

A gradually and steeply increasing air pressure is then applied to the measuring cell. At each step, the system is allowed to equilibrate (determining the capillary pressure) and when the pressure reaches the bubble point of the sample, the air will begin to empty the larger pores of the sample. However, the capillary barrier prevents the air from passing through and what is detected is simply an increase in pressure in the wetting liquid, which, on the other hand, will result in an excess of the wetting liquid in the cell being deposited on a balance and its mass determined. As soon as the equilibrium pressure is reached and the mass dislodged from the pores has been determined, there is no more liquid falling on the balance and the next pressure point can be reached. See Fig. 7.9 for an scheme of the used setup.

7.3.2 Volume-controlled LEIP

In this approach, which is very similar to the wetting liquid pressure controlled porosimetry configuration, the difference is that instead of controlling the applied pressure (usually by the gas pressure supply valve), the volume injected is controlled by means of a syringe or a fairly accurate pump. Discrete amounts of wetting liquid are injected into the sample and the pressure of the liquid is measured at each injection.

An example of a device prepared for the LEIP technique in volume-controlled mode is shown in Fig. 7.10, where two syringes are used to control the volume of water and air injected or expelled from the sample.

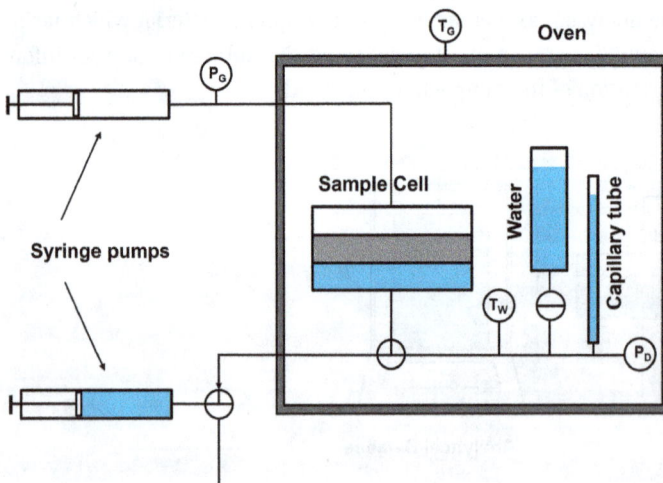

Fig. 7.10: Example of experimental set-up to perform LEIP (volume-controlled) analysis, adapted from [26].

With this procedure and using a sealed system in which evaporation of the air and the pushing fluid is avoided, the saturation of the liquid can be scanned at both increasing (imbibition) and decreasing (drainage) pressure runs of the capillary pressure, and the hysteresis of the capillary pressure can be reported. Therefore, it would perhaps be more correct to call the whole technique LEIP.

Returning to the pressure-controlled porosimetric technique, which is the one commonly used in the determination of the membrane PSD, we have replaced the air flow measurement with a determination of the extruded fluid mass, a measurement that can be equally accurate and above all more static (since it is measured under equilibrium conditions). This makes a complete LEIP experiment take more than 1 h as opposed to 3–5 min for a GLDP porogram.

A typical LEIP experiment (a porogram) plots the mass of wetting liquid collected for each equilibrium capillary pressure value. If at the end of the experiment (when the maximum capillary pressure has been reached) the capillary pressure is slowly reduced, the wetting liquid will flow back into the sample. In this way, a porogram consisting of two curves (an intrusion curve and a subsequent extrusion curve, as those shown in Fig. 7.11) can be determined [20, 21]. Typically, hysteresis occurs between the two curves, since not all the liquid expelled from the pores is reintroduced into them during the reverse process. In any case, the intrusion curve allows us to obtain the PSD as an increase in mass collected on the analytical balance at each stage versus the equilibrium capillary pressure. Using the Young–Laplace equation, we must convert this pressure into the diameter of the pores emptied at each stage (see Fig. 7.12 for the PSD resulting from Fig. 7.11 experiment).

The porogram would consist of collected mass versus equilibrium capillary pressure data. With the help of the Young–Laplace equation, we convert this pressure data into pore diameters. Thus, between two consecutive points a differential amount of mass will have been collected, $\Delta m_i = m_i - m_{i-1}$. At these two points there will have been a change in capillary pressure, Δp_i, which we can convert into a difference in pore diameters, $\Delta d_{p,i} = d_{p,i-1} - d_{p,i}$ (we reverse the sign of the difference to take into account that at each successive step the capillary pressure increases, but consequently, this means that we are analysing smaller pores. Finally, the pore distribution can be presented as a plot of: dm/dd_p ($= \Delta m_i/\Delta d_{p,i}$, for a discrete data set) versus the corresponding mean pore diameter (d_p).

A more elaborate distribution based on the assumption of a capillary pore geometry can be found in the work of Volfkovich and co-workers [17]. However, as we have discussed for other porosimetric techniques, the introduction of geometric or transport models for the analysis of porosimetric results induces errors due to assumptions that cannot always be guaranteed to be fulfilled.

Although the LEP (or LEIP) technique is usually considered one of the existing porometries for the characterization of microporous membranes and as such is referenced in numerous publications [22], few authors have been using it regularly. Among them, apart from several works by Volfkovich, there are some articles by Gostick or a thesis directed by him, focused more on the determination of capillary pressures in gas diffusion layer (GDL) or microporous layer (MPL) of fuel cells [18, 23]. In those works, LEIP has been shown to be clearly superior to HgP, given that the latter (by using a

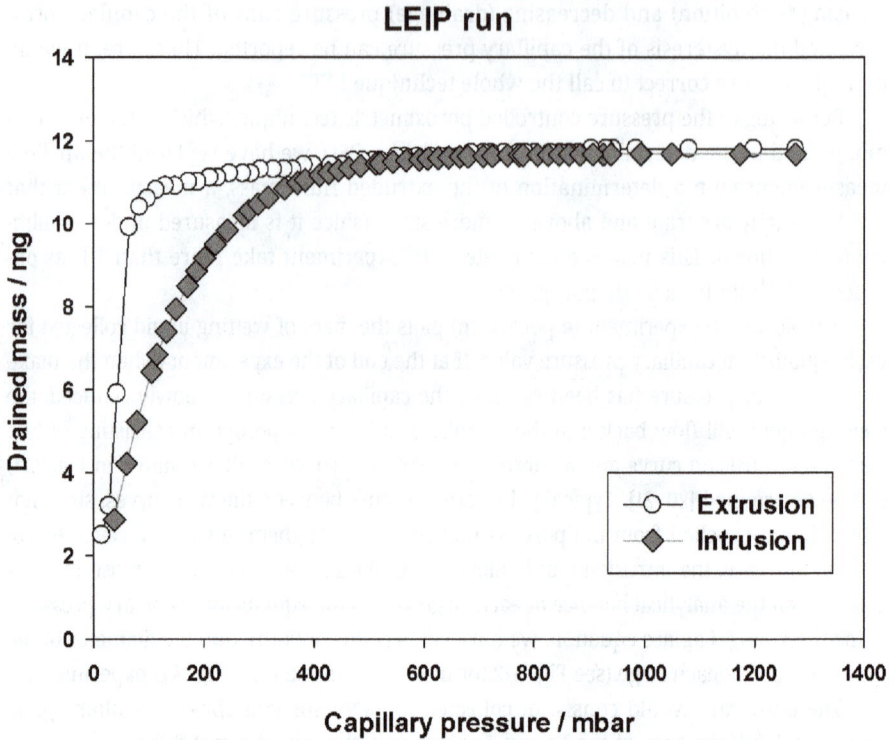

Fig. 7.11: Example of LEIP experimental data for a polymeric MF membrane.

non-wetting liquid for all materials) does not give us information on the wettability of the sample, information that of such devices can be crucial. The LEIP technique, applied to fuel cells, allows us to obtain essential information about the porous system, not only in terms of PSD or porosity, but also important concepts such as breakthrough pressure and wettability of the fluid–solid system.

As far as we have been able to ascertain, the only porometer manufacturer that markets LEIP equipment specifically is PMI, having published several relatively old articles on the subject [24, 25]. PMI markets the LEP-100A model that can go down to 0.06 μm.

Recently, IFTS (Institut de la Filtration et des Techniques Séparatives, France) has designed a new porometer that combines GLDP, LLDP and LEP characteristics with the possibility to perform all three porometries with the same instrument, called fluid–fluid/liquid extrusion porometer [26, 27], becoming one of the most versatile and complete in the market.

As mentioned above, the LEIP technique has been frequently used for the characterization of the GDL or MPL of fuel cells. In this type of materials, the technique gives slightly higher average pore size results than those obtained by similar techniques such as GLDP [27]. This can be attributed to the existence of a high degree of interconnectivity in the pores of such materials. Such interconnectivity would result in a better esti-

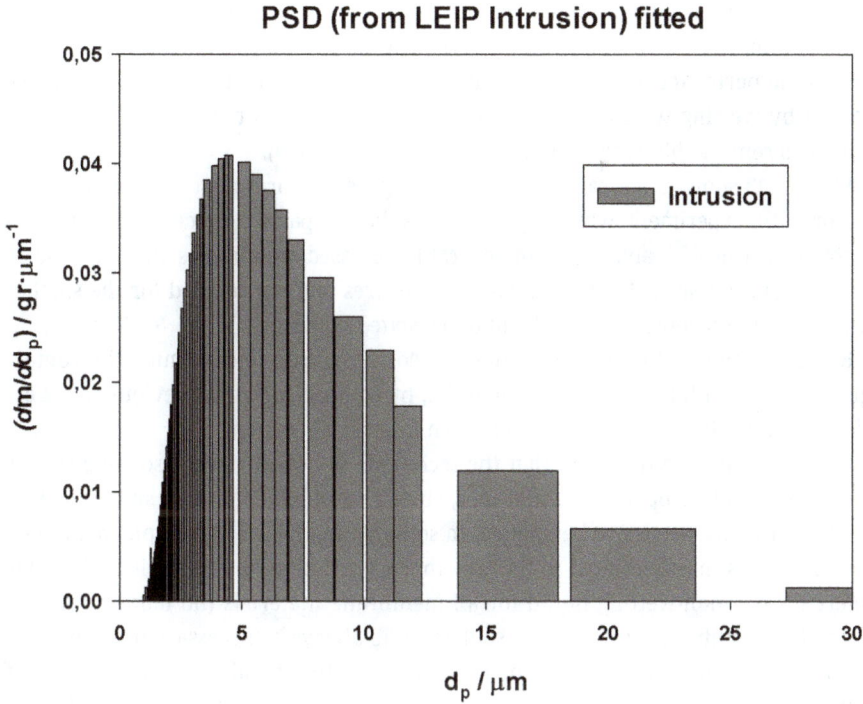

Fig. 7.12: PSD obtained from LEIP experiment shown in Fig. 7.11.

mation of pore sizes using the LEIP technique versus GLDP, as predicted by PNS-type simulations [25, 28].

7.3.3 Liquid entry pressure (LEP)

Finally, we should comment on a variant of this method, also named liquid entry pressure (LEP) in the literature, which presents some differences with the previous technique. In this case, it is used as an external fluid to intrude inside a dry sample. Since the sample is dry, its pores are empty of any external fluid and simply contain air. In a way, we are measuring the reverse of the GLDP technique, since we empty the air inside the pores with fluid (usually water, but other more convenient liquids can be used depending on the hydrophobicity characteristics of the sample), for which we have to apply an external pressure. As the pressure increases, firstly the larger pores and then, smaller and smaller pores are emptied of the air inside and filled with the permeating fluid, and subsequently, flow through the sample can be detected.

The major difference with GLDP is that when applying the Young–Laplace equation, we cannot assume zero contact angle (in this case, the wetting fluid is air and its contact angle with the membrane is not zero). Using this technique, Ailuno measured the maximum, minimum and average pore sizes of various PTFE membranes using water as the permeant fluid ($\gamma = 72$ mN/m, $\theta = 140°$), comparing the results with those obtained by wetting with Porofil and pushing it out with air ($\gamma = 16$ mN/m, $\theta = 0°$), achieving a remarkable agreement between both results [29].

Figure 7.13 shows a typical LEP experiment, quite similar to that found in a typical GLDP or LLDP experiment, while Fig. 7.14 shows the comparison of mean pore size values obtained from LEP and GLDP for several PTFE-based membranes. As can be seen, the agreement is reasonable for the larger pore sizes and not so good for the smaller ones, which are obviously more difficult to be wetted by water.

It is also evident that the technique LEP could be used to determine the contact angle of the air with the membrane material, by comparison with previous GLDP experiences with fully wetting liquids of known interfacial tension.

Finally, it must be indicated that the technique only will give reasonable results when applied to hydrophobic membranes, since hydrophilic ones wet suddenly with almost no pressure needed to be applied. In some sense, the LEP technique mentioned in this section is mostly related to the HgP that we will describe in Section 7.5, as far as mercury is a non-wetting liquid for all membrane materials (no matter whether they are hydrophilic or hydrophobic) so that it will always be necessary to apply pressure to introduce the mercury into the pores. On the other hand, the contact angles of mercury with most materials are, as we shall see, very similar to those used in LEP for clearly hydrophobic membranes ($\theta \sim 135$–$140°$).

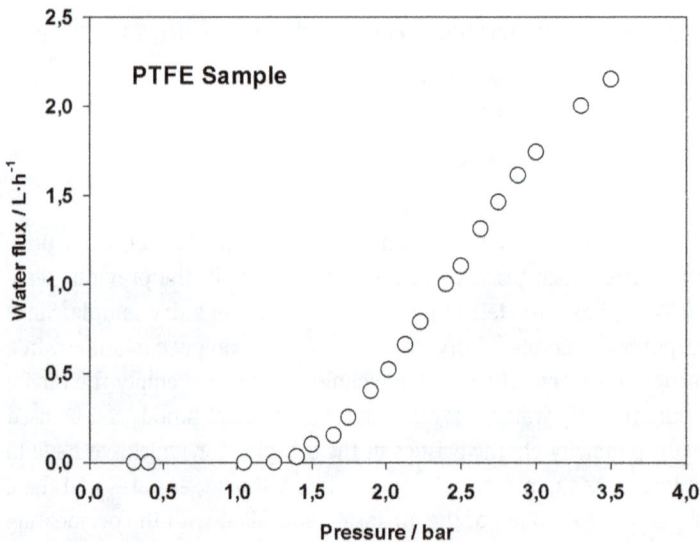

Fig. 7.13: Example of LEP (liquid entry pressure) experimental run for an MF membrane.

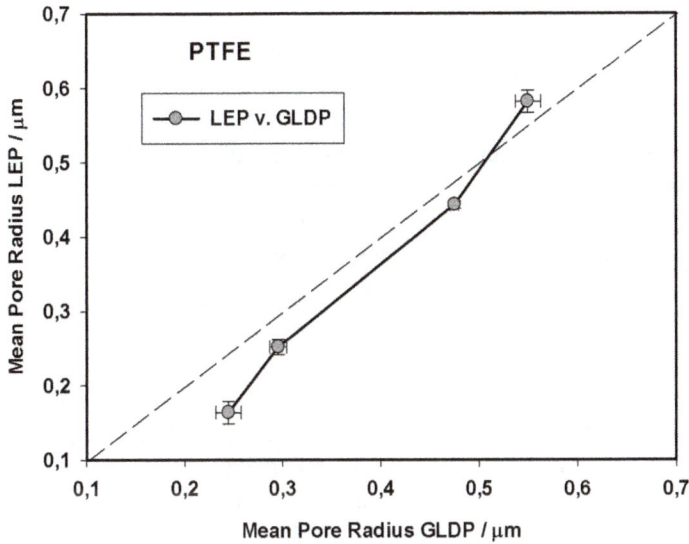

Fig. 7.14: Comparison between mean pore size values obtained by GLDP and LEP for different MF membranes (PTFE).

7.4 Liquid–liquid displacement porosimetry (LLDP)

As already mentioned, when using an air–liquid interface for the characterization of porous materials by means of the extended bubble point technique (GLDP), the high interfacial tensions that air has with most liquids must be considered. Thus, from a value of 73 mN/m at the air–water interface, we can go down to clearly lower values of 16 mN/m for Porofil and similar halogenated liquids. Even so, the pressure required to analyse the pores in the case of UF membranes (pore sizes of less than 0.05 µm) is extremely high and there is a risk of distorting the membrane structure and falsifying the results obtained.

This is why, as early as the pioneering works of Bechhold or Erbe, the use of a liquid–liquid interface was proposed, which significantly reduced the interfacial tension of the membrane. The GLDP technique, when using two liquids (one as a wetting fluid and the other as a pushing liquid), it is called LLDP. Other names found in the literature, referring to the same technique, are "permoporometry" [30], "bi-liquid permporometry" [31]) or "liquid–liquid porosimetry" [32].

In Erbe's original work [33], he indicates that Bechhold, apart from proposing the bubble point method for the determination of the maximum pore present in a membrane, already observed the problem that existed when this method was extended to smaller pores. Bechhold therefore looked for a pair of fluids with the lowest possible interfacial tension, and he found it in the water/isobutyl alcohol pair, which, according to the author himself, had an interfacial tension of only 1.73 mN/m. Erbe pointed out that, at the suggestion of a named Dr H. Karplus (of whose existence and publications we have found no further reference than Erbe's own article), he proposed to extend the bubble point method and combine it with the Hagen–Poiseuille transport model to determine the PSD of a membrane filter. In fact, Erbe himself calls the resulting method the Bechhold–Karplus method, as Ferry also calls it in his 1935 paper [34]. It should also be noted that the equation proposed in the works of Erbe, Bechhold or Ferry, to relate the pore size to the pressure applied to empty it

$$d_p = \frac{4\gamma}{P} \tag{7.20}$$

is referred to in these works as the Cantor equation, being simply the particularization of the Young–Laplace equation to the case in which the contact angle between the wetting liquid and the pore wall is zero, that is, the liquid wets the membrane perfectly. In Section 7.5, we will come across another particular name for the Young–Laplace equation, the so-called Washburn equation, which is normally used to interpret the HgP data.

Returning to Erbe, his main contribution (we have already mentioned that the LLDP method was in fact proposed by Bechhold and the ghost Dr. Karplus) consisted of systematizing the conversion of experimental flux–pressure data into a proper PSD. But before analysing Erbe's proposal, let us discuss what the technique consists of from the experimental point of view.

As already mentioned, the LLDP technique is based on a procedure similar to that of GLDP. The first step consists in soaking the membrane to be analysed in a wetting liquid, making sure that the liquid completely wets all the pores of the membrane. In other words, the wetting liquid must penetrate into all the pores and completely occupy them.

The next step is to dislodge the liquid that occupies these pores by applying pressure to the membrane with a second liquid. It is therefore essential that this second liquid (usually called the pushing liquid) is immiscible with the first (wetting) liquid. Once this fact is assured, we can observe that at low pressures the pushing liquid is not capable of emptying any of the pores of the membrane, so that as we increase the pressure, it will reach a value at which it is sufficient to empty the largest pores present in the membrane. This point (which in the GLDP technique was called the bubble point) will result in the appearance of a first flow of the pushing liquid through the membrane. If, as in the previous technique, we continue to increase the pressure slowly (allowing equilibrium to be reached for each pressure increase), we will empty new pores of the membrane, which will now be filled with the pushing fluid and will give rise to new increases in the flow of this fluid.

Figure 7.15 shows a typical LLDP experiment. In it, the pressure is increased and from a certain value of the pressure, we start to detect a measurable flow. The experiment continues, always obtaining increasing flows, but with a particularity: in the first point, the increase in flow between two consecutive points is greater than expected due to a simple increase in applied pressure. This is because, as we know, new pores previously occupied by the wetting liquid start to open and allow the flow of the pushing liquid. So, in each pressure step there are new pores yet opened contributing to the flow. This increase in permeability results in a slightly convex curve. Whereas at sufficiently high pressures, although new pores open at each stage, there are far fewer of them and the increase in permeability begins to decrease. The appearance of the curve in this part is clearly concave. In between there is a point (called the inflection point) where the curve changes from convex to concave and which corresponds to the maximum value of the pore distribution in terms of flow (i.e., the pore size presenting the higher contribution to the flow, not the one having more population).

The graph shows the permeability lines for representative points (first point, last points as well as the ith and $(i+1)$th points). The permeability is nothing more than the quotient between the flow and the applied pressure at each point. We can see that the permeability (the slope of the line) increases gradually from the initial value to the last points, where the permeability does not vary more (except for slight fluctuations due to experimental errors).

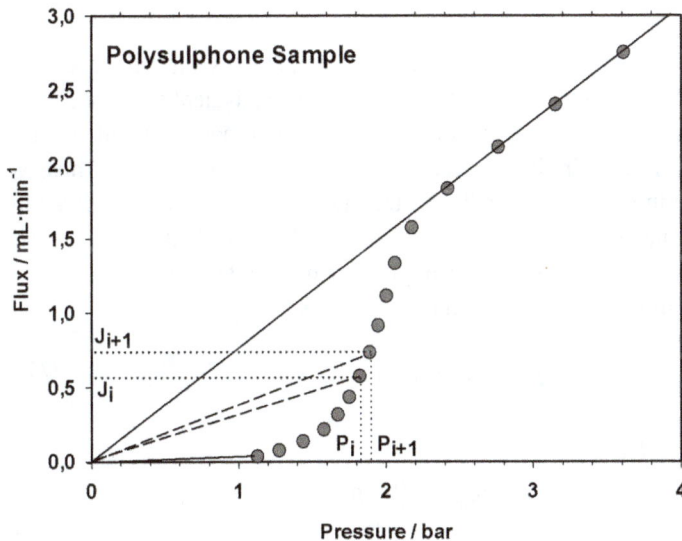

Fig. 7.15: Typical LLDP experiment for a polysulfone UF membrane (unpublished data from authors).

Let us see how to convert this information into the number of open pores, following Erbe's reasoning, which can also be found in Capannelli et al. [35] or Kesting [36]. To better visualize how to do so, in Fig. 7.16 we have enlarged the previous graph, and now we look at a couple of consecutive points of our experience (points i and $i+1$).

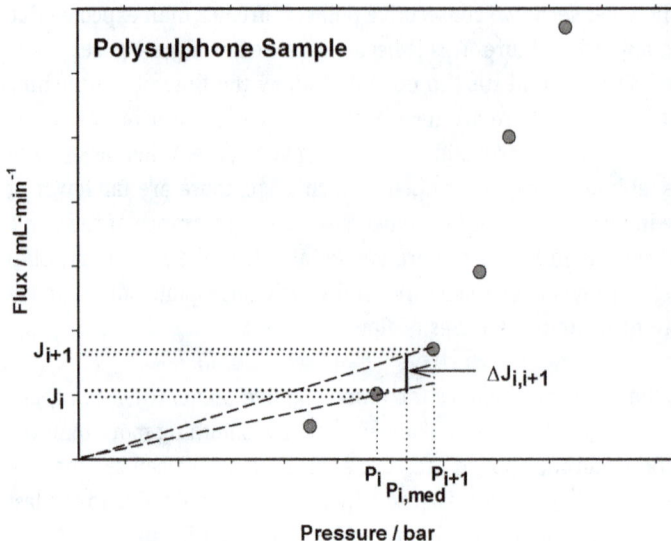

Fig. 7.16: Increase in permeability due to successive pore opening during pressure increase steps in an LLDP experiment (unpublished data from the authors).

If we were to take as the flow increment simply the difference between two consecutive flow readings ($\Delta J_{i,i+1} = J_{i+1} - J_i$) we would be comparing non-equivalent situations, since at point $i+1$ there are more pores open to flow than at the ith point. Instead, we first determine the mean pressure in the interval considered: $P_{i,\text{med}} = (P_i + P_{i+1})/2$. Thus, we see that the increase in flow corresponding to this mean pressure ($\Delta J_{i,i+1}$) is given by the difference in permeabilities (difference in slopes of both straight lines, between point i and point $i+1$, determined by the mean pressure of the interval).

Mathematically, this is expressed as follows:

$$\Delta J_{i,i+1} = J'_{i+1} - J'_i \tag{7.21}$$

with

$$J'_{i+1} = L_{i+1}\, P_{i,\text{med}} = \frac{J_{i+1}}{P_{i+1}}\, P_{i,\text{med}}$$

$$J'_i = L_i\, P_{i,\text{med}} = \frac{J_i}{P_i}\, P_{i,\text{med}} \tag{7.22}$$

where J refers to the experimental flow values, while J' refers to the extrapolated flow values for the mean pressure. In the graph the experimental values are the intersection points of the horizontal dotted lines with the ordinate axis, while the theoretical values are the intersection points of the horizontal dashed lines with this same axis. Finally, we arrive at the following expression:

$$\Delta J_{i,i+1} = \left(\frac{J_{i+1}}{P_{i+1}} - \frac{J_i}{P_i} \right) P_{i,\text{med}} \qquad (7.23)$$

This expression of the flow increment already takes into account the effect of the flow increases due to the pressure increase (since it is calculated for the same pressure point), being only due to the opening of a new set of pores. Thus, from eq. (7.23) we can obtain a first PSD which we will call the flow distribution, and which will be given by each flow increment obtained (from two experimental flow increments) versus the corresponding size of the pores opened in that increment. This pore size is obtained from the Young–Laplace equation (we can actually call it Cantor's equation since we have assumed that the liquid wets the membrane perfectly) so that the PSD will be formed by pairs of values:

$$(\Delta J_{i,i+1}, \ d_{\text{p},i,i+1})$$

being $d_{\text{p},i,i+1} = 2\gamma / P_{i,\text{med}}$.

There is another reference in which the calculations are performed in a similar way [37]. The only difference from the previous procedure is that the flux increment $\Delta J_{i,i+1}$ is calculated from the intersection of the slope lines L_i and L_{i+1} with the vertical at P_{i+1}, instead of the vertical at $P_{i,\text{med}}$. But it is clear that both criteria lead to very similar distributions, at least for experiments with a sufficient number of experimental points.

Before discussing the other possible PSDs that can be obtained from an LLDP porogram, it is worth noting an important comment. When previously discussing the usual criticisms of the GLDP technique, many of which extend to LLDP, it was noted that possibly the most common concern against LLDP is the need to use a transport model to arrive at an absolute distribution of the number of pores of each size present in the sample. It is obvious that these transport models are based on assumptions that we cannot always be sure of, so a good practice would be to present directly as PSD the above flux distribution curve, which does not need any transport model and whose validity is only limited by the validity of the Young–Laplace equation itself. Thus, with the PSD flux, it indicates the contribution to the total flux through the membrane due to each of the pore sizes, information that can be perfectly valid and sufficient to characterize the membrane.

On the other hand, if, as usual, we are interested in PSD in terms of the number of pores, that is, a curve that tells us how many pores there are of each size, we must apply the corresponding transport model. Unlike in the case of GLDP, here there is no possible discussion about the transport model to be applied. Since we are dealing with a liquid flow, the transport will always be convective and will be governed by the Hagen–Poiseuille model for flow through capillaries (or slight modifications of the same model that we will discuss for particular geometries).

Returning to the above graph, to achieve a flow increase given by $\Delta J_{i,i+1}$, at mean pressure $P_{i,\text{med}}$ it is necessary that in steep: $i{\to}i+1$, a number of pores given by the following expression (easily deducible from the Hagen–Poiseuille equation (7.7)) have been opened:

$$n_{i,i+1} = \frac{\eta l}{2\pi\gamma^4} \ P^3_{i,\text{med}} \ \Delta J_{i,i+1} \tag{7.24}$$

Similarly, each of these pores can be considered as a cylinder of diameter $d_{p,i,i+1}$, perpendicular to the membrane surface, so that the area of these pores and their corresponding volume will be given by the following expressions:

$$A_{i,i+1} = \frac{n_{i,i+1} \ \pi \ d^2_{p,i,i+1}}{4}$$

$$V_{i,i+1} = \frac{n_{i,i+1} \ \pi \ l \ d^2_{p,i,i+1}}{4} \tag{7.25}$$

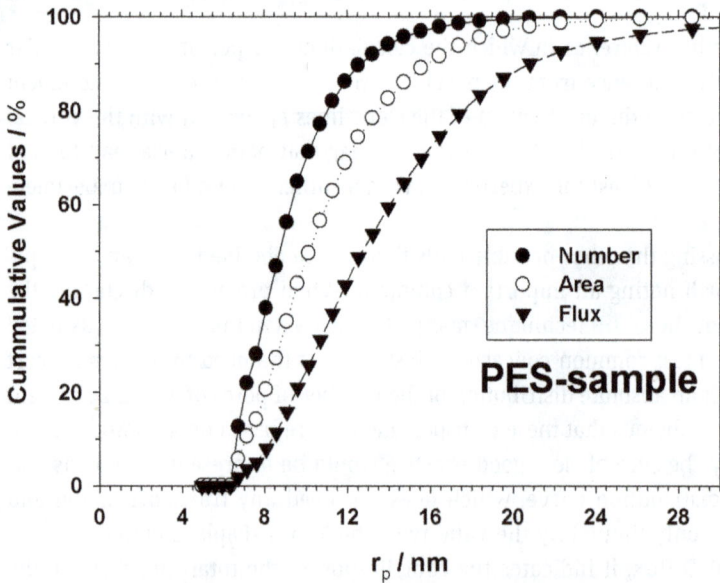

Fig. 7.17: Pore size distributions (cumulative values) in terms of the number of pores, area and flux through such pores, obtained for a typical LLDP experiment on a PES–UF membrane (unpublished data from the authors).

Expressions (7.24) or (7.25) allow us to present a PSD in numbers or in areas or volumes (both absolutely equivalent). In fact, the two most common are the distribution by flow and the distribution by number of pores, although the previous Fig. (7.17) shows an example of the three possible distributions for a given membrane.

Previously, we have seen one way of calculating the pore number. If we look again at Fig. 7.15, the last two points (and certainly some more ones if the experience is sufficiently long in pressures) give rise to a perfectly defined straight line passing through the origin and below which lies the S-shaped curve characteristic of an LLDP experiment. If we call the quotient between the last values of flow and pressure asymptotic permeability (L_{asym}), we can obtain another expression to calculate the pore number from the Hagen–Poiseuille equation (7.7):

$$n'_{i,i+1} = \frac{8\eta l}{\pi \, r^4_{p,i,i+1}} \, L_{asym} \cdot \Delta L_{i,i+1} \tag{7.26}$$

In this expression, the term $\Delta L_{i,i+1}$ has been introduced, and accounts for the contribution of the open pores through an increase in pressure to the total membrane permeability expressed in per cent per one.

Once we have the pore numbers, it is easy to calculate the membrane porosity as follows:

$$\theta = \sum_i A_{i,i+1} = \sum_i n_{i,i+1} \, \pi \, r^2_{p,i,i+1} \tag{7.27}$$

Here the subscript represents the ith, $i+1$th increment of pressure in the experiment. Also, the total number of pores will be

$$N = \sum_i n_{i,i+1} \tag{7.28}$$

7.4.1 Grabar–Nikitine algorithm

Grabar and Nikitine, in a truly pioneering work [38], started from the Hagen–Poiseuille convective flow model to relate the flow through open pores to the number of pores and their size. The main difference between the calculation algorithm that we will now discuss and the previous one proposed by Erbe is that the latter is clearly designed for a discrete set of data, while the Grabar–Nikitine algorithm is deduced for a continuous function of the data (although it can easily be applied to discrete sets of data). Thus, Grabar considers that, for pressure P applied across the membrane, the fluid flux through the pores whose diameters are between $2r$ and $2(r + dr)$ is

$$J \, dr = \frac{\pi \, P \, r^4 \, n(r)}{8 \, \eta l} \, dr \tag{7.29}$$

Given that in addition to the open pores in the interval dr, there are other previously open pores, we will have that at the given pressure, the total flux of liquid that crosses the membrane will be that due to all the pores of sizes between r and r_{max}. This total flux can be obtained by integration of the above equation:

$$J = \frac{\pi P}{8 \eta l} \int_{r}^{r_{max}} r^4 \, n(r) dr \tag{7.30}$$

Deriving this expression and substituting r for p using the Cantor equation (Young–Laplace equation, assumed zero contact angle), we obtain the number of open pores for a given pressure increase:

$$n(r) = \frac{8 \eta l P^5}{\pi (2\gamma)^5} \left[\frac{dJ}{dP} - \frac{J}{P} \right] \tag{7.31}$$

Expression that can be substituted in eq. (7.29) to obtain the total flux and then taking into account the definition of permeability as the ratio between the obtained flux and the applied pressure, we would have that the permeability of the membrane, at pressure P, for which a number of pores $n(r)$ have been opened, will be given by the following expression:

$$L(r) = \frac{\eta P}{\eta' 2\gamma} \left[\frac{dJ}{dP} - \frac{J}{P} \right] \tag{7.32}$$

So the permeability is a function, among other things, of the relative viscosity of the two liquids used.

Obviously, when working with experimental data, the Grabar–Nikitine algorithm cannot be directly applied since it requires a continuous and derivable function of flow–pressure data. There are two possibilities:

– One possibility is to fit the experimental data to some kind of known function and from that fit, apply the Grabar derivative to obtain the PSD. This approach was followed, for example, in Antón et al. [39], where the shape of a porogram assuming a theoretically log-normal PSD was considered. Evidently, fitting the experimental data to known functions or simply to polynomials of sufficient degree allows to obtain much more elegant and clean PSDs, without the typical jumps due to the fluctuation of the experimental data. However, one must be careful with such fits so that the fitted function does not falsify the important parameters of our experimental distribution (mean pore radius and distribution width).

– Another simpler and less laborious possibility is to convert the previous algorithm into a simpler calculation for a discrete data set. For that purpose, we simply convert the pressure and flow differentials, after having been used as infinitesimal expressions, into experimental increments between consecutive data $(dJ/dP \sim \Delta J/\Delta P)$, while P and J are the values of the mean pressure and mean flow in these increments. Both quantities are taken directly from the experimental permeability curve.

In any case, using the Grabar–Nikitine algorithm (modified for discrete data points) or applying a geometrical calculation, both procedures present an important problem related to random fluctuations of experimental data. In actual laboratory measurements, the data deviate from the theoretical behaviour, so that a given pressure increase does not always correspond to the expected flux increase. We define the permeability increase as

$$\Delta L_p = \frac{\Delta J_{i,i+1}}{P_{i,\text{med}}} = \left(\frac{J_{i+1}}{P_{i+1}} - \frac{J_i}{P_i} \right) \tag{7.33}$$

Normally this increment should be positive, as more and more pores are opened with each pressure increase. Nevertheless, as pressure approaches to the last data points (those where it is supposed that all pores are yet emptied from wetting liquid) in Fig. 7.18, it is expected that permeability remains constant and, consequently, the increments given by eq. (7.33) approach zero. But small fluctuations between two consecutive data points could lead to negative increments which make no sense from a theoretical point of view (this permeability decrease should indicate some pores are again closed to flow, which is impossible). Therefore, to solve this problem, several mathematical procedures could be used, being the simplest one to consider $\Delta L_{i,i+1} = 0$, in such cases.

Obviously, in the case of fitting the experimental data to some theoretical or empirical function, which converts the experimental data fluctuations in a continuous function, such experimental errors disappear. As mentioned before, this was the procedure followed, for example, in the work of Antón et al. [39], where a log-normal PSD was assumed for the membrane, and experimental data were fitted to the expected theoretical function of flux–pressure data found for such PSD. The fitting of experimental data gave the values of the mean pore size and standard deviation of the log-normal PSD which best fits to the actual data points. Certainly, when using such fitting procedure (or, really, whatever other fitting procedure, as those quite commonly included in the software of commercial porometers) we must be sure that the fitting is an accurate reflect of the real membrane. This can be achieved by simply requiring that the mean value of the pore size obtained after our fit matches the mean value obtained from the distribution from the direct data without mathematical fit. Given that the permeability at a given pressure corresponds to the slope of the graph, the previous convention cancels out the low permeability values, increasing its difference with respect to the maximum values. These maxima correspond to the working zone of the membrane, that is, the zone where the curve describes an S-curve due to a sudden increase in permeability.

Figure 7.18 shows an example of LLDP experiment comparing original data with the result of fitting to a previously assumed log-normal-based function, with a remarkable agreement between both data sets. Figure 7.19 shows the PSD obtained from fitted data, once Grabar–Nikitine algorithm has been applied. Obviously, it results in a PSD smoother and closer to expected log-normal distribution usual in polymeric UF membranes. While important parameters of the distribution (mean pore size, standard deviation or skewness) does not differ from results obtained by applying conversion algorithm to the original data.

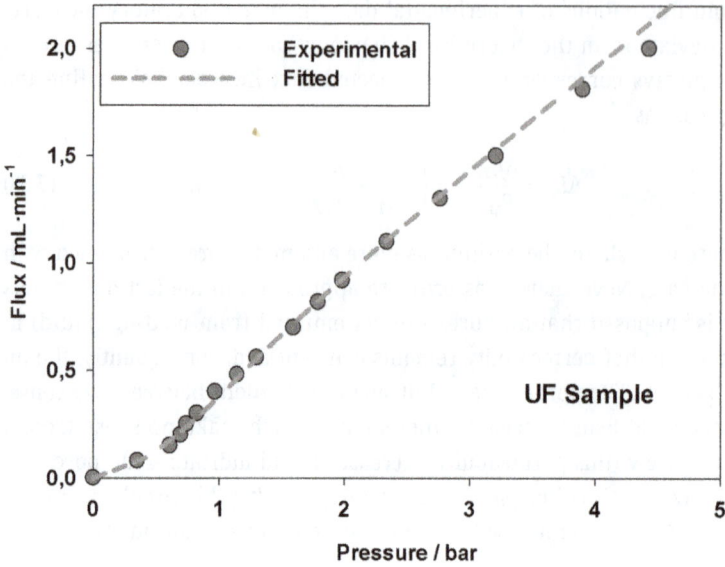

Fig. 7.18: Typical LLDP experiment showing the experimental points along with their fitting.

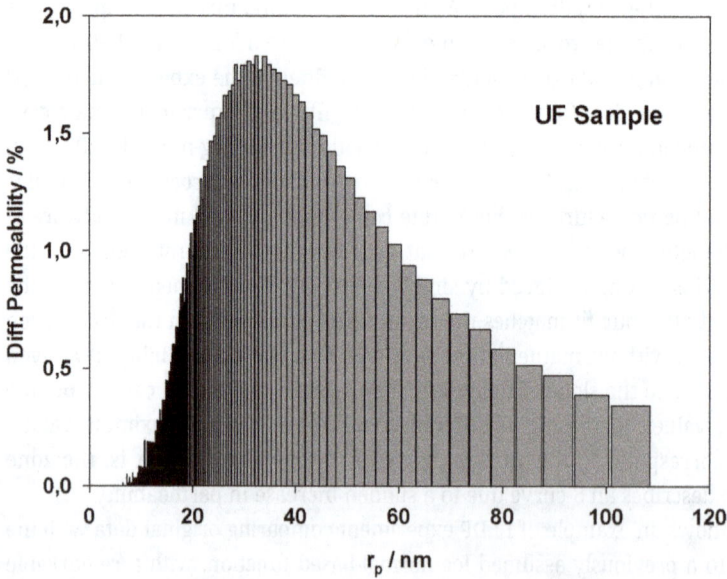

Fig. 7.19: PSD obtained from the treatment of LLDP data fitted as shown in the previous figure.

7.4.2 Liquid mixtures used

As already commented for GLDP, in order to obtain a correct and reliable LLDP experiment, it is necessary to carefully choosing the liquids to be used in the experience. First of all, it must be ensured that they are immiscible liquids. It is also desirable that the interfacial tension between the two liquids is as low as possible in order to be able to analyse the sample using a sufficiently low range of pressures so as not to damage the membrane. Once the pair of liquids that meet these two preconditions has been chosen, one last requirement remains: it is necessary to decide which of the liquids will play the role of wetting liquid (which will be responsible for pre-wetting the membrane by soaking all its pores) and which one will be the pushing liquid (what will be forced by pressure to enter the previously occupied pores, displacing the wetting liquid from them). This assignment of roles to the liquids of the chosen pair does not have to be unique and their roles can be interchanged, if necessary [40]. Basically, the liquid that best wets the membrane will be chosen as the wetting liquid. Given that we can find both hydrophobic membranes (almost all polymeric membranes are hydrophobic in nature) and hydrophilic membranes (either because the material of which they are composed is hydrophilic or because they have been treated to hydrophilize them), generally the most suitable choice as wetting liquid will be an alcohol, whose amphoteric characteristics allow the wetting of the membrane to be optimized.

Since the first published work on LLDP [2], isobutyl alcohol (isobutanol) has been one of the most frequently used liquids in such experiments. When used in combination with water, it has an interfacial tension between 1.7 and 1.9 mN/m depending on the working temperature. In order to ensure immiscibility, in the preparation of porosimetric liquids, the following procedure is usually followed [41]:

– Equal volumes of isobutanol and distilled water (previously degassed) are mixed in a separating funnel. The mixture is shaken vigorously for a few minutes to ensure homogeneous mixing.
– In practise, since miscibility between each liquid is very low and it is constant at ambient conditions, it is enough to mix a gentle amount of water (supposed displacing liquid) with a sensibly lower amount of isobutanol (wetting liquid and more expensive). This will result in lower quantity of isobutanol saturated in water but still enough to wet properly a good number of samples.
– The mixture is then left to stand for a sufficiently long period (usually overnight). As time passes, the mixture will separate into two phases, both transparent but with a clearly distinct interface (see Fig. 7.20).
– When this separation is complete (both phases are clean and shiny with a clear interface between them) they can be collected in separate containers, the lower phase being the aqueous phase and the upper phase the alcoholic phase.

This procedure makes it possible to obtain two phases saturated each one in the other liquid. Thus, when the two liquids, already saturated in each other, come into contact

Fig. 7.20: Phase demixing for a water–isobutanol mixture.

inside the membrane, they will not be soluble in each other and will be completely immiscible.

Since the LLDP analysis is usually performed at temperatures around 20–25 °C, it is convenient to prepare the liquid pairs at ambient temperature close to the operating temperature. Also, this temperature should be very convenient for liquid storage.

The preparation of the porosimetric phases must be carried out each time a measurement or set of measurements is to be made. In the case of the isobutanol/water mixture, it is sufficiently stable to be used for several days without appreciable loss of properties. This is not the case for porosimetric mixtures containing more volatile alcohols (e.g. methanol) which have to be prepared fresh every day, due to the change in composition and properties resulting from the high evaporation of methanol.

In a review by Tanis-Kanbur et al. [42], the main properties of most of the blends that have been used in LLDP porosimetry over time are listed and conveniently referenced. The main properties of the mostly used liquid pairs are given in Tab. 7.1.

Based on our experience, the most stable mixture to work with is the mixture of isobutanol and water, while (among those presented in Tab. 7.1) the mixture with water/isobutanol/methanol, while presenting a clearly lower interfacial tension, is more sensitive to environmental conditions and its properties can vary very quickly. One possibility to avoid the methanol evaporation and consequently the change in the properties of such mixture is to work at sensibly lower temperatures (through an adequate thermostatic bath). Using this approach, we obtained good results compared with usual isobutanol/water mixture, working with the three liquid mixtures at 15 °C [40]. Finally, in Tab. 7.1, the main characteristics (including surface tension) of several commonly used pairs of liquids at a working temperature of 20 °C are summarized [41].

Although the liquid mixtures presented in Tab. 7.1 include the most commonly used ones, there have been other attempts to find liquid mixtures with better properties. It is worth mentioning the work of Philips and DiLeo [43], who proposed a mixture based on water/PEG 8000/ammonium sulfate, on the basis of which Millipore

Tab. 7.1: Most commonly used liquid mixtures on LLDP characterization.

Wetting liquid (water-rich phase)	Displacing liquid (alcohol-rich phase)	Surface tension[1] (mN/m)	Dynamic viscosity[1] (mPa · s)
Water	Methanol/isobutanol	0.35	3.4
Water	Isobutanol	1.7	4.3

[1]$T = 20$ °C.

patented as the CorrTest method of analysis. This mixture has been slightly modified by Bouchiha [44], using water/PEG 1000/magnesium sulfate with interesting results.

7.4.3 Commercial devices

Despite its early introduction as a promising technique, LLDP did not develop rapidly in these early years. Thus, apart from the pioneering work of Bechhold or Erbe, little was published on applications of the technique in later years. We have to go back to the late 1980s and the beginning of the 1990s to find regular publications on the characterization of UF membranes by LLDP [31, 37, 45]. By these years, Capannelli and Bottino's group at the University of Genoa became one of the most active in the improvement and use of the LLDP technique [35, 46, 47].

At the beginning of the 2000s, the authors of this book started a fruitful relationship with the group from Genoa (led by Capannelli and Bottino) focussing on the development and improvement of the LLDP technique. This collaboration led us to jointly design an equipment suitable for the characterization of membranes, whether flat, tubular or hollow fibres. A complete scheme of such device can be seen in Fig. 7.21. The equipment is based on the concept of volume control, using a high-precision syringe pump (ISCO500D) to control the flow through the sample and measuring the equilibrium pressure corresponding to this flow by means of transducers (a schematic of the equipment is shown in the next figure). Although the design has proved to be extremely useful and stable, allowing numerous advances in LLDP research, it presents some problems for its conversion into a commercial equipment.

Firstly, the high cost of the pump (if the required accuracy and reliability are to be achieved) would greatly increase the cost of commercial equipment based on this type of design.

Furthermore, characterization with this system is limited by the capacity of the syringe pump (500 mL in this case). This means that, although the characterization of small membrane samples (flat discs up to 47 mm in diameter, tubular membranes of less than 10 cm in length or small groups of hollow fibres) is granted, when the aim is to analyse larger samples or commercial modules with this device, the measurement capacity is somewhat limited.

For this reason, other devices have opted for a pressure-controlled approach. In this approach, an air pressure is used to push the penetrating liquid through the membrane from a reservoir, and the volume of which can be scaled accordingly. This is the concept behind the equipment patented by Thierry Courtois between 1983 and 1985 (successive French, European and world patents, were issued [48]). Initially, Courtois marketed his own equipment through his company (GEPS) but later joined the French Institute IFTS (Institut de la Filtration et des Techniques Séparatives) which now markets a very complete porometry equipment, including GLDP, LLDP and LEIP options.

Other porometer manufacturers that have introduced options for the LLDP technique in the market are PMI (with its Liquid–liquid Porometer, LLP-1100A, able to work down to 2 nm) and Porometer (whose Poroliq-1000 porometer, developed under the advice of the authors, can also reach 2 nm).

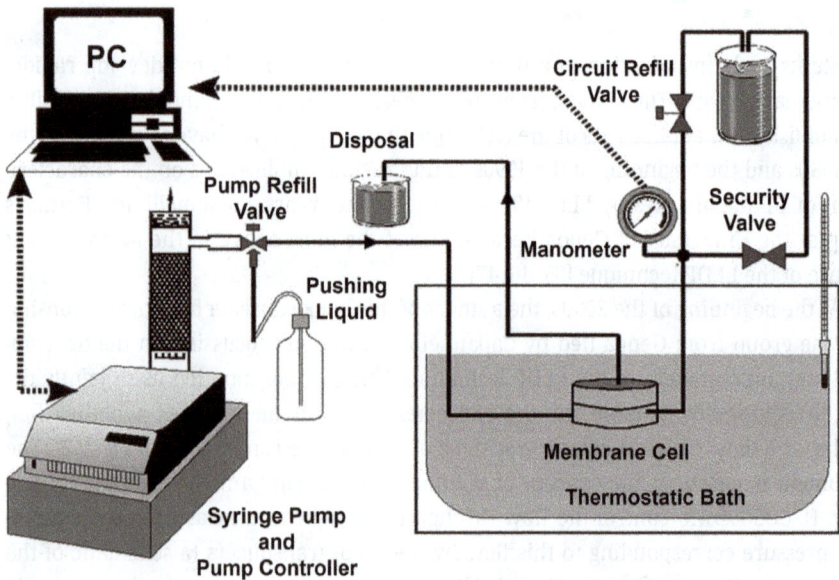

Fig. 7.21: Schematic of a laboratory rig designed to perform LLDP analysis, adapted from [40].

7.4.4 Use of LLDP for estimating MWCO of a membrane

All the porosimetric techniques revised in this chapter, along with those thoroughly discussed across the different chapters of this book (or at least most of them), are able to supply a complete information about the pore size parameters of the studied membranes (mean pore size, maximum and minimum pores, complete PSD, porosity, etc.). This is a very valuable information for selecting a given membrane for a separation purpose, as we commented in the introduction to this text. Nevertheless, especially for most UF membranes and all NF ones, they are marketed using a distinctive characteristic such as their molecular weight cut-off (MWCO). This parameter will be conveniently discussed when pore sizing from solute retention measurements is reviewed in Chapter 8. But, at a first glance, it seems clear that MWCO and pore size refer to related but different concepts.

It seems clear that it would be really interesting to be able to equate cut-off and pore size in some simple and biunivocal way that allows us to quickly know which cut-off corresponds to a given average pore size and vice versa. This relationship between cut-off and pore size is actually implicit in the measurement methodology of both parameters, although its conversion is not so simple.

A few years ago, the authors of this book proposed a method for estimating the MWCO from the PSD obtained for a UF membrane using the LLDP technique [49]. To understand the method, we must remember that the cut-off tells us the molecular weight of the molecule that is 90% retained by our membrane. Our assumption (simple but reasonable) is that the characteristic size of this molecule should be close to the size of 90% of the largest pores in the membrane.

Let us take this intuition as a true basis and define the estimated MWCO as the molecular weight of that molecule, whose size (effective diameter) is equal to the size of the pore that defines the 90% of the largest pores in the membrane. In other words, the molecule whose molecular weight defines the cut-off will be the one whose size is such that only 10% of the pores are larger than it. This means that it will be retained as 90% of the pores of the membrane. Although the idea is simple, there are several preliminary considerations about its validity:

- Firstly, by equating the size of the largest molecule that is retained with the size of the pore that retains it, we are ignoring the important fact that the molecules are elastic. Especially if these molecules are large polymeric chains, their conformation means that they can vary slightly in size under appropriate pressure.
- On the other hand, there is the thorny point of defining the size of the molecule. This size would be clearly well-defined if we are dealing with spherical molecules, but this is a rather particular case, as it is more common for the molecules to have rather linear dimensions that hardly resemble perfect spheres.
- Still, we can consider a certain average diameter (somewhere between the maximum and minimum length of our molecule) as the mean dimension of the molecule, which could be considered as the effective dimension in terms of retention.
- A final point of discussion will be given by the molecule or the type of molecule to which we want to apply this estimation. As we will see in the corresponding chapter, the retention measurements that allow us to define the MWCO of a membrane are usually made from the filtration of a series of dextrans of increasing sizes, although more specific molecules have also been used, such as PEG or different types of sugars [50] or others more specifically related to the application we want to give to our membrane. It is obvious that when a membrane manufacturer offers us a 100 kDa UF membrane, they should indicate with which type of molecule (dextran, PEG, etc.) and under which experimental conditions this value

has been measured. But the truth is that this detailed information rarely accompanies the explanatory pamphlet on the characteristics of the membrane in question.

Faced with these doubts, we opted for a simple model in which we used an empirical correlation valid for dextran between molecular weight and Stokes diameter (obtained from diffusion measurements).

Thus, the diffusion coefficient at infinite dilution of a molecule can be related to its hydrodynamic radius (or Stokes radius) through the Stokes–Einstein expression:

$$r_S = \frac{k_B T}{6 \pi \eta D_\infty} \tag{7.34}$$

where k_B is the Boltzmann's constant, T is the absolute temperature, η is the intrinsic water viscosity and D_∞ is the diffusion coefficient of the molecule at infinite dilution in water.

For the case of the dextran molecule, and assuming they are monodispersed, D_∞ can be correlated with the dextran molecular weight (M), by using the following equation [51]:

$$\log D_\infty = -4.1154 - 0.47752 \log(M) \tag{7.35}$$

Experimentally, our proposal consists of representing the PSD graphically in terms of cumulative pore number (i.e. percentage of pores that have a given size). Such a representation, as shown in Fig. 7.22, gives the cumulative percentage for each size, starting with the minimum diameter pore and ending with the maximum diameter. Assuming that the pores whose size corresponds to 90% of the largest pores in our PSD are those that will dictate the size and, consequently, the molecular weight of the molecule retained at 90%, the intersection of the PSD with the ordinate corresponding to 90%, transferred to the abscissa axis, gives us this limiting size. In the case of the figure, this value corresponds to a pore diameter of 9.30 nm.

Considering this value as the Stokes diameter of the dextran molecule retained at 90%, eq. (7.34) gives us its diffusion coefficient at infinite dilution, which taken to eq. (7.35) gives us a value of the molecular weight that we consider as the MWCO of the given membrane.

In the case of the figure, the result of the estimation is an MWCO of 33.7 kDa, for a membrane whose nominal value (supplied by the manufacturer) is 30 kDa. It is clear that this calculation is only an estimate and that the goodness of the final result (or rather the closeness of this value to the value supplied by the manufacturer) is not assured. Nevertheless, it can be considered that the method gives very close results (always within the order of magnitude of the value sought) with very little experimental work (a good LLDP analysis is sufficient, which can be done in barely 2 h), as opposed to the laborious work of finding a sufficiently large and well-balanced mixture

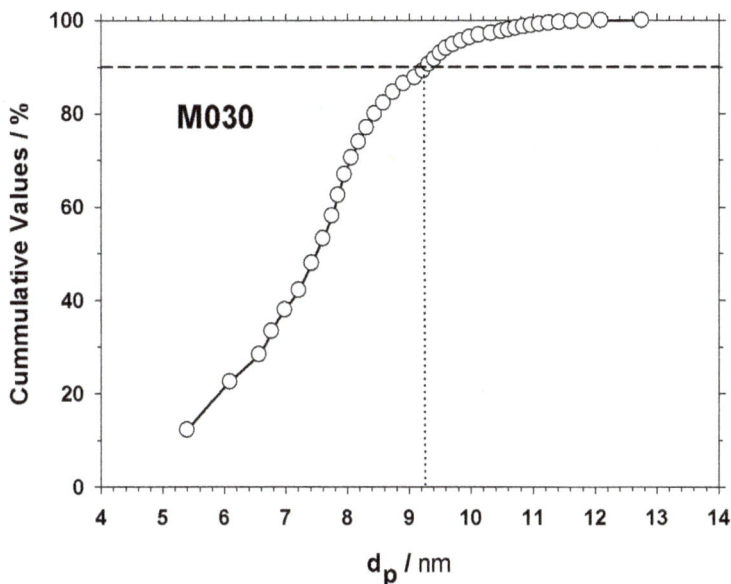

Fig. 7.22: PSD of a polymeric UF membrane (nominal MWCO 30 kDa) as obtained from LLDP, presented in cumulative values of pore numbers. The line corresponding to the 90% biggest pores allows to estimate a value of ~ 34 kDa for that membrane, using the previously mentioned procedure.

of dextrans, determining the most suitable experimental conditions and carrying out the retention experiments, from which the MWCO value is obtained.

Using this methodology, the authors analysed a large set of membranes from the Sartorius R&D laboratory, and from the comparison between the estimates obtained by LLDP and the values measured by Sartorius using dextran retention, a quite remarkable agreement was reached [52], as shown in Fig. 7.23.

Certainly not all membranes are characterized by dextran retention. Thus, PEGs are molecules frequently used in retention experiments. If we want our MWCO estimation to be equally correct, we should correct the empirical expression relating the molecular weight to the diffusion coefficient. Thus, with the same assumptions as in the case of dextrans and assuming the Stokes–Einstein equation to be equally valid, we could replace eq. (7.35) by the following expression, valid for PEGs [53]:

$$\log D_{\infty} = -3.90309 - 0.55 \log(M) \tag{7.36}$$

As an example, Fig. 7.24 presents the comparison between MWCO estimations and nominal values for several UF Biomax membranes (from Millipore) with nominal cut-offs ranging from 5 to 300 kDa. For each membrane, both empirical equations (7.35) and (7.36) have been tested, and results show some best agreement using the PEG valid equation, especially for higher cut-off values where dextran correlation losses somehow validity.

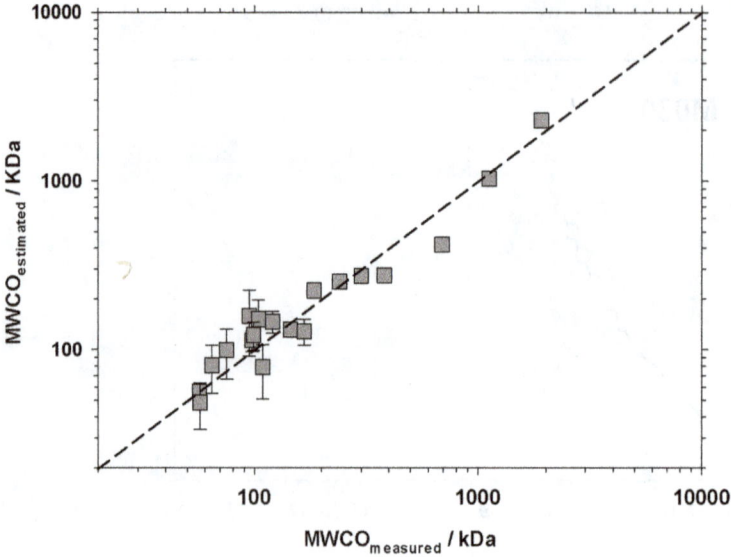

Fig. 7.23: Comparison of estimated MWCO (obtained from LLDP experiments) with experimentally measured values (from dextran retention tests) for several Sartorius polysulfone membranes (adapted from [52]).

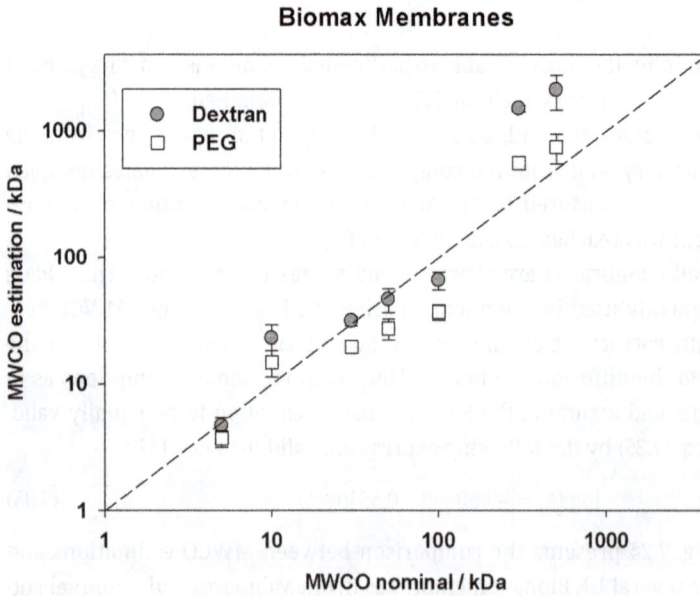

Fig. 7.24: Comparison of MWCO estimations for several Biomax membranes, by using eqs. (7.35) and (7.36) versus the nominal value quoted by the manufacturer (adapted from [120]).

Finally, we should point out that this MWCO estimation procedure does not, of course, require the prior use of LLDP as a characterization technique. It is sufficient that the porometry used to characterize our membrane is sufficiently accurate for the resulting estimate to be close to the real behaviour.

However, as shown clearly in Figs. 7.23 and 7.24, the estimated value for MWCO will be strongly related to how accurate our PSD is in the part corresponding to the largest pores. As we know, from the very foundation of the bubble point-based techniques, these techniques (mainly GLDP and LLDP) are particularly suitable for determining the largest pores of the membrane (those that start the contribution to the total flux). Therefore, both techniques can presumably lead to better estimates of the MWCO than others that are more accurate in the determination of the mean pore size, but may give rise to a greater margin of error in the tails of the distribution. On the other hand, the GLDP technique, due to its experimental set-up, is usually applied to MF membranes, where it does not make sense (at least from a commercial point of view) to talk about MWCO and where any empirical relationship between Stokes diameters and molecular weights will often lose a lot of reliability.

7.4.5 Criticisms of the LLDP technique

As previously discussed for GLDP, much of the criticism of both techniques comes from the assumption of cylindrical pores when applying the Young–Laplace equation. As discussed in Section 7.2.2, some of these objections are not so critical in terms of obtaining reliable results, since the application of the Young–Laplace equation in the usual way allows us to determine the narrowest section of the pore (ultimately, the one that governs the transport and separation in the pore) with little error in most cases, with respect to the actual pore geometry.

The common criticism in using the Hagen–Poiseuille transport model, based on a pore geometry of cylindrical tubes perpendicular to the membrane surface, to obtain the PSD in terms of the number of pores of each size, is more difficult to avoid.

Expressions (7.24) and (7.25) allow us to obtain from the LLDP porogram, the absolute data of the number of pores, area and volume of pores in a sample. Certainly, for these numbers to be realistic, the application of the Hagen–Poiseuille equation should be corrected to account for the particularities of the pores in each case. While the dependence of flux on pore diameter and pressure is correct, the major discrepancy, as has been pointed out by numerous authors [12] comes from the actual geometry of the pores which can rarely be exactly assimilated to a set of cylindrical pores perfectly perpendicular to the membrane surface. Moreover, as the work of Gijsbertsen-Abrahamse and co-workers pointed out, this equation does not account for the fact of pores interconnected, as usually found in many membranes.

These criticisms and other similar ones can be addressed by using a phenomenological approach quite commonly found in many studies dealing with GLDP, mostly, and LLDP. The idea is to include in the Young–Laplace equation a correction factor (usually named as a shape or form factor [54–57]). Thus, the Young–Laplace equation is necessary to convert the applied pressures into the size of the pores concerned, now looks like:

$$P = \frac{2\,K\gamma}{r_p}\cos\theta = \frac{4\,K\gamma}{d_p}\cos\theta \tag{7.37}$$

where K is the so-called shape factor. According to ASTM F-316, and supposing complete wetting ($\theta = 0$), a value of $K = 0.715$ must be used for pore size calculations, but several values can be used depending on the supposed shape of our pores [20]. This shape factor also includes the possible tortuosity of the pores, so, typically values between 0.6 and 1 can be used for K.

Although there seems to be some agreement on the need to include a shape factor in the porosimetric calculations derived from this technique as well as from GLDP, as can be deduced from including such a factor in the recommendations of the ASTM F-316 standard, or in the software implemented in many commercial equipment, it must also be borne in mind that this factor has no theoretical justification, its value being based on purely empirical considerations, validatable only by comparison with the results obtained by other techniques.

On the other hand, using a factor of 0.715 compared to the default value for the ideal case (cylindrical pores, without tortuosity or inclination: $K = 1$) simply shifts the values of the distribution by about 30%. Although in many metrological domains, differences of this order between measurements and actual data may seem excessive; in the case of membrane filters, this discrepancy must be qualified. In general, when characterizing the pore size of a membrane, it is the order of magnitude of the pores that really matters.

An example of this can be found in the filters used in the sterilization of water or other types of solutions. In this regard, the usual recommendation is to use a filter with a nominal pore size of 0.2 µm, although many manufacturers have 0.22 µm filters in their portfolio for the same application, and even 0.45 µm filters can be used successfully. The important thing is to make sure that the pores (at least most of them) are clearly below the size of the substances (bacteria in this case) that we want to retain. This consideration is obviously not only valid to justify slight or not so slight deviations in the results obtained by GLDP/LLDP, but can be extended to any other type of porosimetric method. Such a method will be useful if it allows us to know with certainty the order of magnitude of the pores present in our filter, even if this involves discrepancies of a factor of 2, for example.

In our several published works on LLDP or GLDP, we have never used a shape factor in the conversion of the data, as it is impossible to be sure of the pertinence of the value used. Ultimately, in many studies, it is the comparison between characterizations of similar membranes that is important, in which case it seems good practice to always use the same conversion equation.

Finally, we must emphasize another important factor in ensuring the reliability of LLDP measurements. This is to ensure complete wetting of the sample prior to the liquid displacement experiment. As has been discussed several times, the use of the Young–Laplace equation to convert pressure data into pore sizes assumes a zero contact angle between the wetting liquid and the membrane pores. In the case of MF membranes, with pore sizes of the order of a micron, it is easy to make this assumption. As mentioned above, there are many commercially available liquids, mostly based on halogenated compounds, that easily penetrate all the pores of an MF membrane ensuring complete wetting and a zero contact angle.

But as pore sizes become smaller, this assumption becomes more questionable. Even liquids, such as isobutanol, which readily wets flat surfaces of various materials commonly used in membrane manufacture, can have difficulty entering pore sizes approaching the order of a few nanometres.

To minimize this problem, our experience recommends wetting under vacuum conditions (or a reasonable vacuum) that favours the penetration of the wetting liquid into the smallest pores.

In any case, however, much a liquid proves to be useful in wetting a membrane material, and however much its penetration into the pores is improved, there is an obvious limit to the size of the pores that can be studied by these techniques. This limit is related to the average size of the molecules we use for wetting or emptying the pores. Since the average size of a water molecule is on the order of 0.3 nm and somewhat larger in the case of isobutanol (~ 0.6 nm), it seems clear that it does not make sense to talk about wetting when the order of magnitude of the pores that can be studied by LLDP is close to the order of the nanometre. We can consider, empirically, that the LLDP technique becomes meaningless for pore sizes (diameter) smaller than 2 nm.

Although the various critical aspects mentioned in these sections may lead one to think that the LLDP technique (or GLDP, as the case may be) is unreliable when it comes to characterizing the porous properties of membrane filters, the fact is that, in addition to the disadvantages mentioned, both techniques have undoubted advantages that make them extremely interesting and even a preferred choice over other porosimetric techniques. This is evident in the case of GLDP, whose speed and reliability have elevated it to the standard of reference in the characterization of filters in the micron range. But LLDP, although it has not reached this unanimous consideration, is also one of the most reliable techniques, especially in pore sizes where other techniques have difficulties to reach with sufficient reliability. In any case, in the last chapter of this book, we will attempt an exhaustive review of the comparative advantages and disadvantages of the different porosimetric techniques available.

7.5 Mercury porosimetry

7.5.1 Introduction

Following with the porosimetric techniques based on the Young–Laplace equation, in previous sections we have dealt with those techniques that are based on a wetting fluid filling our membrane. In both liquid displacement porosimetries (GLDP and LLDP), we have selected the fluids (liquids in fact) used to fill the membrane pores to assure complete (or almost complete) wettability. This basically means that the liquid should have a zero (or almost nil) contact angle with the membrane material. But the question now is what happens if the liquid does not wet the membrane, or in other words, if we use a liquid whose contact angle with the material is higher than 90°. This is the case, for example, of mercury, the only metal being in the liquid state at ambient temperature.

The technique, based on the controlled intrusion of mercury inside the pores of a membrane, is called HgP and will be the objective of the following paragraphs.

HgP, also called very often mercury intrusion porosimetry (MIP), is one of the oldest and most widely used porosimetric techniques. The idea for the technique was proposed by Washburn in a short note published in 1921 [58]. However, Washburn never made the experimental measurements to test his hypothesis. It was Ritter and Drake who, taking the idea proposed by Washburn, tackled its experimental realization, in two papers published in 1945 [59, 60]. In the first paper, in addition to describing the experimental procedure necessary to carry out the technique (including a glass dilatometer, a key piece for a correct determination of such small, intruded volumes), Ritter and Drake already emphasize the possible errors induced by the assumption of a circular geometry for the section of the pores (a hypothesis whose validity we have already mentioned in other techniques based on the Young–Laplace equation). In their second work, they applied the technique proposed by Washburn to the characterization of different types of porous materials, such as activated clay, bauxite, diatomaceous earth, activated carbon, silica–alumina (pelleted and gel) or fritted glass. On the other hand, the work of Honold and Skau in 1954 [61] is considered the first application of the technique to the analysis of synthetic membranes. Important works in these pioneering years are the determination of surface areas of porous materials from HgP experiments, by Rootare and Prenzlow [62], or the studies by Liabastre and Orr [63] with track-etched membranes.

Mercury penetration is useful because mercury does not wet almost any materials and will not spontaneously enter into pores by capillary action. Rather, it must be forced into the pores by the action of external pressure. According to the Young–Laplace equation, the required equilibrium pressure is inversely proportional to the radius of the pores. Of course, the required pressure can be quite high in order to penetrate relatively narrow pores. HgP analysis requires a progressive intrusion of mercury into the porous structure under strictly controlled pressures. From pressure versus intrusion data, radius distributions can be easily generated. Clearly, accurate pressure measurements are essential to get reliable pore sizes.

Let us start describing the fundamentals of MIP, starting with the application of the equation governing the mercury penetration inside the membrane pores.

7.5.2 Young–Laplace equation

Firstly, we will go back to the Young–Laplace equation to derive the expression useful for the case of HgP. So, the Young–Laplace equation, in the case of a solid–liquid interface, is

$$p_g - p_\ell = \frac{2\gamma_{g\ell}}{r_p}\cos\theta_{g\ell} \tag{7.38}$$

If we call $\Delta p = p_\ell - p_g$, eq. (7.38) reads

$$\Delta p = -\frac{2\gamma_{g\ell}}{r_p}\cos\theta_{g\ell} \tag{7.39}$$

This equation is also called the Washburn equation. There is another approach to this equation in terms of the work required to force mercury into a pore, which will be described hereafter. In effect, the work required to force mercury into a pore is proportional to the increased surface travelled by the mercury along the pore wall when mercury advances a distance l. Therefore, assuming cylindrical pores:

$$d'W = \gamma'_{g\ell}dS_{g\ell} \Rightarrow W = \gamma_{g\ell}\cos\theta_{g\ell}\int_0^\ell 2\pi r_p d\ell = 2\pi r_p \ell \gamma_{g\ell}\cos\theta_{g\ell} \tag{7.40}$$

$\gamma'_{g\ell}$ is the surface tension along the pore $\gamma'_{g\ell} = \gamma_{g\ell}\cos\theta_{g\ell}$:

$$d'W = -pdV \Rightarrow W = -\Delta p 2\pi r_p^2 \ell \tag{7.41}$$

From eqs. (7.40) and (7.41), we recover:

$$\Delta p = -\frac{2\gamma_{g\ell}}{r_p}\cos\theta_{g\ell} \tag{7.42}$$

Here it is clear that $\Delta p = p_\ell - p_g$.

Equation (7.39) was firstly proposed by Washburn [64] and is the working equation in HgP. For wetting angles less than $\pi/2$, it is $\cos\theta_{g\ell} > 0$ and $\Delta p < 0$ ($p_g > p_j$) indicating that a vacuum should be applied to the liquid (if we assume that $p_g = p_{atm}$) to force it inside a pore of radius r_p. In turn, when $\theta_{g\ell}$ is greater than $\pi/2$, it leads to the requirement of $\Delta p > 0$ ($p_g < p_j$) or to overpressures to force the liquid into the pore.

Mercury tends not to wet most surfaces without being forced into pores. In fact, an arbitrary, uncorroborated contact angle is used customarily, mostly from 130° or 150° [65]. However, there may be significant differences in the wetting angles of mercury on different substances. Angles in the range from 112° to 160° were calculated mostly by measuring mercury intrusion on known porous samples [65]. A value of 140° is frequently used, no matter what solid is under test. The value of pure mercury is assumed for the surface tension $\gamma_{g\ell} = 0.458$ N/m, although it is in fact quite sensitive to contamination from the solid. Uncertainty in $\theta_{g\ell}$ and $\gamma_{g\ell}$ can cause errors that are difficult to quantify. So, a comparison of results obtained with other techniques would be highly advisable if feasible [66].

7.5.3 Porograms

A plot of the intruded (increasing p) and/or the extruded (decreasing p) volume of mercury per unit mass versus pressure is called a porosimetric curve or porogram. Figure 7.25 shows a typical porogram depicting mercury intruded (and then extruded) against applied pressure. All mercury intrusion–extrusion porograms have some common characteristics. These include:

a. Rearrangement of macroscopic portions of the material to optimize their packing with mercury penetrating large interparticle spaces (A).
b. Intrusion at slightly higher pressures into large pores in the material (1) and then to smaller pores (2), in case there were a bimodal PSD. Frequently steps 1 and 2 (and, similarly, 3 and 4) have several more irregular paths corresponding to possible wider multi-peak PSD. Finally, a possible compression might appear.
c. Afterwards when pressure decreases, there is extrusion that does not follow the intrusion path, but rather, extrusion happens at lower pressures (larger radii) (3 and 4), exhibiting a variable degree of hysteresis.
d. Upon completion of the first intrusion–extrusion cycle, some mercury is always retained by the sample, thereby preventing the loop from closing. The dashed line from A to D corresponds to the trapped mercury after a cycle has been completed.

Intrusion–extrusion repeated cycles will continue showing hysteresis and eventually the loop will close, frequently before a third intrusion–extrusion cycle.

Different explanations have been proposed to explain the fact that extrusion curves do not overlap with the intrusion ones. Three main lines of argumentation persist in the literature: the ink-bottle pore contribution; the network contribution due to interconnections of pores; and the contribution of pore–wall–liquid interactions acting solely when extrusion happens. In fact, hysteresis is probably related to different issues, including the difference between dynamic advancing (intrusion) and receding (extrusion) contact angles, and pore network complexity – particularly the presence of "ink-bottle" pores, that is, those having necks that are narrower than the main pore

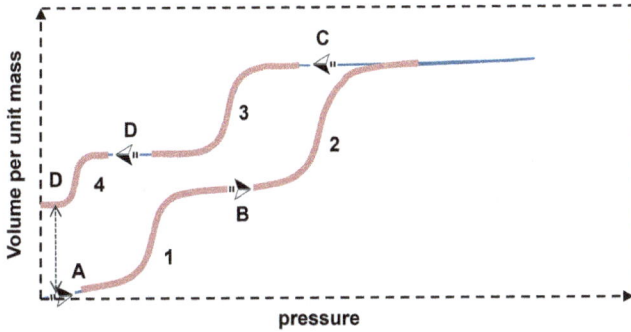

Fig. 7.25: Typical porogram for a bimodal pore size distribution showing hysteresis between intrusion and extrusion.

body, but also including some contribution of compaction [67]. Fig. 7.26 shows the different kind of pores that can be found in a porous material.

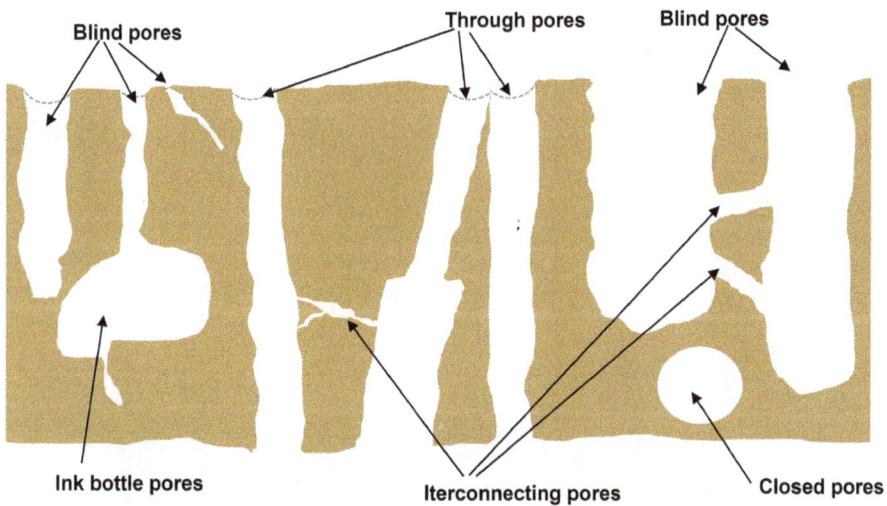

Fig. 7.26: Possible configurations of pores. The different types of pores commonly defined in a porous material are shown.

Of course, the PSD can be obtained from the cumulative volume versus pressure programmes as $(dV/dr_p) = -(\Delta p/r_p)(dV/dp)$ [68, 69]. In effect:

$$\frac{dV}{dp} = \frac{1}{\frac{dp}{dV}} = \frac{1}{\frac{d\left(-\frac{2\gamma_{g\ell}}{r_p}\cos\theta_{g\ell}\right)}{dV}} = \frac{1}{\frac{\left(\frac{2\gamma_{g\ell}}{r_p^2}\cos\theta_{g\ell}\right)dr_p}{dV}} =$$

$$= -\left(\frac{dV}{dr_p}\right)\frac{r_p}{\Delta p} = \left(\frac{dV}{dr_p}\right)\frac{r_p^2}{2\gamma_{g\ell}\cos\theta_{g\ell}} \tag{7.43}$$

7.5.4 Contact angle

In many cases, the extrusion and intrusion curves can be made to coincide by assuming different contact angles for both processes to eliminate the gap between the second intrusion curve and the extrusion curve. This along with lowering pressure until minimizing mercury entrapment should be tried first before trying to infer information on the shapes and interconnectivity of pores by using network models.

The contact angle of a liquid on a solid would be different if it advances over a dry surface or recedes from a wetted surface [70]. The higher contact angle is associated with intrusion into pores while smaller angles should be the receding ones when extruding mercury from pores. This would cause hysteresis in HgP.

The situation when a drop of a liquid is in equilibrium on a solid surface is shown in Fig. 7.27.

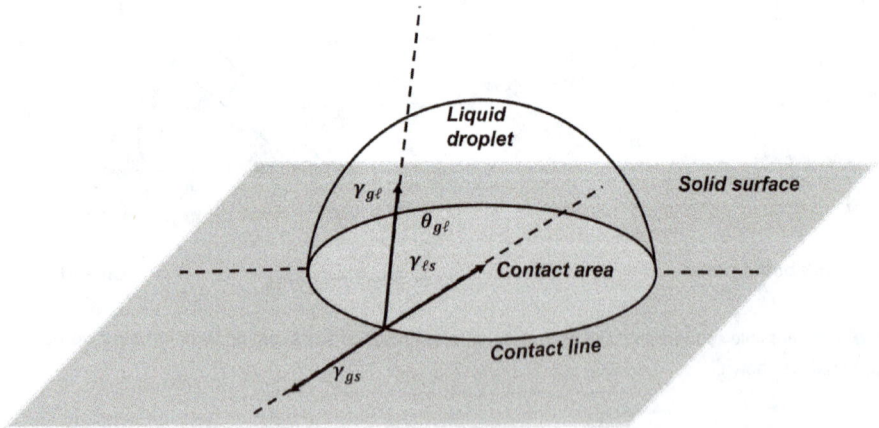

Fig. 7.27: Liquid drop on a solid surface.

Attending to the forces in equilibrium at the gas–solid interface:

$$\gamma_{gs} = \gamma_{g\ell}\cos\theta_{g\ell} + \gamma_{\ell s} \Rightarrow \cos\theta_{g\ell} = \frac{\gamma_{gs} - \gamma_{\ell s}}{\gamma_{g\ell}} \tag{7.44}$$

Fundamentally, when a liquid wets a rough surface, there are two possible wetting states, one with the liquid fully wetting every area of the rough surface and the other with the liquid partially wetting and attached to the peaks of the rough surface. The latter results in the formation of a liquid–solid–air heterogeneous surface. Usually, the first possibility appears for wetting (high contact angle) interfaces, while the second one appear for low wettable (low contact angles) interfaces. These two wetting states have appropriately been recognized as the Wenzel and Cassie–Baxter configurations, respectively, in the literature [70–72]. These two configurations are shown in Fig. 7.28.

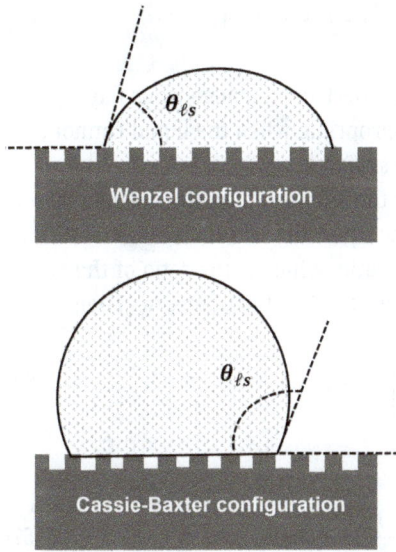

Fig. 7.28: Outline of the Wenzel and Cassie–Baxter configurations of a liquid drop on a rough surface.

Cassie and Baxter [72] considered that the solid surface consists of randomly distributed heterogeneities with fractions:

$$\sum_i f_i = 1 \tag{7.45}$$

and

$$\left. \begin{array}{l} \gamma_{gs} = \sum_i f_i \gamma_{i,gs} \\ \gamma_{\ell s} = \sum_i f_i \gamma_{i,\ell s} \end{array} \right\} \tag{7.46}$$

Thus, eq. (7.44) would lead to

$$\cos \theta_{g\ell}^{C\text{-}B} = \sum_i f_i \frac{\gamma_{gs} - \gamma_{\ell s}}{\gamma_{g\ell}} = \sum_i f_i \cos \theta_{g\ell} \tag{7.47}$$

where $\theta_{g\ell}^{C\text{-}B}$ would be the apparent or measurable contact angle according to Cassie and Baxter, while $\theta_{g\ell}$ would be the Young's contact angle. In the case of Fig. 7.27, there would be two components on the surface of the solid, namely air and the actual solid $f_s + f_g = 1$, and then

$$\cos \theta_{g\ell}^{C\text{-}B} = f_s \cos \theta_{g\ell}^s + f_g \cos \theta_{g\ell}^g \tag{7.48}$$

But assuming that $\theta_{g\ell}^g = \pi$ because the liquid is not supposed to enter the pits or valleys on the solid surface, then

$$\cos \theta_{g\ell}^{C\text{-}B} = f_s \cos \theta_{g\ell}^s + f_s - 1 = f_s \left(1 + \cos \theta_{g\ell}^s\right) - 1 \tag{7.49}$$

where $\theta_{g\ell}^s$ would be the Young's contact angle for a perfectly homogeneous flat surface. The model of Cassie and Baxter is particularly appropriate when the liquid cannot contact the bottom of the valleys appearing on rough surfaces.

On the other hand, when the liquid can wet the solid even on the deep valleys as well as on the steepest peaks of a rough surface, then the actual accessible surface can be taken into account by using the Wenzel's ratio, which is the ratio of the actual solid surface over the apparent surface when projected to a flat plane, r_W. Then

$$\left.\begin{array}{l} \gamma_{gs}^W = r_W \gamma_{gs} \\ \gamma_{\ell s}^W = r_W \gamma_{\ell s} \end{array}\right\} \tag{7.50}$$

and

$$\cos \theta_{g\ell}^W = \frac{\gamma_{gs}^W - \gamma_{\ell s}}{\gamma_{g\ell}} = r \frac{\gamma_{gs} - \gamma_{\ell s}}{\gamma_{g\ell}} = r \cos \theta_{g\ell} \tag{7.51}$$

Note that within the Wenzel's model, roughness would make more hydrophobic the hydrophobic solids ($\theta_{g\ell} > \pi/2$) and more hydrophilic the hydrophilic ones ($\theta_{g\ell} < \pi/2$). While for the Cassie–Baxter regime, the dependence on the proportion of the wetted surface is somehow more variable, in particular it can pass from a hydrophilic Young's angle to a hydrophobic (and vice versa) Cassie–Baxter contact angle [73].

Moreover,

$$\left.\begin{array}{l} f_s = \dfrac{1 + \cos \theta_{g\ell}^{C\text{-}B}}{1 + \cos \theta_{g\ell}^s} \\[3mm] r = \dfrac{\cos \theta_{g\ell}^W}{\cos \theta_{g\ell}} \end{array}\right\} \tag{7.52}$$

We can assume, for example, 140° for the advancing porogram and 102° for the receding one [74]. This gap could be justified in accordance with eq. (7.52), attending the Wenzel regime, by 27% higher rugosity felt when receding. On the other hand, if we assume a Cassie–Baxter regime, we will need to suppose that, again by using eqs. (7.52), receding mercury will feel the material as 29% less heterogeneous. Both these possibilities seem conceivable [75], although difficult to confirm.

Lam [75] showed, by measuring contact angles on quite smooth and homogeneous surfaces, that receding contact angles depend on the solid/liquid contact time. This implies that liquid sorption (and swelling), along with penetration and retention dynamics are the main factors influencing contact angle hysteresis. A relevant contribution can also be related to the irreversibility of the frictional loss [65].

Because these phenomena would affect both the receding but also the advancing contact angles obtained on a wetted surface (after the first cycle), only the advancing contact angle on a dry surface is acceptable to the calculation of pore sizes [75].

7.5.5 Fractal dimension

Fractal theory is currently being used more and more for investigating the multiscale structures of porous materials as far as the micro- and meso-structures of porous materials show extremely complex and irregular features, which are difficult to describe in terms of geometry but that can be studied by the fractal theory [76, 77].

Neimark [78] proposed a correlation of the fractal dimension through the pore surface with the pore diameter relation. The fractal dimension of the pore surface reflects the roughness and irregularities of the internal pore surface of porous materials, and it is commonly acknowledged that it has a typical range of $2 < D_S < 3$. When D_S is equal to 2, it means that the surface is a perfectly smooth plane. The pore surface morphology becomes rougher and more complex when the value of D_S approaches 3. The Neimark's model postulates that energy conservation can be expressed as follows:

$$S = -\frac{1}{\cos \theta_{g\ell}} \int_0^{V_p} p \, dV \tag{7.53}$$

Here the surface extension work is matched with the mercury intrusion work, and S and V refer to pore surface and volume, respectively. Then, by assuming a cylindrical geometry for the pores, it can be shown [79] that

$$\frac{dV}{dr} \propto r^{2-D_S} \tag{7.54}$$

Other models such as the Zhang and Li model [80] do not need any assumptions on a detailed pore geometry.

According to eqs. (7.53), (7.39) and (7.54), log S is proportional to log p:

$$\log S \propto (D_S - 2) \log p \tag{7.55}$$

But $S^{1/D_S} \propto V^{1/3}$ according to Mandelbrot [81] for a fractal surface and the volume inside it. Therefore,

$$\log V \propto 3\left(1 - \frac{2}{D_S}\right)\log p \qquad (7.56)$$

Equation (7.56) can be used to get the fractal dimension of the inner pore surface, which should be different for intrusion and extrusion and for different pressure ranges as well. Using cyclic intrusion–extrusion loops, Villagrán-Zacardi et al. [82] found consistent equal fractal dimensions after the first cycle. Of course, this possible explanation of hysteresis can be complimentary and congruent with the explanations relying on heterogeneity and roughness.

7.5.6 Ink-bottle pores

When the entrance of a pore is narrower than their interior, the pores should be totally penetrated only when a pressure high enough as to wet the entrance is reached giving a pore size versus pressure porogram displaced to the narrower pores (narrower than the actual or inner pore size) or equivalently to higher pressures [66]. On the other hand, during extrusion the converse would be true, because reaching lower pressures should be needed to evacuate the wide inner portion of the pores and some mercury could even remain trapped until very low pressures, due to the breaking of the mercury column in the narrower portions of pores [83], which is made easier due to its rapid emptying.

In this case, the Washburn equation interprets all the intruded volume as corresponding to the size of the width of the narrowest pore neck, and the existence of the larger interior pore bottom will not be detected [84]. Diamond claims as well that when time evolution of penetration is analysed, it is seen that penetration is slower when ink-bottle pores are being filled and faster when they are being emptied. A comparison of first intrusion and first extrusion curves together with the liberation of any trapped mercury by reaching very low pressures could allow to get information on the volume corresponding to ink-bottle pores that, in this way, could only be characterized attending to their total volume.

The PSD does not include ink-bottle pores and should be obtained from subsequent intrusion–extrusion curves because the first cycle cannot be trusted to get the reversibly intrudable or permeability significant PSD. Thus, given that the ink-bottle pores are completely filled during the first intrusion, the ensuing cycles would not show any additional increment of the intruded volume. Once the ink-bottle pores are filled, the process becomes fully reversible, if these ink-bottle pores are not emptied, by avoiding reaching very low pressures, with subsequent cycles almost totally equal. In this case, small residual changes should be attributed to changes in structure, connectivity, or surface features, during subsequent compressions [82].

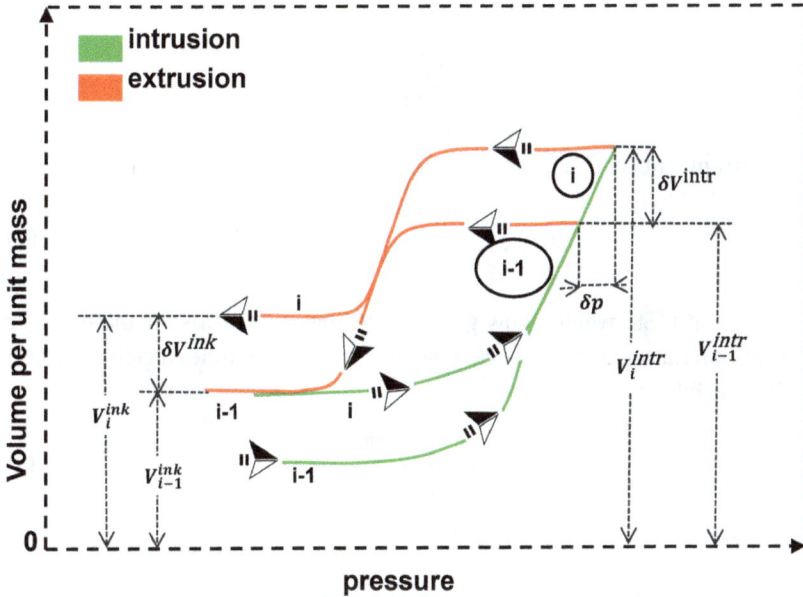

Fig. 7.29: Outline of the PDC-MIP methodology.

An important progress is due to Zhou et al. [85] and was reviewed by Gu et al. [86]. They proposed an alternative method, that is, pressurization–depressurization cycling (PDC) MIC to explore the relevance and contribution of ink-bottle pores. The PDC-MIP test consists in a succession of pressurization and depressurization steps starting from the initial pressure in all the successive steps. The method was reformulated by [87] to the so-called intrusion–extrusion cyclic HgP (IEC-MIP) that according to Fig. 7.29 consists in successive intrusion–extrusion cycles.

The method assumes that from step $i - 1$ to the ith step, the cumulative intruded volume increases by $V_i^{intr} - V_{i-1}^{intr}$, and the cumulative ink-bottle volume should be $V_i^{ink} - V_{i-1}^{ink}$, in such a way that throats of a diameter d_i are filled with mercury volume V_i^{th} correlated with the neighbouring ink-bottle cavities with a volume V_i^{ink} according to

$$\delta V_i^{th} = \left(V_i^{in} + V_{i-1}^{ink} \right) - \left(V_{i-1}^{in} + V_{i-1}^{ink} \right) = \left(V_i^{in} - V_{i-1}^{in} \right) - \left(V_i^{ink} - V_{i-1}^{ink} \right) \tag{7.57}$$

and

$$\delta V_i^{ink} = \left(V_i^{ink} - V_{i-1}^{ink} \right) \tag{7.58}$$

The corresponding δV_i^{th} and δV_i^{ink} correspond to the change in pressure δp and constitute the corresponding volume distributions that could be converted to PSDs by eq. (7.43).

An entrapment coefficient can be defined, for the ith step, as follows:

$$a_i = \frac{\delta V_i^{\text{ink}}}{\delta V_i^{\text{th}} + \delta V_i^{\text{ink}}} \tag{7.59}$$

and a global entrapment coefficient is

$$a = \frac{\sum \delta V_i^{\text{ink}}}{\sum \delta V_i^{\text{th}} + \sum \delta V_i^{\text{ink}}} \tag{7.60}$$

Equations (7.57) and (7.58) would allow getting the differential PSDs for the pores or throats between cavities and for these cavities separately. Sometimes a global connectivity is defined as follows:

$$\eta = 1 - a = \frac{\sum \delta V_i^{\text{th}}}{\sum \delta V_i^{\text{th}} + \sum \delta V_i^{\text{ink}}} \tag{7.61}$$

7.5.7 Canthotaxis, connectivity and non-cylindrical pores

According to Cebeci [88], De Botton [89] and Felipe et al. [90], it can be assumed that there are spherical cavities with direct access to mercury, through cylindrical pores of radius r_p as shown in Fig. 7.30. In this case, if the equilibrium contact angle $\theta_{g\ell}$ between the interface and the pore wall is always attained, then the pressure needed to invade the cavity is not given by eq. (7.39) but by

$$\Delta p = -\frac{2\gamma_{g\ell}}{r_p} \cos \phi = -\frac{2\gamma_{g\ell}}{r_p} \cos(\theta_{g\ell} + \alpha) \tag{7.62}$$

The appearance of the α angle is due to what is called canthotaxis that stands for the contact line pinning at a border [91, 92].

Now following Felipe et al. [90], two cases can be distinguished:

– If $r_p/r_s = |\cos \alpha| < |\cos \theta_{g\ell}|$ (i.e. $\alpha > \pi - \theta_{g\ell}$ or $\phi > \pi$), the radius of curvature starts to increase from $r_p/|\cos \theta_{g\ell}|$ to r_p as pressure increases and then mercury would fill the cavity irreversibly. In this case, because $r_p > r_p/|\cos(\theta_{g\ell} + \alpha)|$, higher pressures should be needed (see Fig. 7.30a, top). Small pores connecting to big voids should give this kind of longer irreversible filling [67], leading to a slower penetration.

– If $r_p/r_s = |\cos \alpha| \geq |\cos \theta_{g\ell}|$ (this happens when $0 \leq \alpha \leq \pi - \theta_{g\ell}$ or $\theta_{g\ell} \leq \phi \leq \pi$), the meniscus at the liquid–vapour interface first anchors at the pore-to-pore edge while its radius of curvature starts to decrease from $r_p/|\cos \theta_{g\ell}|$ until $r_p/|\cos(\theta_{g\ell} + \alpha)|$ as pressure increases. This latter radius of curvature represents the onset of a complete filling of the assumed spherical cavity (see Fig. 7.30b, bottom).

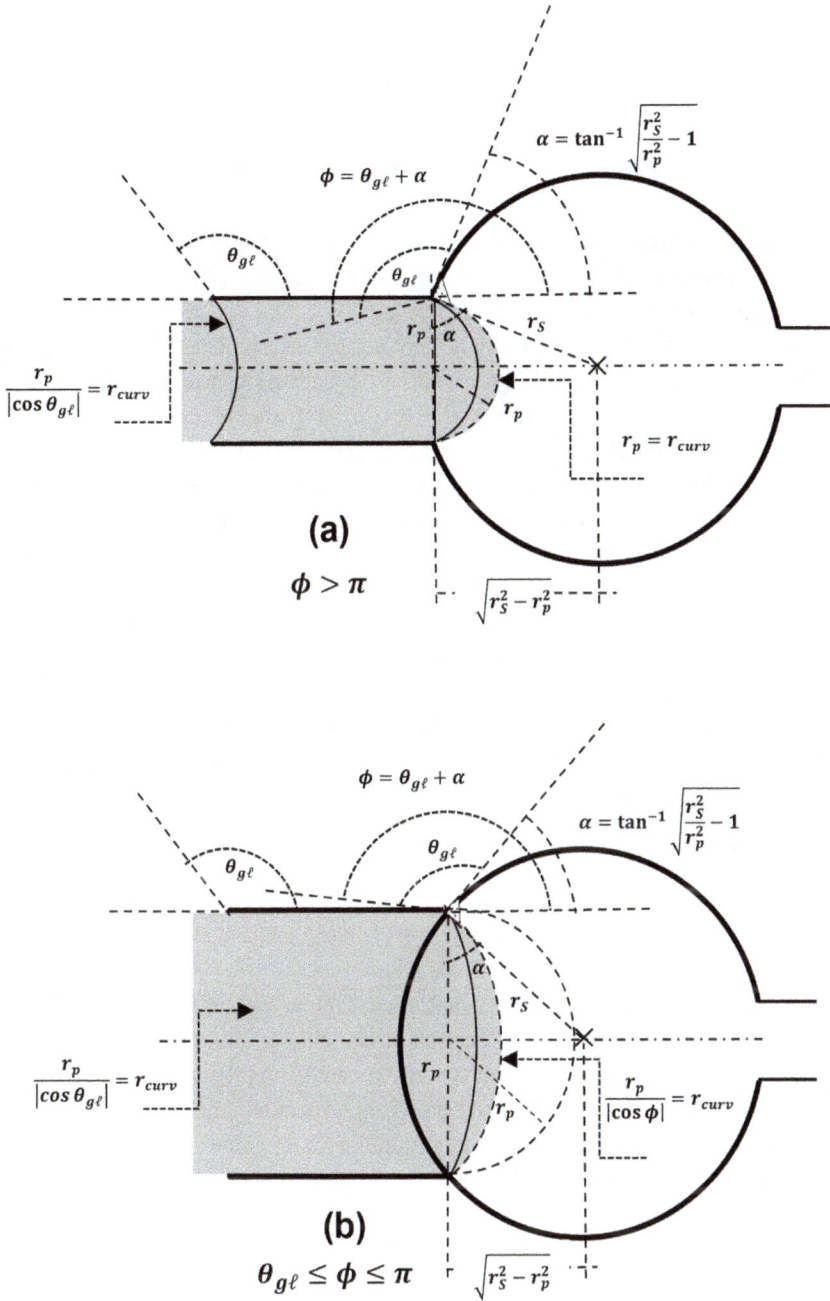

Fig. 7.30: Scheme of a pore entering a spherical cavity.

In summary,

$$\Delta p = \begin{cases} \dfrac{2\gamma_{g\ell}}{r_p} & r_p < r_s |\cos\theta_{g\ell}| \Leftrightarrow |\cos\alpha| < |\cos\theta_{g\ell}| \\[3mm] -\dfrac{2\gamma_{g\ell}}{r_p}\cos\left[\cos^{-1}\left(\dfrac{r_p}{r_s}\right) + \theta_{g\ell}\right] & r_p \geq r_s |\cos\theta_{g\ell}| \Leftrightarrow |\cos\alpha| \geq |\cos\theta_{g\ell}| \end{cases} \tag{7.63}$$

Of course, for an initially empty bond to be permeated by mercury from any one of its two ends, the existence (at its other end) of a continuum liquid path leading to the outer mercury bulk source must also be required.

Canthotaxis is related to the transition between Wenzel and Cassie-Baxter configurations [93]. In any case, if we assume that there is 3D network consisting of bulges and throats, for contact angles in the range $\pi/2 < \theta_{g\ell} < \pi$, mercury intrusion will occur at some value of $\phi > \theta_{g\ell}$ ($\alpha > 0$), whilst partial mercury retraction may occur at some value of $\phi < \theta_{g\ell}$ ($\alpha < 0$). In this case, let us suppose an advancing contact angle of 140° and a receding angle of 102°. If we assume that the advancing angle corresponds to $|\alpha| + \theta_{g\ell}$ while the receding one corresponds to $-|\alpha| + \theta_{g\ell}$, adding both we would get $2\theta_{g\ell}$ that, in our case, would lead to $\theta_{g\ell} = 121°$ and $|\alpha| = 19°$. Now, using $\cos\alpha = r_p/r_s$ we could conclude that the pores could consist of cavities connected to necks with a radius $r_p = 0.95\, r_s$, which seems certainly reasonable.

Felipe et al. [90] tested this procedure with a controlled pore glass reference material in the form of beads, which is a certified reference material for mercury intrusion, designated as CRM, BAM-PM-121. Their results agreed with the results obtained from gas adsorption–desorption.

7.5.8 Connectivity and percolation

At its core, percolation theory examines how a liquid or gas propagates through a network of interconnected pores within a material. A lattice or network is considered where each site may be occupied with a certain probability, leading to the formation of clusters of connected sites. A pivotal concept in the percolation theory is the identification of a critical probability known as the percolation threshold. Below this threshold, the likelihood of a spanning cluster – a continuous path of connected sites from one side of the lattice to the other – is 0, while above the threshold, the probability of a spanning cluster becomes 1. In invasion percolation, an invasion resistance is assigned to each site, and in each step, the liquid invades the site with the smallest resistance among their neighbours [94, 95].

The percolation threshold is defined as the critical concentration or density of occupied sites (or bonds) in a system. Below this threshold, the system consists of isolated clusters, and above it, there is a connected network. Above the threshold, these clusters start to merge, leading to the formation of a continuous network that spans the entire system. The invasion percolation model introduced by Lenormand and Bories [96] and Wilkinson and Willemsen [97] incorporates the "dynamics" of liquid propagation in porous media. In invasion percolation, invasion resistance values, $\mathcal{R}$ from 0 to 1, are allocated to the sites instead of probabilities. The liquid starts to propagate from the initially occupied set. In each step, the invading liquid occupies the interfacial site whose invasion resistance $\mathcal{R}$ is the smallest. In the percolation model, due to Bak and Kalmár-Nagy [98], there are also invasion resistances assigned to the sites, but the occupation of the lattice sites is driven by an external pressure $\mathcal{P}$ normalized to the maximum pressure ($0 \leq \mathcal{P} \leq 1$). For a given pressure, all sites with invasion resistance $\mathcal{R} \leq \mathcal{P}$ that are accessible are occupied. Resistances and pressures should be correlated with pore radii according to the Washburn equation. According to the theory, the presence of a threshold should correspond to an inflection point in the intrusion curve. According to Bak and Kalmár-Nagy [98], the inflection point of the saturation curve is at $\mathcal{P} = 0.59$ for one-sided percolation of a 2D square lattice, and at $\mathcal{P} = 0.31$ for one-sided percolation of a simple cubic lattice. Therefore, the inflection point should give a clue, a rough insight at least, on the morphology of the pore network if it is simple enough.

Pore–network modelling is frequently used to numerically model HgP by using different simulations [99–103] or classical percolation theory [94, 104].

Of course, the initial models assumed that the shapes of real pores within disordered porous structures are regular. The first simulators were based upon two-dimensional pore bonding networks [99, 100], where the individual pores are represented by simple straight cylinders, or slits, arranged in simple lattices, such as square planar ones. Further models used random, three-dimensional pore bond networks [101]. Materials were represented using networks consisting of both pore bodies and pore bonds [90, 91, 105].

The number of pores coordinated at each node in random, three-dimensional pore bond networks can be changed. The pore coordination numbers may vary among nodes, depending on their position, for example, or have a fixed distribution. Each type of size distribution for networks of pore bodies and pore bonds may be different or even depend on the geometry of the sample. For example, these models can assume cylindrical pore bonds with randomly distributed radii between cubic pore bodies with statistically distributed sizes. It is clear that these 3D models can be as complex as needed by adding more and more parameters to be fitted [106–108]. In summary, although these approaches can be helpful, it is quite difficult to model the structure of the pores and to get conclusive information from a complex multifactor model on the actual morphology of the real pore network.

7.5.9 Compressibility

A more down-to-earth factor that must be considered is the expansion of the conductions, the cell, the mercury itself and the sample, under pressure. Intruding mercury without a sample is enough to consider conductions and cell compressibility as a function of pressure [109]. A totally different question is the contribution of the material compressibility and compaction that is impossible to isolate from the changes of volume attributed to the penetration of mercury. Of course, this contribution may vary largely depending on the porous material. The sample without mercury could be compressed up to each pressure to be applied while intruding mercury to characterize the pore volume at this pressure (radius) to isolate the effect of the compressibility of the sample. This procedure should be only meaningful if compression and compaction were totally reversible, which is not the usual case. Therefore, this factor must be taken into account as a potential source of significant errors for very labile and/or high porosity materials.

7.5.10 Application

HgP, developed as a standard method (some of the concerned standards are given in Tab. 7.2) despite its drawbacks, is a vital technique for studying macropores up to 15 μm, which are otherwise impossible to access using other techniques. The pore sizes that can be detected by HgP are shown in Fig. 7.31. When gas adsorption–desorption results are compared with those from MIP, both methods give outcomes that are frequently consistent [66].

Tab. 7.2: Some standards related to mercury porosimetry.

ASTM	D 4404	Determination of pore volume distribution of soil and rock by mercury intrusion porosimetry.
ASTM	D 4284	Determining pore volume distribution of catalyst carriers by mercury intrusion porosimetry
ASTM	D 6761	Determination of the total pore volume of catalysts and catalyst carriers
ISO	15901	Evaluation of pore size distribution and porosity of solid materials by mercury porosimetry and gas adsorption – Part 1: mercury porosimetry
USP	267	Porosimetry by mercury intrusion

ASTM stands for American Society for Testing Materials, ISO for International Organization for Standardization and USP for United States Pharmacopeia.

HgP is very often used to characterize soils, rocks and cement, with many applications of catalyst carriers and, of course, membranes. The annual number of publications found in the WOS web, from Clarivate, on topics such as membranes and HgP are shown in Fig. 7.32.

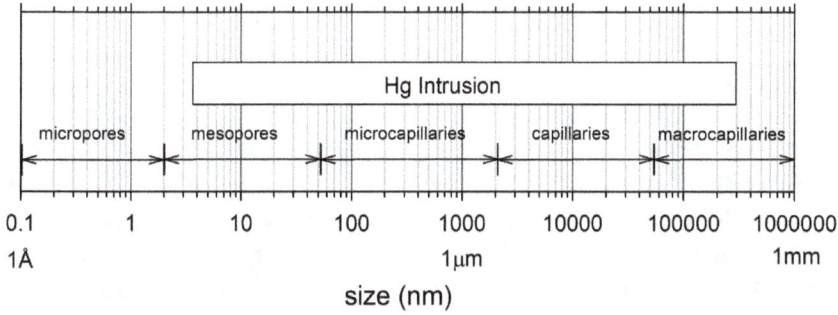

Fig. 7.31: Range of sizes that can be studied by Hg intrusion.

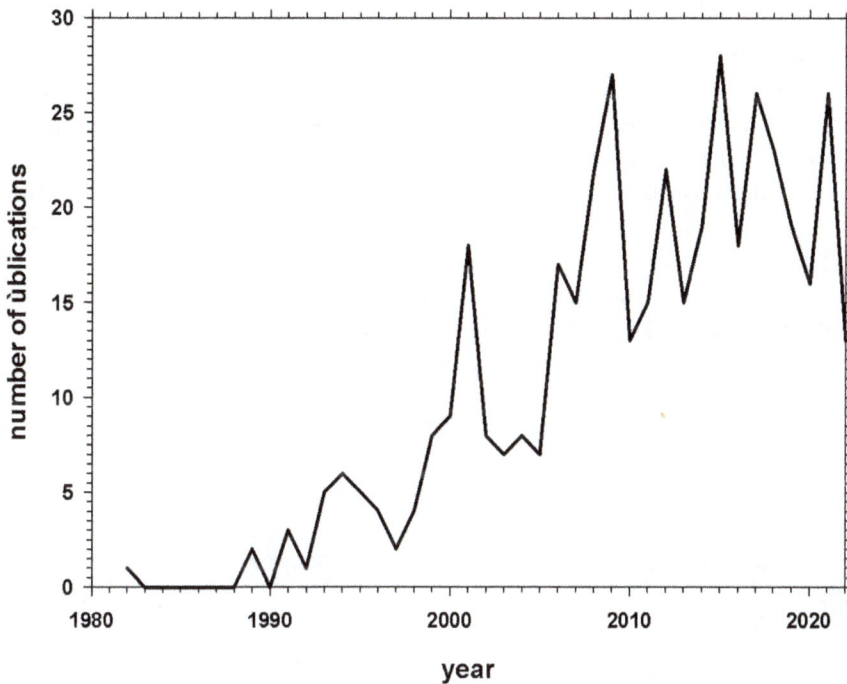

Fig. 7.32: Publications found in the WOS web, from Clarivate, on topics such as membranes and mercury porosimetry.

7.5.11 Commercial devices

The equipment needed to perform an HgP analysis is diverse, depending on the technical solutions provided by their manufacturers, as well as their working ranges and performance. However, all of them have some common characteristics, which we can outline below.

Entering the technical part of HgP, the equipment designed for this type of measurement generally consists of a mercury reservoir, connected by two ports (high-pressure port and low-pressure port, respectively) to the measuring cell. This cell, known as a penetrometer (see Fig. 7.33), consists of an elongated glass vessel, covered in its central part with a conductive material. This conductive coating, usually metal, together with the presence of another, the mercury itself, inside, turns the penetrometer into a capacitor, whose capacity can be measured and related to the volume of mercury inside the glass column.

The sample to be analysed (be it polymeric, ceramic or, as in many applications of the technique, consisting of some porous powder) is introduced into the penetrometer, once it has been previously dried. Subsequently, the penetrometer with the sample inside is degassed by vacuum, and at the low-pressure port, mercury begins to fill the penetrometer. At these low pressures (typically up to 50 psi) the mercury can only enter the interstices between the different pieces of the sample or into the largest pores of the sample (pores of a minimum size of about 4 μm). Subsequently, the penetrometer, insulated and partially filled with mercury, is transferred to the high-pressure port (up to 30–35,000 psi), where it continues to fill the pores with increasing pressures. For each applied pressure, the volume of intruded mercury is measured by determining the new capacity of the penetrometer. Finally, data such as applied pressure and intruded volume are used to construct the porogram.

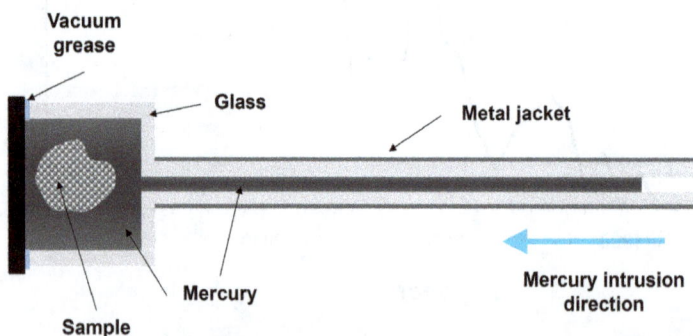

Fig. 7.33: Example of a typical Hg-intrusion penetrometer.

Finally, some short comments must be devoted to the default parameters needed to get reliable results in HgP. Two main values are to be settled when converting a porogram into a PSD.

First is the surface tension of the mercury–air interface. In many of the initial papers related to the application of HgP, a value of 474 mN/m for the surface tension of mercury is quoted, with a value of 480 mN/m also frequently used [63, 110, 111]. Nevertheless, more accurate measurements of this parameter can be found, leading to a value of $\gamma = 0.485$ N/m [112], which can also be adjusted to account for small variations with temperature.

Regarding contact angle, it depends on the material we are analysing, certainly, and so it could change from one material to another material. But it must be noted that, in practice, mercury does not wet any material presenting contact angles ranging from 112° to 150° [113]. Since the most frequently found value corresponds to 130° and given that a precise determination of the exact contact angle for a given material can be very time-consuming and complicated, most of the works published use this value of 130° as a reasonable default [63]. Of course, comparison of the results obtained from HgP with those coming from another porometry could be used to calibrate the technique by adjusting the contact angle closer to the actual values [113].

Several commercial devices designed for HgP are currently available on the market. Among them we can mention:

- Autopore Series from Micromeritics
- Belpore Series from Microtrac MRB
- PoreMaster Series from Quantachrome
- Pascal Series from Thermo Scientific
- Macropore from Carlo Erba
- Porosimeter P2000 from Fison
- Mercury/Nonmercury Intrusion Porosimeter from PMI Inc.

7.5.12 Advantages and disadvantages

Although HgP is a well-established and widely used characterization technique for many types of materials, it is still necessary to consider several peculiarities of the technique that can lead to errors in the interpretation of the results [13].

The biggest problem associated with the use of the HgP technique has to do with the compressibility of the material to be analysed. HgP equipment is designed to apply pressures of up to 450 MPa, which, in theory, allows them to analyse pores of the order of 3–4 nm. However, as the applied pressure increases, the material of which the sample is composed is compressed. This means that at each stage of pressure increase, part of the increase in the volume of intruded mercury is not due to the filling of any pore, but simply to the space left free in the measuring cell as the sample is compressed.

Ideally, the results obtained should be corrected for compressibility by comparison with a non-porous sample of the same material. In this case the compressibility of the material is translated into a hyperbola branch on a graph of Intruded volume vs. applied pressure. The presence of pores in the sample gives rise to deviations in this ideal curve that we can attribute to the presence of the existing pores.

On the other hand, this compressibility depends on the Young's modulus of the material. In the case of ceramic, metallic or sintered materials, this compressibility is small and the corresponding correction to the results is quite simple, although in highly non-compressible materials, negative changes in the intruded volume can be found. In any case, this is not the behaviour in the case of highly compressible polymeric materials, such as those that constitute the majority of commercial membranes, [114]. Thus, for polymeric membranes, the application of the HgP technique is reduced to the study of macropores, if the destruction of the porous structure due to high applied pressures is to be avoided [17].

The only way to circumvent this problem and be able to detect the PSD, within the volume variation due to compressibility, is to have a significant amount of sample in the cell, so that the number of pores to be detected is also significant and stands out from the bottom of the compressibility curve.

Another fact must also be taken into account: many polymeric membranes are asymmetrical, consisting of a support with very large pores, on which a significant number of smaller, truly selective pores have been developed (or joined if it is a composite material). However, although the total number of pores in the active layer (skin) of the membrane is much larger than those present in the support, the total volume of these pores is generally much smaller than the volume given by the support. This is because, in order to maintain a high permeability of the resulting membrane, the skin is as thin as possible.

Thus, in the volume–pressure curve of the joint sample, the contribution of the skin will be small and can easily go undetected. The solution to this problem, if possible, is to separate the support from the active layer and subject enough of the active layer of the membrane to HgP analysis, confident that the pores of the active layer will now be detectable.

The compressibility of samples has often been studied in connection with the fractality or fractal dimension of the sample, especially in the study of coal and similar samples [115].

Another point to note is that mercury intrusion only detects open pores at the mercury inlet, but these pores are not necessarily accessible to the flow of liquid through the sample. This is another problem with the technique, as it accounts for (or does not discriminate against) inkwell-type pores (pores with an entrance to the mercury but whose end is closed, not allowing to pass through). These types of pores will not be of interest in membrane characterization, as only pores that contribute to the flux are of interest to us, mainly for their influence on the permeability of the filter.

Another problem with the technique is the use of a highly polluting fluid, mercury, which must be handled and, of course, recycled with the utmost care to avoid environmental damage. Finally, samples analysed with this technique are contaminated with the mercury trapped in them at the end of the analysis, so HgP must be considered a destructive technique.

Finally, let us not forget that the technique uses the Young–Laplace equation in the interpretation of the results, considering by default cylindrical pores. This is another source of error in the measurement of the real pores that deviate from the ideal model, similar to all the porosimetric techniques that use this equation. The applicability of Young–Laplace, in general, has been discussed at length throughout this chapter.

In any case, the application of the Young–Laplace equation only determines the smallest input size of a pore, but not the actual size of the pore [114]. To understand this, we must consider that, especially in the case of pores of variable cross section, once the narrowest diameter is exceeded (for which it is necessary to wait for a sufficient increase in pressure), the rest of the pore is automatically filled, and this volume is attributable to pores that are narrower than the real value [116]. Similar to other techniques based on the Young–Laplace equation, HgP has difficulties to deal with interconnected pores, where it is not clear which part of the pore refers to such equation, and then some pore–network modelling is needed to correctly interpret those results [114].

Despite these limitations, HgP also has comparative advantages, which we will discuss in detail in Chapter 10 and which have led to the technique being used on more and more types of materials such as membranes, and also ceramics, coals [117], cement [84] or pharmaceutical tablets [118], along with other porous materials containing nanoparticles [117]. Finally, it has recently been used to study bioaerogels [119].

References

[1]   Young T., III. An essay on the cohesion of fluids. Philos Trans R Soc London, 95 (1805) 65–87. https://doi.org/10.1098/rstl.1805.0005.

[2]   Bechhold H. Stoechiom. Verwandtschaftsl.; Z. Phys Chem, 60 (1907) 257–318. Bechhold H. The permeability of ultrafilters. Z Phys Chem, 64 (1908) 328–342.

[3]   Miller D.J., Dreyer D.R., Bielawski C.W., Paul D.R., Freeman B.D. Surface Modification of Water Purification Membranes. Angewandte Chemie (International ed. in English), 56 (2017) 4662–4711. https://doi.org/10.1002/anie.201601509.

[4]   Hayes D.W. An Evaluation of the Bubble Point Method for Determining Pore Size Distributions of Nonwoven Geotextiles. Master's degree Thesis, Auburn University, Auburn, Alabama, USA, (2013).

[5]   Peinemann K.-.V., Pereira Nunes S. (Eds.). Membrane Technology. Volume 1, Membranes for Life Sciences. Wiley-VCH Verlag GmbH, Germany, (2007). ISBN: 13: 978 3527314805.

[6]   Zsigmondy R. Filter for Ultramicroscopic Particles. U.S. Patent, 1(421) 341. (27 June 1922).

[7]   Guo H., Wyart Y., Perot J., Nauleau F., Moulin P. Low-pressure membrane integrity tests for drinking water treatment: A review. Water Res, 44 (2010) 41–57. https://doi.org/10.1016/j.watres.2009.09.032.

[8]   Calvo J.I. Membrane Characterization by Porosimetric Techniques. Lesson Presented at XXIV Summer School on Membranes. Genoa, Italy, (2007).

[9]   Hernández A., Calvo J.I., Prádanos P., Tejerina F. Pore size distributions in microporous membranes. A critical analysis of the bubble point extended method. J Membrane Sci, 112 (1996) 1–12. https://doi.org/10.1016/0376-7388(95)00025-9.

[10]   Li D., Frey M.W., Joo Y.L. Characterization of nanofibrous membranes with capillary flow porometry. J Membrane Sci, 286 (2006) 104–114. https://doi.org/10.1016/j.memsci.2006.09.020.

[11] Gribble C.M., Matthews G.P., Laudone G.M., Turner A., Ridgway C.J., Schoelkopf J., Gane P.A.C. Porometry, porosimetry, image analysis and void network modelling in the study of the pore-level properties of filters. Chem Eng Sci, 66 (2011) 3701–3709. https://dx.doi.org/10.1016/j.ces.2011.05.013.

[12] Gijsbertsen-Abrahamse A.J., Boom R.M., Van der Padt A. Why liquid displacement methods are sometimes wrong in estimating the pore-size distribution. AIChE Journal, 50 (2004) 1364–1371. https://doi.org/10.1002/aic.10124.

[13] Kaneko K. Determination of pore size and pore size distribution: 1. Adsorbents and catalysts. J Membrane Sci, 96 (1994) 59–89. https://doi.org/10.1016/0376-7388(94)00126-X.

[14] Bechhold H., Schlesinger M., Silbereisen K., Maier L., Nurnberger W. Pore diameters of ultrafilters. Kolloid Z, 55 (1931) 172.

[15] Volfkovich Yu M., Bagotzky V.S. The method of Standard porosimetry: 1. Principles and possibilities. J Power Sources, 48 (1994) 327–338. https://doi.org/10.1016/0378-7753(94)80029-4.

[16] Volfkovich Yu M., Bagotzky V.S. The method of Standard porosimetry: 2. Investigation of the formation of porous structures. J Power Sources, 48 (1994) 339–348. https://doi.org/10.1016/0378-7753(94)80030-8.

[17] Kononenko N., Nikonenko V., Grande D., Larchet C., Dammak L., Fomenko M., Volfkovich Y. Porous structure of ion exchange membranes investigated by various technique. Adv Colloid Interface Sci, 246 (2017) 196–216. https://doi.org/10.1016/j.cis.2017.05.007.

[18] Shrestha K.P. Experimental Measurement of Air–Water Capillary Pressure Curves at Elevated Temperatures. Master Thesis, McGill University, Montreal (Canada) (2012).

[19] Gostick J.T., Ioannidis M.A., Fowler M.W., Pritzker M.D. Direct measurement of the capillary pressure characteristics of water–air-gas diffusion layer systems for PEM fuel cells. Electrochem Commun, 10 (2008) 1520–1523. https://doi.org/10.1016/j.elecom.2008.08.008.

[20] Jena A., Gupta K. Advances in pore structure evaluation by porometry. Chem Eng Technol, 33 (2010) 1241–1250. https://doi.org/10.1002/ceat.201000119.

[21] Ozden A., Shahgaldi S., Li X., Hamdullahpur F. A review of gas diffusion layers for proton exchange membrane fuel cells – With a focus on characteristics, characterization techniques, materials and designs. Prog Energy Combust Sci, 74 (2019) 50–102. https://doi.org/10.1016/j.pecs.2019.05.002.

[22] Lee J.-.G., Lee E.-.J., Jeong S., Guo J., Kyoungjin An A., Guo H., Kim J., Leikens T.O., Ghaffour N. Theoretical modelling and experimental validation of transport and separation properties of carbon nanotube electrospun membrane distillation. J Membrane Sci, 526 (2017) 395–408. http://dx.doi.org/10.1016/j.memsci.2016.12.045.

[23] Gostick J.T., Fowler M.W., Ioannidis M.A., Pritzker M.D., Volfkovich Y.M., Sakars A. Capillary pressure and hydrophilic porosity in gas diffusion layers for polymer electrolyte fuel cells. J Power Sources, 156 (2006) 375–387. https://doi.org/10.1016/j.jpowsour.2005.05.086.

[24] Jena A., Gupta K. Characterization of pore structure of filtration media. Fluid/Part Sep J, 14(3) (2002) 227–241.

[25] Jena A., Gupta K. Characterization of Pore Structure of Fuel Cell Components Containing Hydrophobic and Hydrophilic Pores. In: Proceedings of the 41st Power Sources Conference, Philadelphia, PA, USA, (14–17 June 2004).

[26] Peinador R.I., Calvo J.I., Ben Aim R. Comparison of capillary flow porometry (CFP) and liquid extrusion porometry (LEP) techniques for the characterization of porous and face mask membranes. Appl Sci, 10 (2020) 5703. https://doi.org/10.3390/app10165703.

[27] Peinador R.I., Abba O., Calvo J.I. Characterization of Commercial Gas Diffusion Layers (GDL) by Liquid Extrusion Porometry (LEP) and Gas Liquid Displacement Porometry (GLDP). Membranes, 12 (2022) 212. https://doi.org/10.3390/membranes12020212.

[28] Maalal O. Porométrie Liquide–Liquide, Evaporométrie et Simulations sur Réseau de Pores. Ph.D. Thesis, Institut National Polytechnique de Toulouse (Toulouse INP), University of Toulouse, Toulouse, France, (2020).

[29] Calvo J.I., Hernández A., Palacio L., Prádanos P., Ailuno M., Bottino A., Carniglia G., Comite A., Jezowska A., Pagliero M., Firpo R. Characterization of porous PTFE membranes and their potential application in osmotic distillation. Proc. Euromembrane 2018, Valencia (España), (July 2018).

[30] Munari S., Bottino A., Moretti, Capannelli G., Becchi I. Permoporometric study on ultrafiltration membranes. J Membrane Sci, 41 (1989) 69–86. https://doi.org/10.1016/S0376-7388(00)82392-3.

[31] Kim K.J., Fane A.G., Ben Aim R., Liu M.G., Jonsson G., Tessaro I.C., Broek A.P., Bargeman D. A comparative study of techniques used for porous membrane characterization: Pore characterization. J Membrane Sci, 81 (1994) 35–46. https://doi.org/10.1016/03767388(93)E0044-E.

[32] Zeman L.J., Zydney A.L. Microfiltration and Ultrafiltration: Principles and Applications, M. Dekker, New York, USA, (1996). https://doi.org/10.1201/9780203747223.

[33] Erbe F. Die Bestimmung der Porenverteilung nach ihrer Größe in Filtern und Ultrafiltern. Kolloid-Zeitschrift, 63 (1933) 277–285. https://doi.org/10.1007/BF01422935.

[34] Ferry J.D. Ultrafilter membranes and Ultrafiltration. Chem Rev, 18(3) (1936) 373–455. https://doi.org/10.1021/cr60061a001.

[35] Capannelli G., Vigo F., Munari S. Ultrafiltration membranes – Characterization methods. J Membrane Sci, 15 (1983) 289–313. https://doi.org/10.1016/S0376-7388(00)82305-4.

[36] Kesting R.E. Synthetic Polymeric Membranes: A Structural Perspective. 2nd ed., John Wiley & Sons, Inc., New York (USA), (1985). ISBN: 0-471-80717-6.

[37] Kujawski W., Adamczak P., Narebska A. A fully automated system for the determination of pore size distribution in microfiltration and ultrafiltration membranes. Sep Sci Technol, 24 (1989) 495–506. https://doi.org/10.1080/01496398908049787.

[38] Grabar P., Nikitine S. Sur le diamètre des pores des membranes en collodion utilisées en ultrafiltration. J Chim Phys, 33 (1936) 721–741. https://doi.org/10.1051/jcp/1936330721.

[39] Antón F.E., Calvo J.I., Álvarez J.R., Hernández A., Luque S. Fitting approach to liquid–liquid displacement Porosimetry based on the log-normal pore size distribution. J Membrane Sci, 470 (2014) 219–228. http://dx.doi.org/10.1016/j.memsci.2014.07.035.

[40] Peinador R.I., Calvo J.I., Prádanos P., Palacio L., Hernández A. Characterisation of polymeric UF membranes by liquid–liquid displacement porosimetry. J Membrane Sci, 348 (2010) 238–244. https://doi.org/10.1016/j.memsci.2009.11.008.

[41] Calvo J.I., Bottino A., Capannelli G., Hernández A. Pore size distribution of ceramic UF membranes by liquid–liquid displacement porosimetry. J Membrane Sci, 310 (2008) 531–538. https://doi.org/10.1016/j.memsci.2007.11.035.

[42] Tanis-Kanbur M.B., Peinador R.I., Calvo J.I., Hernández A., Chew J.W. Porosimetric membrane characterization techniques: A review. J Membrane Sci, 619 (2021) 118750. https://doi.org/10.1016/j.memsci.2020.118750.

[43] Phillips M.W., DiLeo A.J. A validatable porosimetric technique for verifying the integrity of virus-retentive membranes. Biologicals, 24(3) (1996) 243–253. https://doi.org/10.1006/biol.1996.0033.

[44] Bouchiha S. La caractérisation des membranes d'Ultrafiltration et Microfiltration par les techniques de pénétration FluideFluide (Gaz–Liquide et Liquide–Liquide). In: Rapport de Stage de fin d'études. Université Toulouse III – Paul Sabatier, (2019).

[45] McGuire K.S., Lawson K.W., Lloyd D.R. Pore size distribution determination from liquid permeation through microporous membranes. J Membrane Sci, 99 (1995) 127–137. https://doi.org/10.1016/0376-7388(94)00209-H.

[46] Bottino A., Capannelli G., Petit-Bon P., Cao N., Pegoraro M., Zoia G. Pore size and pore-size distribution in microfiltration membranes. Sep Sci Technol, 26 (1991) 1315–1327. https://doi.org/10.1080/01496399108050534.

[47] Persson K.M., Capannelli G., Bottino A., Trägårdh G. Porosity and protein adsorption of four polymeric microfiltration membranes. J Membrane Sci, 76 (1993) 61–71. https://doi.org/10.1016/0376-7388(93)87005-V.

[48] Courtois T. Device for measuring porosity characteristics of porous media. WO Patent, WO-2005/031318-A2, (April 2005).

[49] Calvo J.I., Peinador R.I., Prádanos P., Palacio L., Bottino A., Capannelli G. Hernández A Liquid–liquid displacement porometry to estimate the molecular weight cut-off of ultrafiltration membranes. Desalination, 268 (2011) 174–181. http://dx.doi.org/10.1016/j.desal.2010.10.016.

[50] Otero-Fernández A., Otero J.A., Maroto-Valiente A., Calvo J.I., Palacio L., Prádanos P., Hernández A. Reduction of Pb(II) in water to safe levels by a small tubular membrane nanofiltration plant. Clean Techn Environ Policy, 20 (2018) 329–343. https://doi.org/10.1007/s10098-017-1474-2.

[51] Cheng L.P., Lin H.V., Chen L.W., Young T.H. Solute rejection of dextran by EVAL membranes with asymmetric and particulate morphologies. Polymer, 39 (1998) 2135–2142. https://doi.org/10.1016/S0032-3861(97)00518-1.

[52] Calvo J.I., Peinador R.I., Thom V., Schleuss T., ToVinh K., Prádanos P., Hernández A. Comparison of pore size distributions from dextran retention tests and liquid–liquid displacement porosimetry. Microporous and Mesoporous Mater, 250 (2017) 170–176. http://dx.doi.org/10.1016/j.micromeso.2017.05.032.

[53] Singh S., Khulbe K.C., Matsuura T., Ramamurthy P. Membrane characterization by solute transport and atomic force microscopy. J Membrane Sci, 142 (1998) 111–127. https://doi.org/10.1016/S0376-7388(97)00329-3.

[54] Giglia S., Bohonak D., Greenhalgh P., Leahy A. Measurement of pore size distribution and prediction of membrane filter virus retention using liquid–liquid porometry. J Membrane Sci, 476 (2015) 399–409. https://doi.org/10.1016/j.memsci.2014.11.053.

[55] AlMarzooqi F.A., Bilad M.R., Mansoor B., Arafat H.A. A comparative study of image analysis and porometry techniques for characterization of porous membranes. J Mater Sci, 51 (2016) 2017–2032. https://doi.org/10.1007/s10853-015-9512-0.

[56] Gustafsson S., Westermann F., Hanrieder T., Jung L., Ruppach H., Mihranyan A. Comparative analysis of dry and wet porometry methods for characterization of regular and cross-linked virus removal filter papers. Membranes, 9 (2019) 1–13. https://doi.org/10.3390/membranes9010001.

[57] Ängeslevä M., Salmimies R., Rideal G., Häkkinen A. Pore diameter measured by capillary flow porometry: The effect of the capillary constant. J Porous Media, 26 (2023) 31–49. https://doi.org/10.1615/JPorMedia.2022043042.

[58] Washburn E.W. Note on a Method of Determining the Distribution of Pore Sizes in a Porous Material. Proc Natl Acad Sci USA, 7(4) (1921) 115–116. https://doi.org/10.1073/pnas.7.4.115.

[59] Ritter H.L., Drake L.C. Pressure porosimeter and determination of complete macropore-size distributions. Pressure porosimeter and determination of complete macropore-size distributions. Ind Eng Chem Anal Ed, 17(12) (1945) 782–786. https://doi.org/10.1021/i560148a013.

[60] Drake L.C., Ritter H.L. Macropore-size distributions in some typical porous substances. Ind Eng Chem Anal Ed, 17(12) (1945) 787–791. https://doi.org/10.1021/i560148a014.

[61] Honold E., Skau E.L. Application of mercury-intrusion method for determination of pore-size distribution to membrane filters. Science, 120(3124) (1954) 805–806. https://doi.org/10.1126/science.120.3124.805.

[62] Rootare H.M., Prenzlow C.F. Surface areas from mercury porosimeter measurements. J Phys Chem, 71(8) (1967) 2733–2736. https://doi.org/10.1021/j100867a057.

[63] Liabastre A.A., Orr C. An evaluation of pore structure by mercury penetration. J Colloid Interface Sci, 64 (1978) 1–18. https://doi.org/10.1016/0021-9797(78)90329-6.

[64] Washburn E.W. The Dynamics of Capillary Flow. Phys Rev, 17(3) (1921) 273–283.

[65] Kloubek J. Hysteresis in Porosimetry. Powder Technology, 29 (1981) 63–73. https://doi.org/10.1016/0032-5910(81)85005-X.

[66] Leofanti G., Padovan M., Tozzola G., Venturelli B. Surface area and pore texture of catalysts. Catalysis Today, 41 (1998) 207–219. https://doi.org/10.1016/S0920-5861(98)00050-9.

[67] Conner W.M., Lane A.M., Hoffman A.J. Measurement of the morphology of high surface area solids: hysteresis in mercury porosimetry. J Colloid Interface Sci, 100(1) (1984) 185–193. https://doi.org/10.1016/0021-9797(84)90424-7.

[68] Moro F., Böhni H. Ink-bottle effect in mercury intrusion porosimetry of cement-based materials. J Colloid Interface Sci, 246 (2002) 135–149. https://doi.org/10.1006/jcis.2001.7962.

[69] Berodier E., Bizzozero J., Muller A.C.A. Mercury intrusion porosimetry. In: A Practical Guide to Microstructural Analysis of Cementitious Materials. CRC Press, Boca Raton, FL, USA, (2016). ISBN: 9781351228497.

[70] Law K.Y., Zhao H. Surface Wetting Characterization, Contact Angle, and Fundamentals. Springer, Heidelberg, Germany, (2016). ISBN: 978-3-319-25212-4.

[71] Wenzel R.N. Resistance of solid surfaces to wetting by water. Ind Eng Chem, 38 (1936) 988–994. https://doi.org/10.1021/ie50320a024.

[72] Cassie A.B.D., Baxter S. Wettability of porous surfaces. Trans Faraday Soc, 40 (1944) 546–551. https://doi.org/10.1039/TF9444000546.

[73] Banerjee S. Simple derivation of Young, Wenzel and Cassie-Baxter equations and its interpretations. arXiv:0808.1460 [cond-mat.mtrl-sci] (2008). https://doi.org/10.48550/arXiv.0808.1460

[74] Salmas C., Androutsopoulos G. Mercury porosimetry: Contact angle hysteresis of materials with controlled pore structure. J Colloid Interface Sci, 239 (2001) 178–189. https://doi.org/10.1006/jcis.2001.7531.

[75] Lam C.N.C. A study of advancing and receding contact angles and contact angle hysteresis. PhD Thesis, University of Toronto, Canada, (2001).

[76] Zeng Q., Li K., Fen-Chong T., Dangla P. Surface fractal analysis of pore structure of high-volume fly-ash cement pastes. Applied Surface Science, 257 (2010) 762–768. https://doi.org/10.1016/j.apsusc.2010.07.061.

[77] Wang L., Zeng X., Yang H., Lv X., Guo F., Shi Y., Hanif A. Investigation and application of fractal theory in cement-based materials: a review. Fractal Fract, 5 (2021) 247. https://doi.org/10.3390/fractalfract5040247.

[78] Neimark A. A new approach to the determination of the surface fractal dimension of porous solids. Phys A: Stat Mech Its Appl, 191 (1992) 258–262. https://doi.org/10.1016/0378-4371(92)90536-Y.

[79] Pfeifer P., Avnir D. Chemistry in non-integer dimensions between two and three: Fractal theory of heterogeneous surface. J Chem Phys, 79 (1983) 3558–3565. https://doi.org/10.1063/1.446210.

[80] Zhang B., Li S. Determination of the surface fractal dimension for porous media by mercury porosimetry. Ind Eng Chem Res, 34 (1995) 1383–1386. https://doi.org/10.1021/ie00043a044.

[81] Mandelbrot B.B. The Fractal Geometry of Nature. WH Freeman and Co, New York, USA, (1983). ISBN: 978-0716711865.

[82] Villagrán-Zaccardi Y., Alderete N., Van den Heede P., De Belie N. Pore size distribution and surface multifractal dimension by multicycle mercury intrusion porosimetry of GGBFS and limestone powder blended concrete. Appl Sci, 11 (2021) 4851. https://doi.org/10.3390/app11114851.

[83] Tsakiroglou C.D., Kolonis G.B., Roumeliotis T.C., Payatakes A.C. Mercury penetration and snap-off in lenticular pores. J Colloid Interf Sci, 193(2) (1997) 259–272. https://doi.org/10.1006/jcis.1997.5058.

[84] Diamond S. Mercury porosimetry. An inappropriate method for the measurement of pore size distributions in cement-based materials. Cem Concr Res, 30 (2000) 1517–1525. https://doi.org/10.1016/S0008-8846(00)00370-7.

[85] Zhou J., Ye G., Van Breugel K. Characterization of pore structure in cement-based materials using pressurization–depressurization cycling mercury intrusion porosimetry (PDC-MIP). Cem Concr Res, 40 (2010) 1120–1128. https://doi.org/10.1016/j.cemconres.2010.02.011.

[86] Gu Z., Goulet R., Levitz P., Ihiawakrim D., Ersen O., Bazant M.Z. Mercury cyclic porosimetry: Measuring pore-size distributions corrected for both pore-space accessivity and contact-angle hysteresis. J Colloid Interf Sci, 599 (2021) 255–261. https://doi.org/10.1016/j.jcis.2021.04.038.

[87] Zhang Y., Wu K., Yang Z., Ye G. A reappraisal of the ink-bottle effect and pore structure of cementitious materials using intrusion–extrusion cyclic mercury porosimetry. Cem Concr Res, 161 (2022) 106942. https://doi.org/10.1016/j.cemconres.2022.106942.

[88] Cebeci O.Z. The Intrusion of Conical and Spherical Pores in Mercury Intrusion Porosimetry. J Colloid Interf Sci, 78 (1980) 383–388. https://doi.org/10.1016/0021-9797(80)90577-9.

[89] De Botton S. Mechanistic Study of Mercury Intrusion and Retraction. M. Sc. Thesis, Universidad Autónoma Metropolitana-Iztapalapa, México, (1985).

[90] Felipe C., Rojas F., Kornhauser I., Matthias T., Zgrablich J.A. Mechanistic and experimental aspects of the structural characterization of some model and real systems by nitrogen sorption and mercury porosimetry. SAGE Publications. Adsorpt Sci Technol, 24(8) (2006) 623–643. https://doi.org/10.1260/026361706781355019.

[91] Tsakiroglou C.D., Payatakes A.C. A new simulator of mercury porosimetry for the characterization of porous materials. J Colloid Interf Sci, 137(2) (1990) 315–339. https://doi.org/10.1016/0021-9797(90)90409-H.

[92] Langbein D. Canthotaxis/Wetting Barriers/Pinning Lines. In: D. Langbein (Eds.), Capillary Surfaces. Springer Tracts in Modern Physics, Vol. 178. Springer, Berlin, Heidelberg. Germany, (2002). https://doi.org/10.1007/3-540-45267-2_7.

[93] Mądry K., Nowick W. Wetting between Cassie-Baxter and Wenzel regimes: A cellular model approach. Eur Phys J E Soft Matter, 44(11) (2021) 138. https://doi.org/10.1140/epje/s10189-021-00140-8.

[94] Sahimi M. Applications of Percolation Theory (Applied Mathematical Sciences, 213). Springer, Berlin. Germany, (2023). ISBN: 978-3031203855.

[95] King P., Masihi M. Percolation Theory in Reservoir Engineering. World Scientific Europe, London, UK, (2018). ISBN: 978-1786345233.

[96] Lenormand R., Bories S. Description of a bond percolation mechanism used for the simulation of drainage with trapping in porous-media. C. R. Hebd Seances Acad Sci Ser B, 291 (1980) 279–282.

[97] Wilkinson D., Willemsen J.F. Invasion percolation: A new form of percolation theory. J Phys, A(16) (1983) 3365. https://doi.org/10.1088/0305-4470/16/14/028.

[98] Bak B.D., Kalmár-Nagy T. Porcolation: an invasion percolation model for mercury porosimetry. Fluctuation Noise Lett, 16(1) (2017) 1750008. http://dx.doi.org/10.1142/S0219477517500080.

[99] Fatt I. The network model of porous media. 1. Capillary pressure characteristics. AIME Petr Trans, 207(7) (1956) 144–159. https://doi.org/10.2118/574-G.

[100] Androutsopoulos G.P., Mann R. Evaluation of mercury porosimeter experiments using a network pore structure model. Chem Eng Sci, 34(10) (1979) 1203–1212. https://doi.org/10.1016/0009-2509(79)85151-9.

[101] Portsmouth R.L., Gladden L.F. Determination of pore connectivity by mercury porosimetry. Chem Eng Sci, 46(12) (1991) 3023–3036. https://doi.org/10.1016/0009-2509(91)85006-J.

[102] Androutsopoulos G.P., Salmas C.E. A simplified model for mercury porosimetry hysteresis. Chem Eng Comm, 176(1) (1999) 1–42. https://doi.org/10.1080/00986449908912144.

[103] Borman V.D., Belogorlov A.A., Byrkin V.A., Tronin V.N., Troyan V.I. Stability of a non wetting liquid in a nanoporous medium. Phys Scr, 89(7) (2014) 075705. http://dx.doi.org/10.1088/0031-8949/89/7/075705.

[104] Wall G.C., Brown R.J.C. The determination of pore-size distributions from sorption isotherms and mercury penetration in interconnected pores: The application of percolation theory. J Colloid Interf Sci, 82(1) (1981) 141–149. https://doi.org/10.1016/0021-9797(81)90132-6.

[105] Matthews G.P., Ridgway C.J., Spearing M.C. Void space modelling of mercury intrusion hysteresis in sandstone, paper coating and other porous media. J Colloid Interface Sci, 171 (1995) 8–27. http://dx.doi.org/10.1006/jcis.1995.1146.

[106] Porcheron F., Monson P.A. Modeling Mercury Porosimetry Using Statistical Mechanics. Langmuir, 20 (2004) 6482–6489. http://dx.doi.org/10.1021/la049939e.

[107] Porcheron E., Monson P.A., Thommes M. Molecular Modelling of Mercury Porosimetry. Adsorption, 11 (2005) 325–329. http://dx.doi.org/10.1007/s10450-005-5945-0.

[108] Porcheron F., Thommes M., Ahmad R., Monson P.A. Mercury Porosimetry in Mesoporous Glasses: A Comparison of Experiments with Results from a Molecular Model. Langmuir, 23(6) (2007) 3372–3380. https://doi.org/10.1021/la063080e.

[109] De With G., Glass H.J. Reliability and Reproducibility of Mercury Intrusion Porosimetry. J Eur Ceramic Soc, 17 (1997) 753–757. https://doi.org/10.1016/S0955-2219(96)00181-1.

[110] Kernaghan M. Surface Tension of Mercury. Phys Rev, 37 (1931) 990–997. https://doi.org/10.1103/PhysRev.37.990.

[111] Rootare H.M. A short literature review of mercury porosimetry as a method of measuring pore-size distributions in porous materials, and a discussion of possible sources of errors in this method. Aminco Lab News, 24(3) (1968) 4A–4H.

[112] Théron F., Lys E., Joubert A., Bertrand F., Le Coq L. Characterization of the porous structure of a non-woven fibrous medium for air filtration at local and global scales using porosimetry and X-ray micro-tomography. Powder Technol, 320 (2017) 295–303. https://doi.org/10.1016/j.powtec.2017.07.020.

[113] Calvo J.I., Hernández A., Prádanos P., Martínez L., Bowen W.R. Pore size distributions in microporous membranes II. Bulk characterization of track-etched filters by air porometry and mercury porosimetry. J Colloid Interface Sci, 176 (1995) 467–478. https://doi.org/10.1006/jcis.1995.9944.

[114] Giesche H. Mercury Porosimetry: A General (Practical) Overview. Part Part Syst Charact, 23 (2006) 9–19. https://doi.org/10.1002/ppsc.200601009.

[115] Li Y.-.H., Qing Lu G., Rudolph V. Compressibility and fractal dimension of fine coal particles in relation to pore structure characterisation using mercury porosimetry. Part Part Syst Charact, 16 (1999) 25–31. https://doi.org/10.1002/(SICI)1521-4117(199905)16:1<25::AID-PPSC25>3.0.CO;2-T.

[116] Rigby S.P. Recent Developments in the Structural Characterisation of Disordered, Mesoporous Solids. Johnson Matthey Technol Rev, 62 (2018) 296–312. https://doi.org/10.1595/205651318X696710.

[117] Calvo J.I., Bottino A., Prádanos P., Palacio L., Hernández A. Porosity. In: E.M.V. Hoek, V.V. Tarabara (Eds.), Encyclopedia of Membrane Science and Technology. Wiley Intersci. Pub, New York, USA, (2013) Pages: 1062–1086.

[118] Dees P.J., Polderman J. Mercury porosimetry in pharmaceutical technology. Powder Technol, 29 (1981) 187–197. https://doi.org/10.1016/0032-5910(81)85016-4.

[119] Horvat G., Pantić M., Knez Ž., Novak Z. A brief evaluation of pore structure determination for bioaerogels. Gels, 8 (2022) 438. https://doi.org/10.3390/gels8070438.

[120] Imbrogno A., Calvo J.I., Breida M., Schwaiger R., Schäfer A.I. Molecular weight cut off (MWCO) determination in ultra- and nanofiltration: Review of methods and implications on organic matter removal. Sep. & Purif. Technol., 354 (2024) 128612. https://doi.org/10.1016/j.seppur.2024.128612

Chapter 8
Other techniques for porosimetric studies

8.1 Pore size distribution by solute retention

As mentioned in the introduction to this book, there are several characterization methods (among which those revised in this text) that provide information on membrane structure, while others focus on aspects or parameters related to the functional performance of the membrane. In this sense, a fundamental aspect in the performance and usefulness of membranes is their selectivity towards certain solutes or mixtures of solutes.

This type of information is usually obtained by means of solute retention tests, that is, filtration experiments for our membrane in which the membrane filter is placed in contact with solutions containing certain solutes of chosen sizes or molecular weights, and the percentage of these solutes retained by the membrane is then determined.

From such experiments, we can define the observed retention coefficient, R_O, for a given solute as the following relationship:

$$R_O = 1 - \frac{c_p}{c_f} \tag{8.1}$$

where c_p and c_f are the solute concentrations in the permeate and in the feed, respectively.

This retention coefficient is called "observed" because it does not take into account the accumulation of the retained material on the membrane surface. This phenomenon, called concentration polarization, is inevitable and inherent to the filtration process itself. If we could determine the concentration of solute in contact with the membrane (which is obviously much higher than in the feed), we could replace this observed (or experimental) retention coefficient with the true retention coefficient, R, as follows:

$$R = 1 - \frac{c_p}{c_m} \tag{8.2}$$

Now c_m is the concentration of the solute in contact with the membrane ($c_m > c_0$). See Fig. 8.1 for a clearer definition of these concepts.

Experiments on filtration of solutes with increasing molecular weights are fundamental in determining a key parameter in the description of a membrane, the molecular weight cut-off (MWCO). When purchasing a membrane for a specific application, the cut-off value is often the first criterion for the decision. It is obvious that this value alone does not ensure the suitability of a particular filter for a specific industrial application. Nevertheless, the MWCO value is essential for a first selection among the most suitable filters.

https://doi.org/10.1515/9783110792195-008

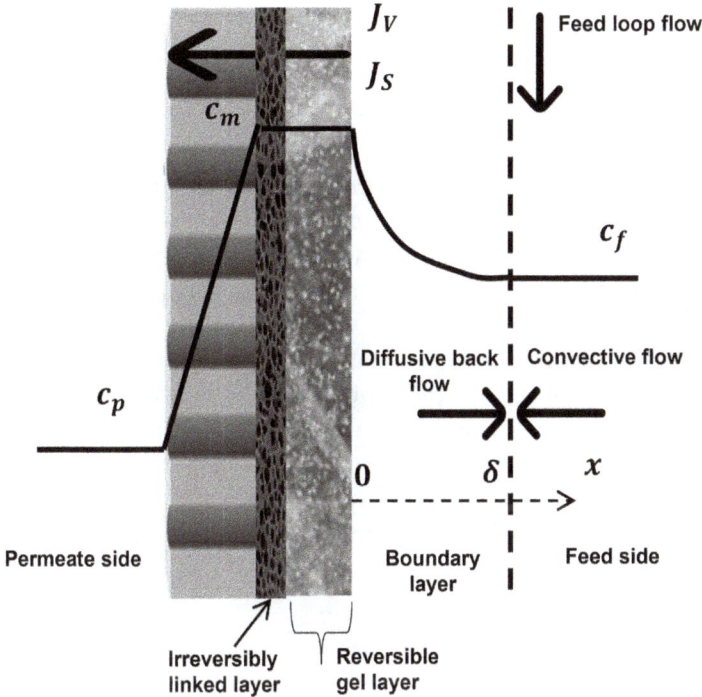

Fig. 8.1: The concentration profile through a membrane system including the concentration polarization boundary layer.

Although it is not the purpose of this book, it is useful to briefly recall the definition and how the cut-off is usually determined.

First of all, we would say that: "MWCO is defined as the lowest molecular weight of that solute which is retained by 90% when filtered through the membrane under study".

The first thing to bear in mind is that this definition is not univocal, as it depends on the type of molecule used to determine the cut-off value. Thus, MWCO is typically from retention experiments using molecules as diverse as dextrans, polyethylene glycols (PEGs), globular proteins and so on.

Furthermore, this definition and the conditions under which this value is to be determined are not standardized, despite its importance and several attempts at standardization or at least homogenization [1]. Thus, the conditions of the filtration cell (size and shape of filtration channels, use of spacers, etc.), the recirculation speed, the applied pressure and even the working temperature are issues that affect the MWCO value obtained (obviously apart, as yet commented, from the type of molecule considered in the filtration experiments).

Membrane manufacturers usually carry out experiments on the retention of solutions containing solutes of various molecular weights. Generally, molecules such as dextrans or PEGs are used, which can be found commercially in a wide range of molecular weights and with a certain homogeneity for each molecular weight. By performing the corresponding retention tests, it is determined which of these molecular weights is retained at 90%, and this value is assigned as MWCO.

It seems reasonable to assume that the molecular weight (and consequently) of a retained molecule is related to the average size of the retaining pores. It is based on this assumption that we can consider the next porosimetric method that will be discussed in this text, the solute retention test. But first we will discuss the thermodynamic basis of the process that will allow us to understand the equations from which to interpret the retention data and to be able to convert these data into a pore size distribution (PSD).

Finally, we will also dedicate a few pages to comment on various empirical correlations that have been presented over the years, which allow us to relate, in an obviously approximate way, the size of a pore with the molecular weight of the molecule that is retained in it. These correlations will allow us to propose a method for estimating the MWCO, based on PSD data, obtained by any valid porometry, which we have already discussed when introducing liquid–liquid displacement porosimetry, on the basis of which this estimation method was proposed [2].

8.1.1 Solute retention test fundamentals

Based on the linear thermodynamics of irreversible processes, K.S. Spiegler, O. Kedem and A. Katchalsky, in 1958 to 1965, developed equations characterizing the transmembrane transport of homogeneous solutions [3]. These equations have been firmly established as a powerful tool to study the transport of non-electrolytes through porous membranes.

A comprehensive non-equilibrium thermodynamic formalism, suitable for many actual membrane applications, was developed from the Spiegler, Kedem and Katchalsky equations [4–6].

For a non-equilibrium spontaneous system, each flux is in general dependent on the same number of forces as the number of fluxes. Forces and fluxes grouped in factors multiplying a force and a flux give the energy dissipation. The Onsager reciprocity laws, however, reduce the number of independent phenomenological coefficients when force–flux linear correlations exist by introducing the so-called phenomenological correlations [7, 8].

In any case, here we will discuss the approach to the interpretation of solute retention test based on the use of hindered transport models.

Efforts to elucidate the selectivity of porous membranes using models of hindered transport date back almost a century. One of the earliest influential studies was due to Ferry [9] who derived hindrance factors for convective transport from steric considerations. Hydrodynamic terms to describe reductions in the solute mobility within the pores were introduced by Pappenheimer et al. [10] and Renkin [11], among others.

The concept of steric and/or hydrodynamic hindrances in fine pores has since then found the extensive use in characterizing membrane based on their functional performances [12].

The easiest strategy is to assume that once a molecule is smaller than a pore, solvent convection will carry it readily through [13, 14]. The study based on this supposition determines the percentage of total flux transporting molecules smaller than the pore for each pore size. This method was applied, among others, by Michaels [15], who suggested a log-normal distribution to suit the form of sieving curves as a function of the molecular radius. Nevertheless, Munch et al. [16] and Zeman and Wales [17] have carried out tests utilizing membranes with track-etched pores, which have a narrow range of pore sizes. Their research amply supports the theoretical models that take into consideration hydrodynamic and steric interactions in the movement of molecules through pores. In such cases, the retention coefficient would be based on the relationship between molecules and pore sizes, as proposed first by Ferry [9]. Afterwards, Nobegra et al. [18] and Aimar et al. [19] contributed significantly to the standardization of such procedure in the 1990s and beyond.

8.1.2 Retentive and non-retentive pores

To evaluate the PSD of an incompletely retentive membrane, we can presume that for each molecular size present in the feed, there is a percentage of the pores that should be totally retentive while the rest of them would allow the passage of such molecules without any restriction.

We can write the mass balance as

$$J_V c_p = J_V^t c_m \tag{8.3}$$

Here c_p and c_m are the concentrations in the pore and at the pore entrance in the feed side (on the membrane surface). J_V is the total volume flow trespassing the membrane, and J_V^t is the flow through the transmitting pores. Note that these flows correspond to volumes per unit of the membrane area and time (m/s).

The ratio of the transmitted volume flow and the transmitted solvent (water) flow is the ratio between the viscosity of the solution and the viscosity of the solvent (water):

$$\frac{J_V^t}{J_W^t} = \frac{\eta_W}{\eta_s} \tag{8.4}$$

This is because these flows can be considered as given by the Hagen–Poiseuille law and thus inversely proportional to viscosity:

$$J_W^t = \frac{n_p \pi r_p^4}{8\eta_W} \left(\frac{\Delta p}{\Delta x}\right) \tag{8.5}$$

Here n_p is the number of pores per unit area of the membrane. For diluted solutions, the ratio in eq. (8.4) is close to 1, $\eta_w/\eta_s \approx 1$.

The flow of solvent can be divided into two portions as follows:

$$J_W = J_W^t + J_W^r \tag{8.6}$$

J_W^t is the fraction of the solvent passing through the transmitting pores and J_W^r is the fraction passing through the retaining pores. Note that we can define a retention coefficient R (the one we defined as true retention coefficient in (8.2)) as

$$R = 1 - \frac{c_p}{c_m} \tag{8.7}$$

Note that c_p/c_m is the fraction of mass transmitted into the pores and thus R is a good measure of the retained fraction of mass. In terms of this retention coefficient:

$$\left. \begin{array}{l} J_W^t = J_V(1-R) \\ J_W^r = J_V R \end{array} \right\} J_W^t + J_W^r = J_V \tag{8.8}$$

This is only valid if the solvent does not contribute appreciably to the volume permeated which is approximately true for diluted solutions. These equations allow the evaluation of J_W^t and J_W^r (and $J_W = J_W^t + J_W^r$) once J_V is known for each R. The quotient J_W^t/J_W versus the solute molecular weight, by using different solutes or wide molecular size distributions, corresponds to the cumulative of flux passing through the pores allowing the passage of increasing molecular weights. If molecular weights can be correlated with the sizes of their molecules, the cumulative distribution of pores versus their size would be obtained, assuming that permeation is only restricted by size:

$$\frac{d\left(n_p^t/n_p\right)}{dr_p} = \frac{1}{r_p^2} \frac{d\left(J_W^t/J_W\right)}{dr_p} \tag{8.9}$$

8.1.3 Concentration polarization

In a steady state, the amount of solute convectively transported to the membrane surface is equal to the solute that permeates through the membrane minus the contribution of back-diffusion to the bulk feed solution according to the commonly admitted film theory [20]:

$$J_V c = J_V c_p - D\frac{dc}{dx} \tag{8.10}$$

The boundary conditions are (see Fig. 8.1):

$$\left. \begin{array}{l} x = 0 \rightarrow c = c_m \\ x = \delta \rightarrow c = c_f \end{array} \right\} \tag{8.11}$$

Here δ is the thickness of the concentration polarization layer where concentration decreases from c_m to the feed concentration c_f. The integration of eq. (8.10) then leads to

$$J_v = \frac{D}{\delta} \ln\left(\frac{c_m - c_p}{c_f - c_p}\right) = K_m \ln\left(\frac{c_m - c_p}{c_f - c_p}\right) \tag{8.12}$$

where $K_m = \frac{D}{\delta}$ is the mass transfer coefficient. Equation (8.12) can be written as

$$c_m = c_p + (c_f - c_p)e^{J_v/K_m} \tag{8.13}$$

If we define an observed retention coefficient as

$$R_O = 1 - \frac{c_p}{c_f} \tag{8.14}$$

Equation (8.13) can be written as

$$\ln\left(\frac{1 - R_O}{R_O}\right) = \ln\left(\frac{1 - R}{R}\right) + \frac{J_v}{K_m} \tag{8.15}$$

Equation (8.15) correlates R, which is not directly accessible, with J_v in terms of R_O and the mass transfer coefficient K_m.

8.1.4 Mass transfer coefficient

The Sherwood correlation links the Sherwood number with the Reynolds and the Schmidt dimensionless numbers [21–23]:

$$Sh = A \, Re^{\alpha} \, Sc^{\beta} \tag{8.16}$$

where these dimensionless numbers are for a stirred cell:

$$\left. \begin{array}{l} Sh = K_m r_c / D \\ Re = \rho \omega r_c^2 / \eta \\ Sc = \eta / \rho D \end{array} \right\} \tag{8.17}$$

Here, K_m is the mass transfer coefficient, r_c is the radius of the circular cell, ω is the stirring speed, ρ is the density, η is the viscosity and D is the diffusion coefficient. Isolating the mass transfer coefficient, we obtain

$$K_m = \left\{ A \left[r_c^{2\alpha-1} \right] \left[\left(\frac{\rho}{\eta} \right)^{\alpha-\beta} \right] \left[D^{1-\beta} \right] \right\} \omega^\alpha = B\omega^\alpha \tag{8.18}$$

Then, eq. (8.14) leads to

$$\ln\left(\frac{1-R_0}{R_0} \right) = \ln\left(\frac{1-R}{R} \right) + \frac{J_V}{B\omega^\alpha} \tag{8.19}$$

When J_V increases, c_m also does (and consequently R) until a stable value is achieved. When this maximum retention, R_{max}, is obtained – for high J_V values – eq. (8.19) gives a straight line, as the first summand of the right-hand side would also be constant, $\ln[(1-R_{max})/R_{max}]$, because the true retention coefficient R would be the maximum one, R_{max}. The fit of a set of experimental data, for a constant applied pressure gradient, of $\ln[(1-R_0)/R_0]$ versus J_V/ω^α, for high J_V, to eq. (8.18), will provide B as the inverse of the slope. In addition, R_{max} can be obtained from the ordinate intercept. Once K_m is known (by eq. (8.18)) for each ω, the true retention coefficient can be calculated (by eq. (8.15)) for every J_V. Actually, the mass transfer coefficient evaluated by the method outlined is only strictly valid for high J_V in conditions where convective flux is predominant, leading to an approximately constant c_p/c_m (and R approaching a constant R_{max}). This procedure to find the maximal true retention coefficient and the mass transfer coefficient is called the velocity (ω) variation method.

In the case of a tangential speed sweeping the membrane surface along a given hydraulic channel, eq. (8.19) should be substituted by

$$\ln\left(\frac{1-R_0}{R_0} \right) = \ln\left(\frac{1-R}{R} \right) + \frac{J_V}{B\upsilon^\alpha} \tag{8.20}$$

where B is

$$B = \left\{ A \left[d_{Ch}^{\alpha-1} \right] \left[\left(\frac{\rho}{\eta} \right)^{\alpha-\beta} \right] \left[D^{1-\beta} \right] \right\} \tag{8.21}$$

Because $Sh = K_m \, d_{Ch}/D$ and $Re = \rho \upsilon \, d_{Ch}/\eta$ in this case, d_{Ch} is the equivalent diameter of the channel on the membrane and υ is the linear velocity through this channel.

The textbook by Welty et al. [24] provides a variety of correlations for the Sherwood number for mass transfer to different geometries. Specifically focusing on membrane transport, Mason and Malinauskas [25] studied the different Sherwood correlations to be used. There is a huge number of different correlations [26] proposed for the mass transfer coefficient with turbulent flow [27]. The most endorsed version is the Dittus–Boelter correlation with $\alpha = 0.8$ and $\beta = 1/3$ [28]. Richardson et al. [29] proposed the Dittus–Boelter correlation for turbulent flow considering $A = 0.023$, $\alpha = 0.8$ and $\beta = 0.33$.

8.1.5 True retention and pore size

It could be assumed that all the molecules of the solute should be retained by all the pores of size below that of the solute molecule [30]. Thus, R expressed as per cent would correspond to the percentage of pores with radii below that of the molecule. But this percentage would depend on the applied pressure gradient in a way that we need to know. Here we will follow the lines developed by García-Martín et al. [26].

The Nernst–Planck equation for the transport says that the solute flux is the sum of a diffusion term and a convective one [31, 32] according to

$$\vec{j}_s = -\frac{D'c'}{RT}\left(\frac{d\mu'}{dx}\right) + c'\vec{j}_v \tag{8.22}$$

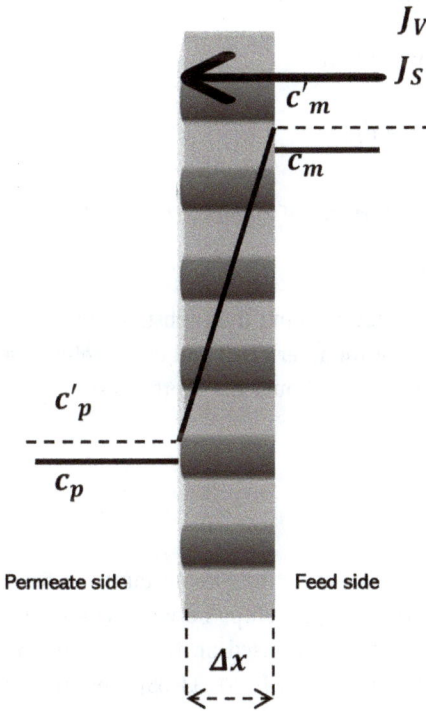

Fig. 8.2: Partition at the interfaces of the membrane.

Here primes refer to the membrane side at the interfaces and the fluxes are defined per unit area of the pores (see Fig. 8.2 for clarity). The chemical potential would include the contribution of electric potential if the solute was charged; otherwise, this would be unnecessary. According to thermodynamics, chemical potential under isothermal conditions is given by

$$\mu' = \mu'^0 (T, p'_0) + \int_{p'_0}^{p'} \bar{V}'(p') dp' + RT \ln \frac{a'}{a'_0} =$$

$$= \mu'^0 (T, p'_0) + \int_{p'_0}^{p'} \bar{V}'(p') dp' + RT \ln \frac{\gamma' c'}{\gamma'_0 c'_0} \tag{8.23}$$

where $\mu'^0 (T, p'_0)$ is the chemical potential in the reference state, a'_0 is the activity in this reference state, p_0 is the corresponding pressure, γ_0 is the activity coefficient, c'_0 is the concentration and $\bar{V}'$ is the molar volume. Then

$$\frac{d\mu'}{dx} = \bar{V}' \frac{dp'}{dx} + RT \left(\frac{d \ln c'}{dx} + \frac{d \ln \gamma'}{dx} \right) \tag{8.24}$$

If we neglect both dp'/dx and $d \ln \gamma'/dx$, then we obtain:

$$j'_S = -\frac{D'}{RT} \left(\frac{dc'}{dx} \right) + c' j'_V \tag{8.25}$$

Assuming that there is equilibrium at both the interfaces of membrane and solution:

$$\mu' = \mu \tag{8.26}$$

By considering eq. (8.23) and assuming diluted solutions (and thus substantially equal molar volumes at the bulk and membrane sides at the interfaces) and equal reference states, we would get $c'_m / c_m = c'_p / c_p = 1$. Certainly, this should not be the case in general but:

$$\frac{c'_m}{c_m} = \frac{c'_p}{c_p} = \phi \tag{8.27}$$

The steric partition coefficient, ϕ, can be evaluated by taking into account the fluid mechanics at the pore–bulk interfaces. Different relationships between ϕ and the ratio of molecular and pore radii $\lambda = r_S/r_p$ can be used based on the shape of the pores. For cylindrical pores $\phi = (1 - \lambda^2)$ is customarily used [8, 11, 33–35]. Note that it has been assumed that the same equilibria appear at both sides of the membrane.

The hindrance coefficients correlate diffusivities and solute speed inside, D' and J'_V, and outside, D and J_V, the membrane:

$$\left. \begin{aligned} K_d &= \frac{D'}{D} \\ K_c &= \frac{J'_V}{J_V} \end{aligned} \right\} \tag{8.28}$$

K_d is the diffusive and K_C is the convective hindrance factors. Consequently:

$$j'_S = -K_d D \frac{dc'}{dx} + K_C c' j_V \tag{8.29}$$

and

$$j'_S = -K_d \phi D \frac{dc}{dx} + K_C \phi c j_V \tag{8.30}$$

But the flux per unit of membrane area j_S and j_V are

$$J_S = j_S \Theta = j'_S \phi \Theta$$
$$J_V = j_V \Theta \tag{8.31}$$

If Θ is the pore area per unit of membrane area (the surface porosity), then eq. (8.30) is

$$\frac{J_S}{\Theta} = -K_d \phi D \frac{dc}{dx} + K_C \phi c \frac{J_V}{\Theta} \tag{8.32}$$

$$J_S = -K_d \phi \Theta D \frac{dc}{dx} + K_C \phi c J_V \tag{8.33}$$

Hindrance factors are related to fluid mechanics for the motion problem of a molecule through a cylindrical or a slit pore of infinite length [36, 37]. The hindrance coefficients depend on the solute to pore–size ratio, $\lambda = r_s/r_p$, like the steric partition coefficient. Several relationships have been proposed to calculate the hindrance for cylindrical as well as slit geometry, considering different locations of a spherical molecule within the radial coordinates of the pore (centre line approximation, cross-sectional average, solvent velocity parabolic profile, etc.). The most comprehensive and acknowledged relationships are due to Bungay and Brenner [38]:

$$\left. \begin{aligned} K_d &= \frac{6\pi}{K_t} \\ K_C &= \left(1 + 2\lambda - \lambda^2\right) \frac{K_s}{2K_t} \end{aligned} \right\} \tag{8.34}$$

$$\left. \begin{aligned} K_t &= \frac{9}{4}\pi^2 \sqrt{2}(1-\lambda)^{-5/2} \left(1 + \sum_{n=1}^{2} a_n(1-\lambda)^n\right) + \sum_{n=0}^{4} a_{n+3}\lambda^n \\ K_s &= \frac{9}{4}\pi^2 \sqrt{2}(1-\lambda)^{-5/2} \left(1 + \sum_{n=1}^{2} b_n(1-\lambda)^n\right) + \sum_{n=0}^{4} b_{n+3}\lambda^n \end{aligned} \right\} \tag{8.35}$$

In Fig. 8.3, the corresponding hindrance factors are represented as a function of $\lambda = r_s/r_p$, according to the Bungay and Brenner parameters given in Tab. 8.1.

Tab. 8.1: Constants in eq. (8.35) for the hindrance coefficients according to Bungay and Brenner [38].

a_1	−1.2167	b_1	0.1167
a_2	1.5336	b_2	−0.0442
a_3	−22.5083	b_3	4.018
a_4	−5.6117	b_4	−3.9788
a_5	−0.3363	b_5	−1.9215
a_6	−1.216	b_6	4.392
a_7	1.647	b_7	5.006

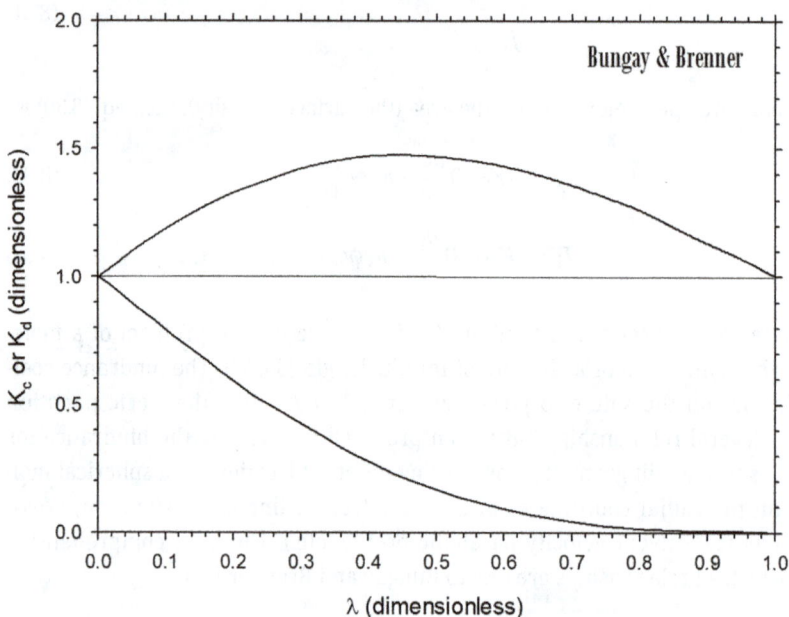

Fig. 8.3: Hindrance factors as a function of λ according to Bungay and Brenner.

A term considering the effect of pressure [39] should be added as follows:

$$\left.\begin{aligned} K'_c &= K_c + K_d \frac{16\lambda^2}{9}(2-\phi)\\ K'_d &= K_d \end{aligned}\right\} \tag{8.36}$$

There have been several attempts to include non-rigid, non-spherical solutes in the theoretical framework of Bungay and Brenner and connected relationships [35] to non-spherical [40] molecules, such as those that resemble a rigid rod [33], and non-rigid molecules, such as chains that can join freely [41] or coiled polymers [42]. Yet, in the case of small enough test solutes as for example in the case of nanofiltration, the molecules of the solutes can be thought of as being very nearly stiff and spherical.

Equation (8.33), considering that $J_S = c_p J_V$, can be written as

$$\frac{dc}{dx} = \frac{1}{\Theta D}\left[\left(\frac{K_C}{K_d}\right)c - \left(\frac{1}{\phi K_d}\right)c_p\right]J_V \tag{8.37}$$

This equation can be integrated from the concentration at the pore entrance c_m to that at the pore exit c_p:

$$R = 1 - \frac{c_p}{c_m} = 1 - \frac{K'_c\phi}{1-(1-K'_c\phi)e^{-Pe}} \tag{8.38}$$

The Peclet number Pe, corresponding to the ratio of convective to diffusive transport, is

$$Pe = \frac{K'_c}{K'_d}\left(\frac{J_V}{D}\right)\frac{\Delta x}{\Theta} \tag{8.39}$$

where $\Delta x/\Theta$ can be evaluated from measurements of pure water permeability P_w that, according to eq. (8.5), is

$$P_w = \frac{J_w^t}{\Delta p} = \frac{n_p\pi r_p^4}{8\eta_w}\left(\frac{1}{\Delta x}\right) = \frac{r_p^2}{8\eta_w}\left(\frac{\Theta}{\Delta x}\right) \tag{8.40}$$

and

$$\frac{\Delta x}{\Theta} = \frac{r_p^2}{8\eta_w P_w} \tag{8.41}$$

Thus, the Peclet number is

$$Pe = \frac{1}{8\eta_w D}\left(\frac{K'_c}{K'_d}\right)\frac{r_p^2 J_V}{P_w} \tag{8.42}$$

Equations (8.38) and (8.39) state that R is a function of J_V with r_p (which is included in K'_c/K'_d and ϕ through $\phi = r_p/r_s$). Therefore, a representation of R as a function of J_V would allow to get r_p for each r_s (each solute).

Note that according to eqs. (8.38) and (8.39):

$$\lim_{J_V\to\infty} R = 1 - K'_c\phi \tag{8.43}$$

But eq. (8.87) can be written as

$$J_S = -K_d\phi\Theta D\frac{\Delta c}{\Delta x} + K_c\phi c J_V \tag{8.44}$$

Now it can be compared with the corresponding equation of Spiegler–Kedem–Katchalsky equations [43–46]:

$$J_S = \bar{c}(1-\sigma)J_V + \omega\Delta\Pi \tag{8.45}$$

With $\Delta\Pi = R\,T\Delta c$ and $c \cong \bar{c}$, we can conclude that

$$\omega = -\phi\frac{K_d D}{RT}\left(\frac{\Theta}{\Delta x}\right) \tag{8.46}$$

and

$$\sigma = 1 - K'_c\phi \tag{8.47}$$

Or $\sigma_p = 1 - K'_c\phi$ to reflect that the effect of pressure has been considered.

Therefore, because we know from the Spiegler–Kedem–Katchalsky that

$$\sigma = \left(\frac{\Delta p}{\Delta\Pi}\right)_{J_V=0} = \left(\frac{\Delta p}{\Delta\Pi}\right)_{J_d=0} \leq 1 \tag{8.48}$$

These definitions could be used to measure $\sigma_p = R_{max}$ (according to eq. (8.19)) as well. Of course [35]

$$\lambda \to 0 \Rightarrow \begin{cases} \begin{rcases} \phi \to 1 \Rightarrow \sigma_p \to 1 - K'_c \\ \sigma_p \to 0 \end{rcases} \Rightarrow K'_c \to 1 \\[2mm] \begin{rcases} \phi \to 1 \Rightarrow \omega \to -\dfrac{K'_d D}{RT}\left(\dfrac{\Theta}{\Delta x}\right) \\[2mm] \omega \to -\dfrac{D}{RT}\left(\dfrac{\Theta}{\Delta x}\right) \end{rcases} \Rightarrow K'_d \to 1 \end{cases} \tag{8.49}$$

$$\lambda \to 1 \Rightarrow \begin{cases} \phi \to 0 \Rightarrow \sigma_p \to 1 \Rightarrow K'_c \to 1 \\ \omega \to 0 \Rightarrow K'_d \to 0 \end{cases} \tag{8.50}$$

These requirements are clearly accomplished by the Bungay and Brenner correlations for the hindrance factors.

Then, if we accept that $\sigma_p = R_{max}$ and the Bungay and Brenner correlation, then we can plot it as a function of $\lambda = r_s/r_p$ as shown in Fig. 8.4 to get the cumulative PSDs that after differentiation would get the actual PSD. Note that $R_{max} > 0.90$ for $\lambda = r_s/r_p > 0.75$ approximately.

The procedure consists of using a certain solute with a given size r_s, because the membrane has a distribution of pore sizes r_p that would give a range of λ. Then the maximal retention or reflection coefficient is measured by the variable speed method, for example, and the cumulative distribution is built and, from it, the differential distribution as well. Of course, several solutes or a polydisperse solute can be used to get

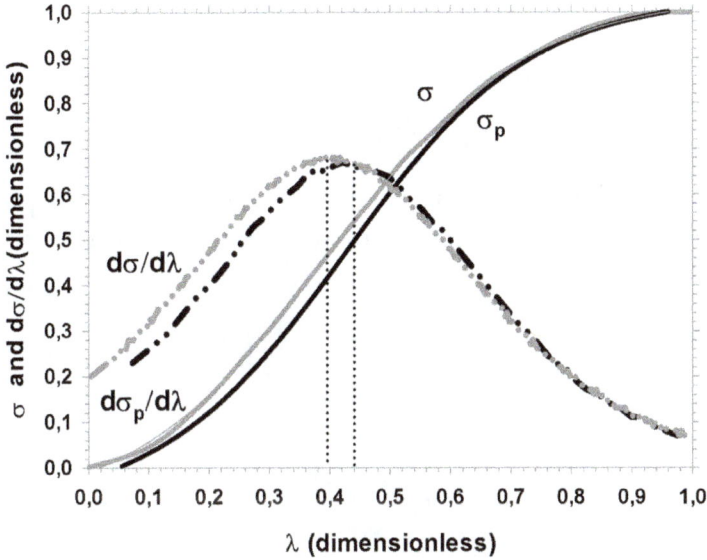

Fig. 8.4: Reflection coefficient with σ_p and without the effect of the pressure gradient $\sigma = 1 - K_C \, \phi$. The corresponding differential distributions are shown too, in both cases for the Bungay and Brenner model.

more data on retention to define better the cumulative distribution by varying λ simultaneously by changing r_s [47].

8.2 Molecular weight cut-off and pore size

A parameter frequently used to characterize the membrane PSD and retention properties is the MWCO. It is arbitrarily defined as the molecular weight in Dalton (g/mol) at which 90% of the macromolecular solute is rejected by the membrane. According to the predictions in Fig. 8.4, the corresponding pore sizes could be over 75% of the pores present.

To measure the MWCO of membranes, dextran, PEG and proteins with different molecular weights are frequently employed. In order to transform molecular weights to sizes, it is mandatory to obtain a relationship between both these parameters. Obviously, there is no universal correlation [48, 49].

The Stokes radius of a macromolecule is correlated with its diffusivity in a solution by using the Stokes–Einstein equation:

$$D = kT6\pi\eta r_S \tag{8.51}$$

This equation gives the diffusivity in terms of viscosity η and the Stokes radius r_S with k being the Boltzmann constant. The diffusivity is also

$$D = 2.5 \times 10^6 kT\eta (M\bar{\eta})^{1/3} \tag{8.52}$$

in terms of the molar mass M and the intrinsic viscosity $\bar{\eta}$. If we merge them, we get

$$r_S = 2.122 \times 10^{-8} M\bar{\eta}^{1/3} \tag{8.53}$$

where r_S is in cm, M is in Dalton and $\bar{\eta}$ is in dL/g. For PEG and a polyethylene oxide (PEO) of known molecular weight, the intrinsic viscosity can be calculated as follows:

$$\bar{\eta} = 4.9 \times 10^{-4} M^{0.672} \tag{8.54}$$

$$\bar{\eta} = 1.192 \times 10^{-4} M^{0.76} \tag{8.55}$$

respectively.
　　Therefore,

$$r_S = 16.73 \times 10^{-10} M^{0.557} \tag{8.56}$$

for PEGs [50, 51], and

$$r_S = 10.44 \times 10^{-10} M^{0.587} \tag{8.57}$$

for PEOs [52].
　　Similarly, other correlations have been proposed for other solutes. In any case, it is still convenient to use solutes as similar as possible to those of interest.

8.2.1 Applications

Numerous parameters must be taken into account when selecting the right MWCO for a given application, including sample concentration, molecular shape, and operating conditions such as temperature, pressure and cross-flow velocity. The transport of molecules would be influenced by other additional factors. For instance, the passage of the solute can be increased when dealing with linear molecules, high transmembrane pressure (TMP) and low sample concentration, whereas it can be decreased by low temperature and membrane fouling. Additionally, it is critical to remember that different manufacturers' qualifying procedures for molecular weight are not necessarily equivalent. There might be certain restrictions on this measurement method because there is no agreed-upon industry standard for the MWCO assessment. It is therefore recommended to choose an MWCO that is at least 2 times smaller than the molecular weight of the solute that should be retained.

8.3 Mean pore size from gas permeability measurements

Among the techniques that can be used to determine information about the porosity and pore size of a membrane sample, we can include those based on determinations of the permeability of our sample to various fluids.

In particular, it is very common to determine the permeability of the membrane to pure water. This measurement is always important in the characterization of a sample because of its high permeability. In applications treating feed solutions in the liquid state mean an important capacity of the membrane to treat large volumes of fluid in the shortest possible time, reducing the operating costs of the process. Practically, any researcher or technologist who starts working with a membrane of unknown properties will first perform a liquid permeation test (with ultrapure water) in order to get an idea of the aqueous permeability of the membrane.

Thus, if we define the aqueous permeability, L_p, as the quotient between the volume flux (J_v) that passes through the membrane in the unit of time and the applied TMP (ΔP):

$$L_p = \frac{J_v}{\Delta P} \tag{8.58}$$

We can apply the convective transport model (Hagen–Poiseuille) to relate this data to the structural characteristics of the membrane. To do this, we will consider a simple model of a membrane formed by a number N of pores, all of equal size, d_p, which cross the surface of the membrane. Thus, the flux would be given by the following expression:

$$J_v = \frac{N\pi d_p^4}{128\,\eta\tau}\frac{\Delta P}{l} \tag{8.59}$$

where η is the viscosity of the permeating liquid (pure water in most cases), τ is the tortuosity of the pores (to include the fact that they may not be perpendicular to the surface) and l is the sample thickness (or pore length).

If we assume that all pores have a circular normal section, then the above equation simplifies to

$$J_v = \frac{\Theta\,d_p^2}{32\,\eta\tau}\frac{\Delta P}{l} \tag{8.60}$$

where Θ is the porosity of the membrane (percentage of the membrane area occupied by pores).

Thus, the permeability would be finally

$$L_p = \frac{\Theta\,d_p^2}{32\,\eta\tau l} \tag{8.61}$$

Therefore, by measuring the aqueous permeability of the membrane and introducing into the above equation reasonable estimates of the porosity and thickness of the membrane (values that, for example, can be obtained from suitable microscopic images), we would obtain the value of the mean pore diameter of the sample analysed.

Obviously, this radius value is only an average value that does not take into account the existence of a possible pore distribution. Even so, this value can be a quite reasonable and interesting indication to determine the possible applications of our membrane.

However, there is a method, also based on permeability measurements, which allows us to determine both parameters, the average pore radius and the porosity of the membrane. This method, known as the gas permeation test [53, 54], consists of performing several permeation measurements of an inert gas through the membrane, at different TMPs. In fact, gas permeation measurements have often been used to check the absence of pinholes in commercial membranes.

From the gas permeance data, the "dusty gas model" (DGM) can be applied to these values [55, 56]. DGM describes the gas transport through the membrane in terms of convective (Poiseuille) and diffusive (Knudsen) flows [57, 58].

Thus, the permeance P (defined as the ratio of the gas flux to the TMP) plotted against the mean gas pressure in the measuring cell (P_m is the average of the readings of two pressure gauges located at the inlet and outlet of the gas in the cell) results in a straight line, where we can evaluate with simplicity its slope (β) and its ordinate at the origin (α):

$$P = \alpha + \beta \cdot P_m \qquad (8.62)$$

Once these two parameters are obtained from the fitting of experimental data, DGM allows us to obtain the average pore size based on the known data, as follows:

$$r_p = \frac{16}{3} \left(\frac{\beta}{\alpha}\right) \left(\frac{8RT}{\pi M}\right)^{0.5} \eta \qquad (8.63)$$

where r_p is the average pore radius, R is the ideal gas constant, T is the temperature, M is the molecular mass of the permeant gas and η is its viscosity.

Likewise, the porosity of the membrane (actually the porosity per unit pore length) is calculated from the following expression:

$$\frac{\Theta}{L} = \frac{8RT\eta}{r_p^2} \beta \qquad (8.64)$$

As we can see, the model does not allow us to get the actual value of the porosity, but the ratio of porosity–membrane thickness. In any case, the porosity value can be known, provided that we can reasonably estimate (again cross sections of SEM images are a fine and precise way to obtain such a value) the thickness of the membrane or rather its active layer.

Certainly, the information obtained in this type of measurement (essentially the average pore size) is not, as we have indicated above, as complete as a porosimetric analysis that allows us to obtain the entire PSDs in a membrane. But it is also true that the mean pore value, when the size distribution is reasonably narrow, is a sufficiently reliable indicator of the filtration capacity of the membrane under study.

This method has been successfully used to characterize polymeric [59, 60] or ceramic [54, 61] membranes, with various configurations, generally using pure nitrogen as permeant gas (a gas that is readily available, cheap and inert with membrane materials). Another advantage is that it allows simple information to be obtained for tubular samples or hollow fibres, for which the most common porosimetric techniques present some technical difficulty.

The technique is generally used in combination with other porosimetric techniques in order to compare or refine the results of these techniques [54, 59].

DGM can be modified to take into account certain non-linearities in the behaviour of the gas during permeation, such as those due to the compressibility of the gas itself or to significant temperature variations. In this way, the model can be used to analyse certain membrane processes such as membrane distillation, where such non-linear conditions frequently occur [62].

References

[1]   CHARMME Network "Harmonization of characterization methodologies for porous membranes",
      EC Contract SMT4-CT 98-7518. Web page: http://www.charmme.livstek.lth.se

[2]   Calvo J.I., Peinador R.I., Prádanos P., Palacio L., Bottino A., Capannelli G., Hernández A. Liquid–liquid
      displacement porometry to estimate the molecular weight cut-off of ultrafiltration membranes.
      Desalination, 268 (2011) 174–181. http://dx.doi.org/10.1016/j.desal.2010.10.016.

[3]   Nagy E. Equations of Mass Transport through a Membrane Layer. Elsevier, Amsterdam,
      The Netherlands, (2019). ISBN: 9780128137222.

[4]   Essig A., Caplan S.R. The use of linear nonequilibrium thermodynamics in the study of renal
      physiology. Am J Physiol, 236(3) (1979 Mar) F211–9. https://doi.org/10.1152/ajprenal.1979.236.3.F211.
      PMID: 371416.

[5]   Jarzyńska M., Pietruszka M. Derivation of Practical Kedem – Katchalsky Equations for Membrane
      Substance Transport. Old and New Concepts of Physics, 5 (2008) 459–474. https://doi.org/10.2478/
      v10005-007-0041-8.

[6]   Rubinstein I., Schur A., Zaltzman B. Artifact of "Breakthrough" osmosis: Comment on the local
      Spiegler-Kedem-Katchalsky equations with constant coefficients. Nature Sci Rep, 11 (2021) 5051.
      https://doi.org/10.1038/srep45168.

[7]   Lebon G., Jou D., Casas-Vázquez J. Understanding Non-equilibrium Thermodynamics: Foundations,
      Applications, Frontiers. Springer–Verlag, Berlin, (2008). ISBN: 978-3-540-74252-4.

[8]   Kondepudi D. Introduction to Modern Thermodynamics. Wiley, Chichester UK, (2008) 333–338.
      ISBN: 978-0-470-01598-8.

[9]   Ferry J.D. Statistical evaluation of sieve constants in ultrafiltration. J Gen Physiol, 20 (1936) 95–104.
      https://doi.org/10.1085/jgp.20.1.95.

[10] Pappenheimer J.R., Renkin E.M., Borrero L.M. Filtration, diffusion and molecular sieving through peripheral capillary membranes. Am J Physiol, 167 (1951) 13–46. https://doi.org/10.1152/ajplegacy. 1951.167.1.13.

[11] Renkin E.M. Filtrate, diffusion and molecular sieving through porous cellulose membranes. J Gen Physiol, 38(2) (1954) 225–243. https://doi.org/10.1085/jgp.38.2.225.

[12] Deen W.M. Hindered transport of large molecules in liquid-filled pores. AIChE J, 33(9) (1987) 1409–1425. https://doi.org/10.1002/aic.690330902.

[13] Green D.M., Antwiler G.D., Mancrief J.W., Dercherd J.F., Popovitch R.P. Measurements of the transmittance coefficient spectrum of Cuprophan and RP 69 membranes: Application to middle molecule removal via ultrafiltration. Trans Am Sot Artif Intern Organs, 22 (1976) 627.

[14] De Balmann H., Nobrega R. The deformation of dextran molecules: Causes and consequences in ultrafiltration. J Membr Sci, 40 (1989) 311–327. https://doi.org/10.1016/S0376-7388(00)81153-9.

[15] Michaels A.S. Analysis and prediction of sieving curves for ultrafiltration membranes: A universal correlation?. Sep Sci Technol, 15 (1980) 1305–1322. https://doi.org/10.1080/01496398008068507.

[16] Munch W.D., Zestar L.P., Anderson J.L. Rejection of polyelectrolytes from microporous membranes. J Membr Sci, 5 (1979) 77–102. https://doi.org/10.1016/S0376-7388(00)80439-1.

[17] Zeman L., Wales M. Steric rejection of polymeric solutes by membranes with uniform pore size distributions. Sep Sci Technol, 16 (1981) 275–290. https://doi.org/10.1080/01496398108068519.

[18] Nobrega R., de Balmann H., Aimar P., Sanchez V. Transfer of dextran through ultrafiltration membranes: A study of retention data analysed by gel permeation chromatography. J Membr Sci, 45 (1989) 17–36. https://doi.org/10.1016/S0376-7388(00)80842-X.

[19] Aimar P., Meireles M., Sanchez V. A contribution to the translation of retention curves into pore size distributions for sieving membranes. J Membr Sci, 54 (1990) 321–338. https://doi.org/10.1016/S0376-7388(00)80618-3.

[20] Sablani S.S., Goosena M.F.A., Al-Belushi R., Wilf M. Concentration polarization in ultrafiltration and reverse osmosis: A critical review. Desalination, 141 (2001) 269–289. https://doi.org/10.1016/j. memsci.2020.118199.

[21] Cussler E.L. Diffusion: Mass Transfer in Fluid Systems. Cambridge U. Press, Cambridge, UK, (1997). ISBN: 9780521871211.

[22] King C.J. Separation Processes. Dover, New York, USA, (2013). ISBN: 978-0-486-49173-8.

[23] Ramachandran P.A. Advanced Transport Phenomena. Analysis, Modeling, And Computations. Cambridge University Press, Cambridge, United Kingdom, (2014). ISBN: 978-0-521-76261-8.

[24] Welty J.R., Rorrer G.L., Foster D.G. Fundamentals of Momentum, Heat, and Mass Transfer, 6th ed., Wiley, New Jersey, (2014). ISBN: 978-0470504819.

[25] Mason E.A., Malinauskas A.P. Gas Transport in Porous Media: The Dusty-gas Model. Elsevier, Amsterdam, (1983).

[26] García-Martín N., Silva V., Carmona F.J., Palacio L., Hernández A., Prádanos P. Pore size analysis from retention of neutral solutes through nanofiltration membranes. The contribution of concentration-polarization. Desalination, 344 (2014) 1–11. https://doi.org/10.1016/j.desal.2014.02.038.

[27] Gekas V., Hallström B. Mass transfer in the membrane concentration polarization layer under turbulent cross flow: I. Critical literature review and adaptation of existing Sherwood correlations to membrane operations. J Membr Sci, 30 (1987) 153–170. https://doi.org/10.1016/S0376-7388(00)81349-6.

[28] Dittus F.W., Boelter L.M.K. Heat transfer in automobile radiators of the tubular type. Int Commun Heat Mass Transfer, 12 (1985) 3–22. https://doi.org/10.1016/0735-1933(85)90003-X.

[29] Richardson J.F., Harker J.H., Backhurst J.R. Coulson & Richardson's Chemical Engineering, 5th ed., Butterworth Heinemann, Oxford Univ. Press, Oxford, UK, (1995). ISBN: 978-0-08-041865-0.

[30] Edward J.T. Molecular volumes and the Stokes–Einstein equation. J Chem Ed, 47(4) (1970) 261–270. https://doi.org/10.1021/ed047p261.

[31] Peppin S.S.L., Nonequilibrium thermodynamics of concentration polarization, PhD Thesis, Univ. Alberta, Canada, (1999).

[32] Déon S., Dutournié P., Bourseau P. Modeling nanofiltration with Nernst–Planck approach and polarization layer. AIChE J, 53 (2007) 1952–1969. https://doi.org/10.1002/aic.1120.

[33] Giddings J.C., Kucera E., Russell C.P., Myers M.N. Statistical theory for the equilibrium distribution of rigid molecules in inert porous networks. Exclusion Chromatography, J Phys Chem, 72 (1968) 4397–4408. https://doi.org/10.1021/j100859a008.

[34] Dechadilok P., Deen W.M. Hindrance factors for diffusion and convection in pores, Ind. Eng Chem Res, 45 (2006) 6953–6959. https://doi.org/10.1021/ie051387n.

[35] Silva V., Prádanos P., Palacio L., Hernández A. Alternative pore hindrance factors: What one should be used for nanofiltration modelization?. Desalination, 245 (2009) 606–613. https://doi.org/10.1016/j.desal.2009.02.026.

[36] Bandini S., Vezzani D. Nanofiltration modeling: The role of dielectric exclusion in membrane characterization. Chem Eng Sci, 58(159) (2003) 3303–3326. https://doi.org/10.1016/S0009-2509(03)00212-4.

[37] Bandini S., Bruni L. Transport Phenomena in Nanofiltration Membranes. In: E. Drioli, L. Giorno (Eds.), Comprehensive Membrane Science and Engineering. Vol. 2 (2010) 67–89. https://doi.org/10.1016/B978-0-08-093250-7.00006-2. ISBN: 978-0-08-093250-7.

[38] Bungay P.M., Brenner H. The motion of a closely fitting sphere in a fluid-filled tube. Int J Multiph Flow, 1 (1973) 25–56. https://doi.org/10.1016/0301-9322(73)90003-7.

[39] Silva V., Prádanos P., Palacio L., Calvo J.I., Hernández A. Relevance of hindrance factors and hydrodynamic pressure gradient in the modelization of the transport of neutral solutes across nanofiltration membranes. Chem Eng J, 149 (2009) 78–86. https://doi.org/10.1016/j.cej.2008.10.002.

[40] Opong W.S., Zidney A.L. Diffusive and convective protein transport through asymmetric membranes. AIChE J, 37 (1991) 1497–1510. https://doi.org/10.1002/aic.690371007.

[41] Priest R.G. Integral-equation method for calculating entropy of confined chains. J Appl Phys, 52 (1981) 5930–5933. https://doi.org/10.1063/1.328521.

[42] Davidson M.G., Deen W.M. Hydrodynamic theory for the hindered transport of flexible macromolecules in porous membranes. J Membr Sci, 35 (1988) 167–192. https://doi.org/10.1016/S0376-7388(00)82442-4.

[43] Kedem O., Katchalsky A. Thermodynamic analysis of the permeability of biological membranes to non-electrolytes. Biochimica et Biophysica Acta, 27 (1958) 229–246. https://doi.org/10.1016/0006-3002(58)90330-5.

[44] Kedem O., Katchalsky A. A Physical Interpretation of the Phenomenological Coefficients of Membrane Permeability. J Gen Physiol, 45(1) (1961) 143–179. https://doi.org/10.1085/jgp.45.1.143.

[45] Katchalsky A., Curran P.F. Nonequilibrium Thermodynamics in Biophysics, Volume 4 in the Series Harvard Books in Biophysics. Harvard Univ. Press, Cambridge, Massachusetts, USA, (1965). ISBN: 978-0674494114.

[46] Kedem O., Spiegler K.S. Thermodynamics of hyperfiltration (reverse osmosis): Criteria for efficient membranes. Desalination, 1(4) (1966) 311–326. https://doi.org/10.1016/S0011-9164(00)80018-1.

[47] Ochoa N.A., Prádanos P., Palacio L., Pagliero C., Marchese J., Hernández A. Pore size distributions based on AFM imaging and retention of multidisperse polymer solutes – Characterisation of polyethersulfone UF membranes with dopes containing different PVP. 187(1–2) (2001) 227–237. https://doi.org/10.1016/S0376-7388(01)00348-9.

[48] Singh S., Khulbe K.C., Matsuura T., Ramamurthy P. Membrane characterization by solute transport and atomic force microscopy. J Membr Sci, 142(1) (1998) 111–127. https://doi.org/10.1016/S0376-7388(97)00329-3.

[49] Gumí T., Valiente M., Khulbe K.C., Palet C., Matsuura T. Characterization of activated composite membranes by solute transport, contact angle measurement, AFM and ESR. J Membr Sci, 212 (2003) 123–134. https://doi.org/10.1016/S0376-7388(02)00490-8.

[50] Hsieh F.U., Matsuura T., Sourirajan S. Reverse osmosis separations of polyethylene glycols in dilute aqueous solutions using porous cellulose acetate membranes. J Appl Polym Sci, 23 (1979) 561–573. https://doi.org/10.1002/app.1979.070230226.

[51] Meireles M., Bessieres A., Rogissart I., Aimar P., Sanchez V. An appropriate molecular size parameter for porous membranes calibration. J Membr Sci, 103 (1995) 105–115. https://doi.org/10.1016/0376-7388(94)00311-L.

[52] Nabi G. Light-scattering studies of aqueous solutions of poly(ethylene oxide). Pakistan J Sci, 20 (1968) 136–140.

[53] Khayet M., Feng C.Y., Khulbe K.C., Matsuura T. Preparation and characterization of polyvinylidene fluoride hollow fiber membranes for ultrafiltration. Polymer, 43 (2002) 3879–3890. https://doi.org/10.1016/S0032-3861(02)00237-9.

[54] Calvo J.I., Bottino A., Capannelli G., Hernández A. Pore size distribution of ceramic UF membranes by liquid–liquid displacement porosimetry. J Membr Sci, 310 (2008) 531–538. https://doi.org/10.1016/j.memsci.2007.11.035.

[55] Evans R.B., Watson G.M., Mason E.A. Gaseous diffusion m porous media at uniform pressure. J Chem Phys, 35 (1961) 2076–2083. https://doi.org/10.1063/1.1732211.

[56] Mason E.A., Wendt R.P., Bresler E.H. Similarity relations (dimensional analysis) for membrane transport. J Membr Sci, 6 (1980) 283–298. https://doi.org/10.1016/S0376-7388(00)82170-5.

[57] Schofield R.W., Fane A.G., Fell C.J.D. Gas and vapour transport through microporous membranes. I Knudsen–Poiseuille transition. J Membr Sci, 53 (1990) 159–171. https://doi.org/10.1016/0376-7388(90)80011-A.

[58] Datta R., Dechapanichkul S., Kim J.S., Fang L.Y., Uehara H. A generalized model for the transport of gases in porous, non-porous, and leaky membranes. I. Application to single gases. J Membr Sci, 75 (1992) 245–263. https://doi.org/10.1016/0376-7388(92)85067-S.

[59] Khayet M., Matsuura T. Determination of surface and bulk pore sizes of flat sheet and hollow–fiber membranes by atomic force microscopy, gas permeation and solute transport methods. Desalination, 158 (2003) 57–64. https://doi.org/10.1016/S0011-9164(03)00433-8.

[60] Deshmukh S.P., Li K. Effect of ethanol composition in water coagulation bath on morphology of PVDF hollow fibre membranes. J Membr Sci, 150 (1998) 78–85. https://doi.org/10.1016/S0376-7388(98)00196-3.

[61] Pakizeh M., Omidkhah M.R., Zarringhalam A. Study of mass transfer through new templated silica membranes prepared by sol–gel method. Int J Hydrogen Energy, 32 (2007) 2032–2042. https://doi.org/10.1016/j.ijhydene.2006.10.004.

[62] Gao F., Chena X.C., Yua G., Asumanaa C. Compressible gases transport through porous membrane: A modified dusty gas model. J Membr Sci, 389 (2011) 200–206. http://doi.org/10.1016/j.memsci.2011.05.064.

Chapter 9
Porosity measurement techniques

When talking about porosity, referring to the structural study of membranes, it is very common to use this term as a synonym for the knowledge of the whole PSD of the given membrane, or at least, of the mean pore size. However, as should have become clear throughout this text, knowledge of the PSD is much more complete in the sense that it allows us to know how many pores there are of each size in our membrane. It is clear that from the PSD, it is easy to determine the porosity of the membrane. But it can also be interesting to know the overall porosity of the filter without the need to acquire complete knowledge of each of the pores in the filter. To do this, we must look at the definition of porosity, a term that, to be precise, has a more rigorous definition that we will consider in this chapter as well as the various existing methods for determining this porosity.

Generally speaking, we can define porosity as the percentage of a porous material occupied by pores. However, we can talk about the percentage of pores in a surface area (surface porosity) or in the total volume of the material (volume porosity). Surface porosity can be determined quickly and reliably by using microscopic images of the membrane surface (provided they are acquired with sufficient definition and adequate illumination) and applying the image analysis methodologies discussed in Chapter 2.

But this surface porosity can be quite different from the total or volumetric porosity, depending on the conformation and tortuosity of the pores within it.

Thus, in this chapter, we will briefly review some of the techniques that can be used to determine the volumetric porosity of our membrane, defined as follows:

$$\Theta = \frac{V_e}{V_t} = \frac{V_t - V_s}{V_t} = 1 - \frac{V_s}{V_t} \tag{9.1}$$

where V_e is the empty volume of the membrane (i.e. the volume occupied by the pores), V_s is the solid volume and V_t is the total volume.

In order to do so, we will go from less to more difficult (and consequently from less to more accurate) in the methods for determining the volume porosity.

Recall that density is obtained as the quotient between the mass and the volume of a material. Thus, when calculating the density of a porous material, we can calculate or define different densities depending on the volume considered.

In the previous figure (Fig. 9.1) we have represented a porous material as formed by a solid matrix (in grey) and a set of pores inside it. We distinguish between open pores (white), those that communicate with the outside of the material (or pass through it, such as the pores in a membrane that give rise to flow through it) and closed pores (cyan), which have no communication with the outside.

https://doi.org/10.1515/9783110792195-009

Porous material

Skeletal volume (V_{sk})

True volume (V_t)

Envelope volume (V_{en})

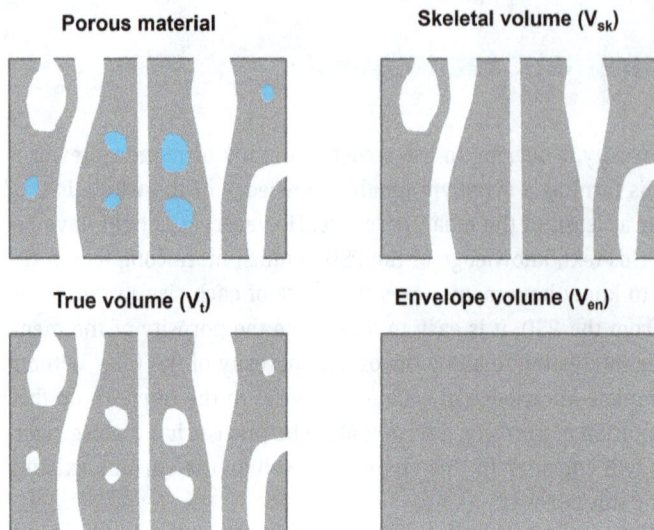

Fig. 9.1: Different volumes that can be defined in a porous material.

Thus, the true volume can be defined as the total or real volume of the solid matrix minus all the pores, both open and closed. On the other hand, if we do not discount the closed pores (which are not accessible to our measurement), the resulting volume will be the skeletal volume. Finally, we speak of envelope volume when we only look at the limits of the solid matrix (including in this volume all the pores, again both open and closed).

Thus, depending on which of these volumes we are determining, we will speak of real density, skeletal density or envelope density (often called bulk density) [1].

9.1 Apparent density method

We will, therefore, begin with the simplest porosity measurement method, the bulk density method. Strictly speaking, this method cannot be considered a measure (not even an indirect or imprecise one) of the porosity, but an estimate based on simple assumptions, which allows us to get an approximate idea of the average porosity of the sample.

For this, we need to know the density of the material (ρ_{true}) of which it is composed, i.e. the density of a non-porous sample composed of the same material as the membrane under analysis. If we also determine the geometric dimensions occupied by the membrane (for a flat membrane its envelope volume, V_{env}, would simply be the area of the membrane multiplied by its thickness), and then weigh on a balance a dry sample of the membrane whose dimensions we have previously determined (m_s), we can finally estimate the porosity with the following expression:

$$\Theta = 1 - \frac{V_{true}}{V_{env}} = 1 - \frac{m_s}{\rho_{true} \, V_{env}} = 1 - \frac{\rho_{env}}{\rho_{true}} \tag{9.2}$$

where ρ_{env} is the so-called bulk density of the membrane (the quotient of its mass m_s and its envelope volume V_{env}).

It should be clear to the readers that this estimation method is quite imprecise in most cases. This is primarily due to the difficulty of knowing the real density of the material from which the membrane is made (in many cases, a commercial membrane is made of several materials of different densities). Also, the determination of the volume of the sample can be quite imprecise, as the thicknesses are usually very small (tens to hundreds of microns).

But the main difficulty lies in the fact that for asymmetric membranes (often composed of two materials or at least two clearly differentiated layers, support and active layer) the method will give an average porosity between the two layers, which can present very remarkable differences in their respective densities.

Actually, this problem for the determination of porosity will be encountered in asymmetric membranes regardless of the measurement method used. In some cases, it will be easy to separate the active layer from the support and then determine the porosity of both parts separately. However, in general, this will not be possible (e.g. for a ceramic tubular membrane, the support and the active layer are really a transition between more or less large granules, which have undergone simultaneous sintering, and it is not possible to separate both parts accurately).

9.2 Pycnometric method

However, the most commonly used method of measuring porosity is based on pycnometric determination (i.e. by using a pycnometer). Looking back in history, the first record of the use of a pycnometer can be found in the story of Archimedes.

As is well known, Archimedes had to resolve the doubts of King Heron of Syracuse about the possible deception in the manufacture of his crown. To do so, he used a very simple method: he filled a container with water up to the rim, inserted the crown and determined the amount of water that overflowed from the container. He then did the same with a quantity of gold whose mass coincided with that of the crown itself. The equality of the two excesses of water would mean equality in the densities of the crown and the uncut gold. Legend says that the goldsmith had deceived the king and that this deception cost him his head, but we gained Archimedes' principle of hydrostatic buoyancy.

Basically, the pycnometric method is based on the same idea. Following Archimedes' principle, we can determine the mass of the dry membrane (m_1), the mass of the pycnometer filled with water (m_2) and finally the mass of the same pycnometer flush with water but now with the membrane inside (m_3):

$$\Theta = 1 - \frac{m_1 + m_2 - m_3}{\rho_1 \, V_{\text{env}}} \tag{9.3}$$

where ρ_1 is the density of the liquid (water in this case), at the working temperature and V_{env} the apparent total volume of the sample (envelope volume).

The pycnometry thus described (using water as the filling liquid) presents several problems. First, water does not easily enter the pores of a membrane, unless the membrane is very hydrophilic. If part of the pores is not filled with water, the porosity obtained from the above calculation will be underestimated.

On the other hand, the volumes of water intruding into the pores will generally be very low, making the method inaccurate. Still, it can be useful to determine a porosity value for symmetrical microfiltration membranes (most of them hydrophilic in nature or rendered hydrophilic by some treatment).

Another possibility is to replace water with a liquid which, having a high contact angle for any material, forces us to fill the pores by pressure. This can be done with mercury, so the mercury porosimeters (whose characteristics and fundamentals were discussed in Chapter 8) include, as a common feature, the possibility of determining a porosity value based on a pycnometric determination.

In this case, we start by knowing the exact volume of the penetrometer (V_p) and then introduce mercury under pressure inside the pores of the membrane (previously placed inside the penetrometer), determining the volume of intruded mercury (V_{int}). Finally, weighing the penetrometer with the sample and the mercury ($m_{p,\text{Hg,s}}$), as well as the mass of the empty penetrometer (m_p) and finally the mass of the dry sample (m_s), we determine Θ by the following expression:

$$\Theta = 1 - \frac{V_{\text{int}}}{V_p - \rho_{\text{Hg}} \left(m_{p,\text{Hg,s}} + m_p - m_s \right)} \tag{9.4}$$

where ρ_{Hg} is the density of mercury.

Mercury pycnometry is obviously more accurate than water pycnometry (the measurement of the volumes in the penetrometer is based on highly accurate conductivity measurements) and on the other hand, we ensure that mercury penetrates into all pores, regardless of whether the material is hydrophilic or hydrophobic. Furthermore, the porosity value is a simple calculation that we can easily perform as a last step in using HgP to determine the PSD of the membrane.

However, HgP has certain limitations [2] that should be accounted for. In particular, mercury is a highly polluting substance and strict environmental management rules must be observed. Also, the sample analysed with HgP is destroyed and cannot be used later.

There is another pycnometric technique based on a gas instead of a liquid, proposed in 1971 by Feldman, for the study of cements [3]. This gas is generally helium, as it is very poorly absorbent in any type of material. The resulting technique is called helium pycnometry, which has reached a commercial development and a multitude of applications in the study of porous materials of all types [4–7].

Helium atoms, practically spherical, and with a diameter between 0.20 and 0.23 nm, do not present Van der Waals forces, which translates into practically no adsorption capacity on the surface of solids. As a result, helium atoms can penetrate pores smaller than 0.3 nm [8], which makes this gas particularly suitable for calculating the true volume of the solid skeleton of porous materials.

In helium pycnometry, the pycnometer (of stable and calibrated volume) is filled with a gaseous fluid of known density (He that behaves as an ideal gas). When the sample whose density or specific weight is to be known is subsequently introduced, it displaces a certain volume of fluid equivalent to the volume of the sample under study. Once the mass or weight of the sample is known exactly, it is possible to calculate its density or specific gravity, regardless of whether the morphology of the sample is more or less regular or irregular, or whether it is in block or powder form.

Both the mercury and helium pycnometric methods actually determine the skeleton volume of the sample, as they are unable to enter the closed pores (in order to determine the real or true volume). However, helium pycnometry can penetrate pores where mercury cannot reach, so it will usually give a higher value of porosity [9]. In any case, none of these pycnometric techniques can determine the porosity due to all pores (open and closed).

In any case, the technique has acquired the status of standard ASTM/D/5550–94: "Standard test method for specific gravity of soil solids by gas pycnometer", and there are several commercial devices on the market (AccuPyc II 1345 from Micromeritics or G-DenPyc 2900 from GoldAPP Inst.) that allow the determination of porosity and specific gravity in a simple way.

The authors proposed in 1998 [10], a porosity determination method for membranes based on the measurement of the pressure resulting from imposing a constant gas flow in the vessel containing the sample. The method is essentially an adaptation of helium pycnometry (it uses, in fact, this same gas because of its ideal behaviour and low adsorption capacity) and is designed to be performed with conventional GAD equipment.

From the equation of state of ideal gases and assuming, as we have said, a constant gas flow, we arrive at the following expression:

$$P = P_i + \left(\frac{k}{V_g}\right) t \tag{9.5}$$

where P_i and k are constants to be determined and V_g is the volume occupied by the helium in the container, also exactly unknown. By means of three consecutive measurements: one with the empty container, another with a non-porous material of known volume and finally with the sample to be determined, the constants are obtained and the porosity is determined, with a fair degree of accuracy. The following figure (Fig. 9.2) compares various porosity measurements using some of the techniques indicated here, as well as other indirect measurements, with the value obtained by the helium flow method mentioned above.

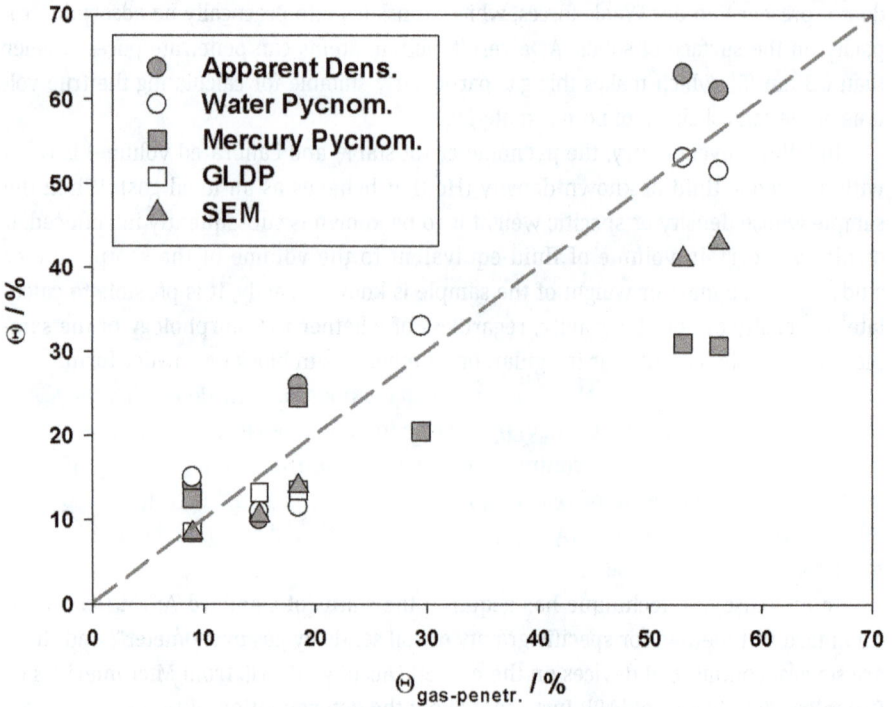

Fig. 9.2: Comparison of surface and volume porosities for several membranes measured by means of several porosity determination techniques (adapted from [10]).

Thus, the air–liquid displacement data are obtained from the GLDP technique by summing the porosity contribution of all the pores present in the PSD. Similarly, SEM-CIA data refer to the result of image analysis determination from SEM images of the membrane surface. Obviously, the resulting porosity in this case will be a surface porosity and not a volume porosity, being only comparable for materials with very regular pores.

It is worth noting that an indirect method (not aimed at porosity determinations), such as the calculation obtained from the PSD coming from a GLDP experiment, could lead to reasonable porosity estimations, at least, for those membranes where cylindrical capillary pores do not deviate too far from real porous structure.

References

[1]   Orsolini P., Michen B., Huch A., Tingaut P., Caseri W.R., Zimmermann T. Characterization of Pores in Dense Nanopapers and Nanofibrillated Cellulose Membranes: A Critical Assessment of Established Methods. ACS Appl Mater Interfaces, 7(46) (2015) 25884–25897. https://doi.org/10.1021/acsami.5b08308.

[2]   Dos Santos Macedo R., Ulsen C., Jacomo A.Y., Figueiredo P.O., Nery G.P., Uliana D., Müller A. Pycnometry for assessing porosity of fine recycled aggregates. Constr Build Mater, 308 (2021) 125091. https://doi.org/10.1016/j.conbuildmat.2021.125091.

[3]   Feldman R.F. The flow of Helium into the interlayer spaces of hydrated Portland cement paste. Cem Concr Res, 1 (1971) 285–300. https://doi.org/10.1016/0008-8846(71)90004-4.

[4]   Ayral A., Phalippou J., Woignier T. Skeletal density of silica aerogels determined by helium pycnometry. J Mater Sci, 27 (1992) 1166–1170. https://doi.org/10.1007/BF01142014.

[5]   Semel F.J., Lados D.A. Porosity analysis of PM materials by helium pycnometry. Powder Metall, 49 (2006) 173–182. https://doi.org/10.1179/174329006X95347.

[6]   Yang X., Sun Z., Shui L., Ji Y. Characterization of the absolute volume change of cement pastes in early–age hydration process based on helium pycnometry. Constr Build Mater, 142 (2017) 490–498. https://doi.org/10.1016/j.conbuildmat.2017.03.108.

[7]   Zheng H., Yang F., Guo Q., Pan S., Jiang S., Wang H. Multi-scale pore structure, pore network and pore connectivity of tight shale oil reservoir from Triassic Yanchang Formation, Ordos Basin. J Pet Sci Eng, 212 (2022) 110283. https://doi.org/10.1016/j.petrol.2022.110283.

[8]   Van Krevelan D.W. Coal. Typology-Chemistry-Physics-Constitution. Elsevier, Amsterdam, Netherlands, (1961). ISBN: 9780444895868.

[9]   Beaudoin J.J. Porosity measurement of some hydrated cementitious systems by high pressure mercury intrusion-microstructural limitations. Cement & Concrete Res, 9 (1979) 771–781. https://doi.org/10.1016/0008-8846(79)90073-5.

[10] Palacio L., Prádanos P., Calvo J.I., Hernández A. Porosity measurements by a gas penetration method and other techniques applied to membrane characterization. Thin Solid Films, 348 (1999) 22–29. https://doi.org/10.1016/S0040-6090(99)00197-2.

Chapter 10
Conclusions

Throughout the previous chapters we have described and analysed various porosimetric techniques currently available for determining the number and size of pores present in a membrane.

In this review, which we believe is reasonably exhaustive, we have started with the microscopic techniques, following the simple maxim "you cannot measure what you cannot see". We have devoted Chapter 2 to various microscopic techniques that have been used in the analysis of membranes and other types of porous surfaces.

In Chapter 3, we have considered the techniques that we have included under the name of spectroscopic, where we have included positron annihilation lifetime spectroscopy (PALS), synchrotron radiation (SR) or ellipsometry.

The rest of the techniques analysed in this book base the interpretation of the experimental data on a set of three equations, well known in the scientific literature and related to each other, such as the Kelvin, Gibbs–Thomson and Young–Laplace equations. These three equations, as we have mentioned, have a common base and have been developed to treat from the thermodynamic point of view, diverse behaviours of curved interfaces. For this reason, we have dedicated Chapter 4 to the thermodynamic basis of the three equations.

Chapters 5–7 are devoted to analysing the different techniques that can be deduced from the three equations mentioned above, with a detailed analysis of them including the simplifications and assumptions that must be made to use each one of them. Likewise, of all the techniques we have sought to comment briefly on their advantages and disadvantages as well as the particularities and manufacturers of the different commercial equipment that can be used (when such equipment exists).

Finally, we have dedicated Chapter 8 to other porosimetric techniques that cannot be included in the previous sections, and finally, Chapter 9 has been dedicated to those measurements in which we are only interested in the total porosity of the sample and not in the complete pore size distribution (PSD) of the sample.

However, the reader may miss some of the other porosimetric methods that have been proposed in the scientific literature by hard-working researchers. Nevertheless, the ones collected and analysed here are the main porosimetric methods among which a researcher can choose when he begins to face the structural knowledge of his membranes or of those others, commercial or modified, that are part of his study.

The truth is that when deciding which porosimetric technique is the most suitable, the novice researcher or the company that wishes to start a filter manufacturing process for a specific application may have reasonable doubts in all these techniques. Which is the most convenient technique and which one will give me more reliable, faster, more accurate results?

https://doi.org/10.1515/9783110792195-010

This is a question to which, unfortunately (or not so much, it depends), there is no single answer:

- A first decision will come from the laboratory itself. Certainly, if a researcher works in a laboratory where a porosimetric method already exists and has been used previously, it seems reasonable for him to use the technique for his own studies.
- Another important reason may be the type of membranes we want to study. Thus, for MF membranes, for which the use of the gas–liquid displacement porosimetry (GLDP) technique is already standardized, this seems a suitable choice.
- Another case may be the study of membranes whose pores closely resemble the capillary pore model (perpendicular to the surface). Although these membranes are not the most extensively marketed membranes, we can find the case of track-etched membranes (Nuclepore or Cyclopore are the best-known examples) or those obtained by anodic deposition (the most obvious example would be Anopore, made of aluminium oxide and working range MF). In these cases, a surface study can be easily extrapolated to the behaviour of the pore along the entire membrane so that microscopic techniques (SEM would be sufficient for the range of pores in question) will give us a very valid estimate of the complete PSD of the filter.
- Likewise, the range of pores in the sample (or rather expected, if we do not have prior knowledge of them) is a major reason to choose one technique over the other. Techniques that give accurate results for MF filters may be quite inaccurate (or not accessible at all) for UF membranes.

As a general starting point, we could say that a good porosimetric characterization combines information from several complementary techniques so that the advantages and disadvantages of these techniques are adequately compensated. Thus, it is very common to use a microscopic technique for a superficial observation of the sample (perhaps together with the visualization of a cross section to determine how the morphology of the pores varies along the filter), accompanied by a porosimetric technique that gives us the PSD of the complete filter or preferably of its active layer.

In any case, in order to make a decision on the most appropriate porosimetric technique or techniques in each case, it is worth bearing in mind the advantages and disadvantages of different existing techniques. Although we have already discussed most of these advantages in the chapters dedicated to each technique, this is a good time to make a comparative analysis that will allow us to have a clearer vision of the suitability of each technique for our particular study. We will devote the rest of this chapter for this comparison.

10.1 Comparison between different techniques

One of the most important points to consider when selecting the most suitable porosimetry to characterize the membranes we are interested in is the range of pores that the membrane will presumably present. Figure 10.1 presents most of the techniques reviewed in this book along with their various ranges of detectable pore sizes.

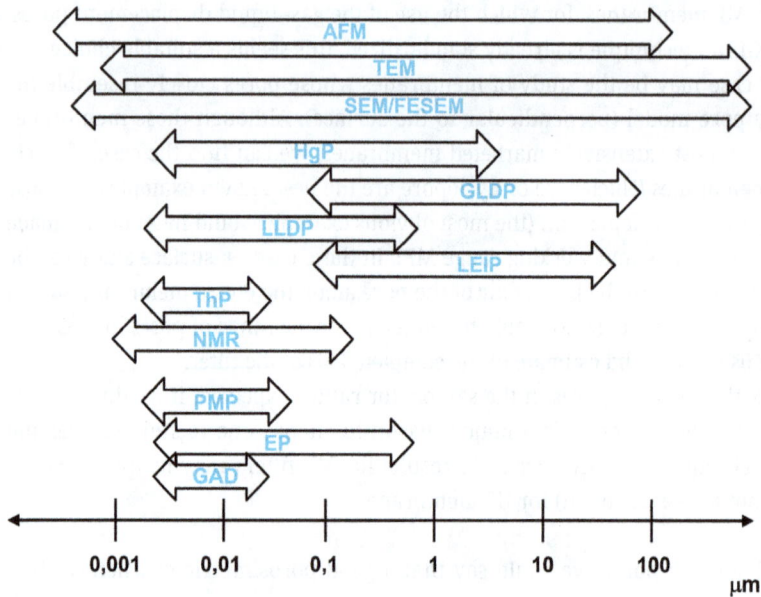

Fig. 10.1: Range of pores to be detectable with the different porosimetries studied in this book (spectroscopic techniques are not included).

Likewise, Fig. 10.2 shows the ranges typically found in the membranes used in the main membrane processes under pressure gradient, which, as can be seen, cover the whole range of pore sizes possible in a filter.

Tung et al. [1], in a review of characterization methods, or Zamani et al. [2], in a research on evapoporometry (EP), published in 2017, compared the advantages and disadvantages of different porosimetric methods that can be used to determine the PSD of membrane filters. In the following paragraphs we will discuss the features presented in both the works based on the conclusions arising from this book. Moreover we'll complete their discussion with some methods not considered by those authors that, as we have seen across these chapters, can also be used to determine the porosity and pore size characteristics of membranes.

Proteins — Divalent Ions
- Bacteria — Viruses — Sugars
- Suspended — Colloids — Antibiotics — Monovalent
  particles — Polypeptides — Dyes — Ions

MF   **UF**   **NF**   **R**
                                                O

10 – 0.1   0.1 – 0.01   0.01 – 0.001 μm 0.001 – 0.0001 μm
μm   μm

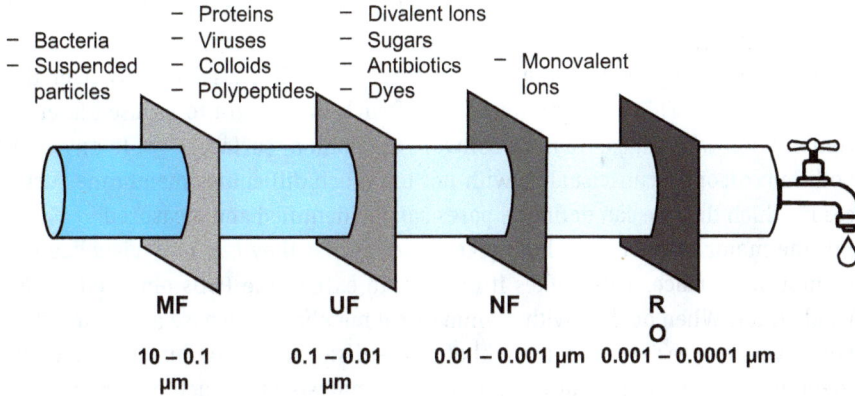

Fig. 10.2: Range of particles to be retained in the most frequently used membrane processes (those governed by a pressure difference across the membrane) along with their characteristic pore sizes.

10.1.1 Microscopic methods

Apart from the particular features of the different microscopic techniques, which were conveniently discussed in Chapter 2, including the advantages and disadvantages of the different working modes, here we will like to provide a brief general overview of the possibilities and benefits of microscopic methods when searching for information about the pore characteristics of our membranes.

The first distinction when discussing the benefits of porosimetric techniques has to do with the difference between direct and indirect techniques. As we have been discussing in these chapters, the so-called direct techniques refer to microscopic methods, which are the only ones that, depending on the existing pore size range, are able to visualize them directly:

– Both electron microscopies (mainly SEM but also TEM) and atomic force microscopy (AFM) allow pores to be visualized in ranges from the MF to the NF. In fact AFM and even a well-operated SEM can visualize the interstices of RO membranes, although in this case it is difficult to speak of pores per se. Both techniques are very precise and can be used to study the layers deposited onto the membrane surface and/or the inner pores because of fouling (especially biofouling). If we are asked to make a distinction, it could be that in the case of AFM, the sample preparation is nil, as opposed to SEM, which usually requires working with metallic coatings or carbon replicas. However, current FESEM microscopes eliminate the need for such sample preparation. Finally, AFM has an important advantage over SEM in that it can work both in a gaseous environment and with liquid-embedded samples (even at the cost of losing some resolution). This can be particularly important when analysing samples with biological content (viruses or bacteria retained in the membrane). On the other side, SEM is able to envisage a large membrane area with maximum resolution, while for AFM, too large scanning areas usually mean reduced resolution when tip passes rapidly over the pore entrances. This could be important if such images are to be used later in combination with Image Analysis software to obtain a "surface" PSD. In any case, both microscopies can visualize with not too much difficulties membrane surfaces in which that we can define as pores can be identified and measured.

– But the major drawback of both techniques is that they can only visualize the membrane surface. This makes it difficult to extend the PSDs obtained to the whole filter. When dealing with asymmetrical membranes, whose active and support layers are easily separable without pore distortion at the junction between them, we can make a separate microscopic analysis of the active and support layers, resulting in a more complete knowledge of the interior of the membrane. Obtaining cross sections of the sample (by cryogenic fracture, for example) will help us to visualize the interior structure of the sample, but even then, we can never be sure how far the validity of the PSD obtained from the surface analysis extends. This is why we commented in the previous pages that, for a good structural characterization of the membrane, it is very convenient (and very visual on the other hand) to combine the information obtained from an indirect porosimetric technique with a good microscopic image of the membrane, completed with a good image analysis.

– Finally, a last comment is about the subjectivity of the image analysis of the micropictures obtained from microscopic methods. Usually, the definition of what is a pore and what is not (decision that you should take when using an image analysis software) is non-univocal and then subjected to interpretation. The fact is clearly true at the pore borders. Only when high-quality and high-definition images have been obtained, the subjectivity margin of the operator is clearly minimized. So, as commented in Chapter 2, we can conclude that it would be very convenient to define having a common and universally accepted standard about the use of image analysis for membrane characterization, which unfortunately is not available at present [3].

10.1.2 Kelvin equation-based methods

Moving on to indirect porosimetric techniques, we have already seen that most of these techniques are based on the use of three equations that refer to different aspects of the behaviour of curved interfaces. Starting with the Kelvin equation, we have reviewed three porosimetric techniques based on this expression: GAD (gas adsorption–desorption), PmP (permporometry) and EP. Let us discuss their respective advantages and disadvantages:

10.1.2.1 Gas adsorption–desorption

– Starting with GAD, it is applicable to the study of membranes from MF to NF, with minimal sample preparation (drying and degassing), does not require a large laboratory and can analyse relatively large samples. In fact, when dealing with asymmetric membranes, the GAD technique needs sufficient pore volume of the active layer to distinguish its contribution from that of the support (with much larger pores). This requires either using a large amount of sample or separating the active layer from the support, whenever possible.

– One of its disadvantages is that, as an indirect technique, it is necessary to assume a certain pore geometry in order to apply the Kelvin equation, a geometry whose pertinence cannot always be guaranteed, and which presents particular problems when dealing with membranes with strongly interconnected pores. The measurement of the pressure and volume of adsorbed gas, essential for correct results, has a limited precision in most commercial equipment, and the determination becomes less precise as the pore size to be determined increases. As mentioned in the previous paragraph, GAD needs sufficient pore volume to obtain a correct PSD, which presents problems for low porosity membranes. The results of GAD experiments are also influenced by a correct estimation of the t-layer thickness. Finally, the technique is not suitable for discriminating structural changes in the membrane due to fouling or biofouling.

10.1.2.2 Permporometry

– As for PmP, we have a flow-based technique, which makes the technique interesting as it is more like normal membrane operation. It requires little laboratory space, minimal sample preparation and can determine continuous pores both in the UF and MF range.

– On the downside, it must also assume a pore geometry. The accuracy of the measurements (flow and partial pressure) and the control of their values can be limited, with the error increasing with pore diameter. It is also inaccurate with low-porosity samples and sensitive to the existence of interconnected pores. The results are also influenced, as in the case of GAD, by a correct estimation of the t-layer thickness. Finally, the technique is not suitable for discriminating structural changes in the membrane due to fouling or biofouling.

10.1.2.3 Evapoporometry

– Finally, for the EP technique, it is characterized by mass-based measurements and does not need to assume pore geometry, and gravimetric measurements can be highly accurate. On the other hand, it allows the study of MF and UF membranes, with inexpensive equipment (consisting of elements that normally already exist in the laboratory or are easy to replicate), which occupies little space, requires minimal sample preparation, works in ambient conditions and has a simple experimental procedure. It is also capable of studying fouling and biofouling.

– Among its drawbacks are its accuracy decreases with pore size (larger pores are measured less accurately, being considered inadequate above 150 nm, which greatly limits its application in MF), it is unable to isolate continuous pores, it needs corrections to deal with asymmetric membranes (corrections that imply assuming a pore geometry model) and it has problems with low porosity membranes or interconnected pores. Finally, it also needs corrections to account for the thickness of the adsorbed t-layer. Mention should also be made of the long duration of an EP experiment (typically 10–14 h, which can distort the results obtained if the environmental conditions are not carefully controlled).

10.1.3 Gibbs–Thomson equation-based methods

Based on the Gibbs–Thomson equation, we have analysed in this text two techniques: nuclear magnetic resonance (NMR)-based porosimetry and thermoporometry (ThP).

10.1.3.1 Thermoporometry

– With respect to the ThP technique, it is a technique based on mass measurements, suitable for the UF and MF ranges, with a small amount of equipment and minimal sample preparation.

– Its disadvantages include the fact that it is an indirect technique that must assume a pore geometry in the interpretation of the results. Its accuracy is limited by the precision in the measurement of the heat involved, an error that increases as the pores become larger. It has problems in determining the information for low porosity samples and is not suitable for assessing the effect of biofouling. Finally, it requires a correction of the results to take into account the thickness of the unfrozen layer in contact with the pore walls.

10.1.3.2 Nuclear magnetic resonance

– The NMR technique has the advantage of being non-destructive, with a measurement range between 2 nm and 1 μm, depending on the absorber used. NMR cryoporometry has an advantage over DSC-ThP, as the latter measures transient heat fluxes and therefore has a minimum speed at which the measurement can be made. The NMR method returns an absolute signal that can be measured arbitrarily slowly, or in discrete steps, to obtain a better resolution or signal-to-noise ratio. On the other hand, the method is not too sensitive to detect interconnected pores or the existence of pore constrictions, which otherwise can be an advantage in samples of complex structure. Finally, NMR applications include studies of silica gels, bones, cements, rocks and many other porous materials.

– Among the disadvantages of the technique, we would have to highlight that,
   given that it measures the surface-to-volume ratio of the sample, the technique
   needs to make prior assumptions about the geometry of the sample. Also, commercial NMR equipment is quite expensive, and a specific probe is needed to
   apply cryoporosimetry, which is an excessive expense for a characterization laboratory. However, many university departments have NMR equipment purchased
   for other applications that can be used for this technique at a reasonable cost. Finally, the typical duration of an experiment is 24 h, which is excessive. However,
   the major disadvantage of the technique is its current lack of development and research compared to other more firmly established porosimetry techniques [4–6].

10.1.4 Young–Laplace equation-based methods

Finally, based on the Young–Laplace equation, we have revised in this text the mercury
porosimetry (HgP) and the two liquid displacement techniques (GLDP and liquid–liquid
displacement porosimetry (LLDP)) with their variants (liquid entry pressure (LEP) and
liquid intrusion/extrusion porosimetry (LEIP)).

10.1.4.1 Mercury porosimetry

– Starting with HgP, possibly its greatest advantage comes from a wide range of
   pores it cover. In most commercial equipment performing this technique, the
   range of pores that can be analysed (combining both high- and low-pressure
   ports) varies from about 1 mm to nearly 3 nm. This requires, at the lowest limit of
   the range, the use of pressures up to 60,000 psi (414 MPa). Obviously, this pressure may be too high for the membrane material, which may be deformed, and
   the results distorted. However, this problem is not encountered with ceramic-
   type membranes, which can withstand these pressures without any problems
   other than rupture or breaking, which does not affect their internal structure.
   Other notable advantages of the technique are its proven speed and accuracy. In
   addition, the technique provides important information about the sample, apart
   from its PSD, such as total pore volume, total pore surface area and sample densities, and then porosities (volume and skeleton values).
– In the disadvantages or drawbacks, we have already mentioned the high pressures that HgP may need to reach small pores. But it is also difficult to extract the
   information corresponding to the PSD of the active layer in the case of asymmetric membranes. The problem is that the total pore volume in the active layer, by
   its own desirable characteristics, is much smaller than that of the support so the
   PSD of the pores in the active layer will be a small contribution to the total PSD,
   sometimes difficult to separate. Whenever possible, the solution, as is also recommended in the case of GAD, is to separate the active layer and the support and

analyse them separately. If this is not possible, make sure that enough of the active layer is included in the penetrometer so that the resulting PSD is not masked in the error range of the equipment. Like all Young–Laplace-based techniques, it needs a geometrical model of the pores to interpret the results. Another point to note is that the technique, given its working method, is not able to discriminate between pores that contribute to the flow and closed pores (inkwell type). The method will count all of them and it will be up to us (by means of complementary permeability measurements) to check what percentage of the measured pores is open to flow. Last but not least, mercury is a very polluting liquid, which must be handled with the utmost caution in the laboratory. This means that samples, once analysed by HgP, must be discarded with the appropriate control, and this technique cannot be used for online production line control or pre-use testing.

10.1.4.2 Gas–liquid displacement porosimetry/liquid–liquid displacement porosimetry

–   Liquid displacement techniques, as thoroughly discussed in Chapter 7, share many of their advantages and disadvantages since they are based on the same operating principle and use the same assumptions to interpret their results. Thus, one of their major advantages is that they provide a flow-based PSD, closer to the usual use of membranes. Remember that, in most membrane applications, driving force is still a transmembrane pressure which leads to convective flow between both membrane sides. Regarding variable section pores, the information obtained always refers to the narrowest area of the pore, which is the one that governs the retention of solutes. This is another point favourable to the use of these techniques. The working range goes from conventional filters and MF (GLDP) to UF and the upper limit of NF membranes (LLDP). Sample preparation is simple, and measurement is remarkably fast (from a few minutes in the case of GLDP to about 1 h for LLDP). In the case of GLDP it is a widely used and recommended technique in MF membranes, with a wide range of commercial equipment available to the researcher. This is not the case for LLDP, for which there is currently little commercial equipment capable of performing it with sufficient rigour (possibly the equipment marketed by IFTS is the most suitable and reliable of the existing commercial ones). The cost of this commercial equipment is certainly high, although reliable versions can be assembled in the laboratory itself at a much lower cost.
–   Among the disadvantages attributable to both techniques is the fact that they are indirect techniques, in which a pore geometry needs to be assumed in order to interpret the results. However, as discussed in Chapter 7, if we limit ourselves to the PSD obtained directly from the measured fluxes (without converting these distributions into absolute numbers of pores in each size), this limitation is very diluted and can be perfectly ignored. In any case, a shape factor can be included in the Young–Laplace equation but, in our opinion, as we discussed in Chapter 7,

this factor is based on a detailed knowledge of the actual pore structure. And, in most cases, such previous knowledge is not possible or just difficult to convert into an adequate number. In our opinion, the best procedure (except if reliable information is previously known about the porous structure) consists of assuming no shape factor ($K = 1$) which avoids subjectivity.

– The required pressures are high when approaching the range between UF and NF (in the order of 50 bar) but generally these membranes are manufactured to adequately withstand these pressures which are similar to the usual working pressures. There is a certain discrepancy in the results when the membrane has a high proportion of interconnected pores, although this is a disadvantage common to practically all porometries. Finally, changes in PSD induced by irreversible fouling or aggressive cleaning can be studied with both techniques [7–9], although it is not able to discriminate the existence of biofouling. In commercial equipment, membranes are analysed with flat samples (extracted from their corresponding cartridges) or with shortened tubular samples or small bundles of fibres. This can lead to some differences between these short-scale results and the results of real modules. Recently [10] the LLDP technique has been improved to adapt it to the study of membrane cartridges (of small dimensions) which can be studied directly without the need to remove the membrane from its interior. This makes it possible to envisage the possibility of periodic inline monitoring of commercial cartridges in their own working situation. A final negative point of these techniques, especially LLDP, is the difficulty of wetting the membrane pores. It should be remembered that a correct and complete wetting of the inside of the pores is necessary to obtain reliable results. While in the case of MF membranes, where GLDP is applied conventionally, wetting is largely guaranteed with the various liquids used commercially, in the case of UF membrane pores, especially when approaching the nanometre limit, it can be difficult to ensure. Performing this wetting under vacuum can significantly improve the results, and in case you are not sure that the smallest pores are properly wetted, you can pre-filter the wetting liquid under pressure so that in the end, you can be sure that all pores are filled with the wetting liquid.

10.1.4.3 Liquid entry pressure/liquid intrusion/extrusion porosimetry

Techniques derived from the bubble point such as LEP and LEIP are, in their operational principle, quite similar to the aforementioned techniques, with which they largely share their advantages and difficulties. Although the LEIP technique is appraised to be reliable in determining the PSD of membranes, obviously if we manage to find a capillary barrier suitable for the pore size of our membrane under analysis, it does not seem that the technique has found great diffusion among researchers (apart from the circle of researchers directly related to the introducer of the technique himself).

On the other hand, the LEP technique has been mostly used to test the hydrophobicity/hydrophilicity of filters as well as to determine contact angles with different liquids [11], and it does not seem easy to adapt it to obtain the complete PSD of commercial membranes.

10.1.5 Spectroscopic methods

In this review of the various advantages of each of the porosimetric techniques included in this text, we have left aside the spectroscopic techniques that we analysed in Chapter 3. Both PALS and, to a lesser extent, SR and ELLP, are techniques that offer interesting information about our membranes and that, as we have seen, can also be used to obtain their PSD. However, none of the cases is obtaining specific information on pore size, the most outstanding feature of these techniques, so they are only considered when complementary or comparative information to that obtained by other porosimetric methods is sought, for example, to validate the reliability of one of these methods.

10.2 Literature review

In this section we are going to make a simple (non-exhaustive) compilation of the number of articles published on each of the techniques presented in the book. Although some techniques have been around longer than others, this review can give some idea of the degree of scientific interest that each technique has generated.

Throughout this book, we have included several hundred references to support what is contained therein. In particular, for each technique, we have referenced the first works that proposed, developed and helped to extend it. Additionally, numerous examples of later applications in the field of characterization of all types of membranes are also given. All these works and many others not referenced can be found in the main scientific publication databases such as the Web of Science (from Clarivate) or Scopus (from Elsevier). These databases and several others have been developed and trained in the last years and they include a broad range of publications along with several features of searching.

In these pages we will summarize the type of information that can be found in both databases related with porosimetric techniques.

Thus, a WoS search allows us to get a reasonable idea of the research carried out on these porosimetric techniques over the years. Figure 10.3 shows the results of articles and research reviews that WoS refines in a search with the terms porometry or porosimetry, as topics. Overall, 1,001 articles or works appear over these years. It is noteworthy that the first works appear at the beginning of the 1980s. It is true that some techniques (e.g. GAD, HgP or those based on the bubble point) had already laid their foundations much earlier (remember, for example, that mercury porosimetry was proposed by Washburn in 1921, developed by Ritter & Drake in 1945 and applied for the first time to membranes by Honold & Skau in 1954). However, these techniques did not receive the joint name of porosimetry in those years, probably because the importance of the determination of the PSD of synthetic membranes was not yet clear.

The appearance of Loeb and Sourirajan's synthetic asymmetric cellulose acetate membrane in 1960 prompted the search for new membranes and industrial applications for them. Although it is true that in Loeb's membrane (which quickly found an outstanding application in seawater desalination by reverse osmosis) it does not make much sense to talk about pore size, as more membranes designed for different applications were obtained, it became necessary to know the PSD of these filters in order to be able to adjust their properties and potential in different applications. It can be considered that the first work in which porosimetry is specifically mentioned is a brief note published in the *Colloid Journal of the USSR* in 1982 by Dytnerskii et al. [12], in which the porous characteristics of membranes obtained by plasma polymerization are analysed.

In any case, after this first milestone, the number of publications associated with the term porosimetry has grown steadily over the years, reflecting the growing importance of porosimetric characterization in parallel to the rapid growth of membrane science and technology.

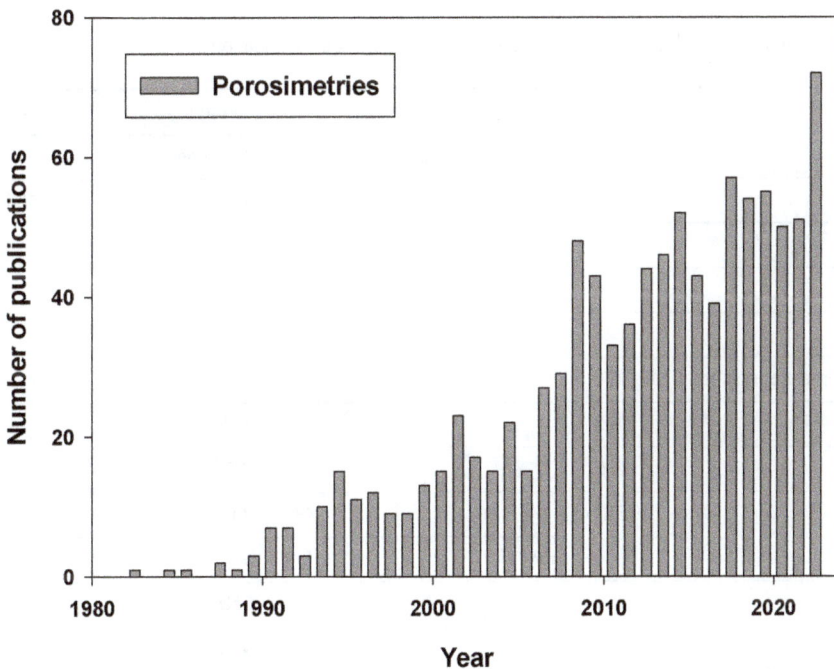

Fig. 10.3: Publications containing porosimetry or porometry in their subject, according to WoS, from Clarivate.

Similarly, this study could have been accomplished by using the Scopus database as a source, with very similar results, as expected. By way of example, the results obtained by these latter databases, depending on the name of the porosimetric technique entered in the search terms (title, abstract, keywords), are given in Tab. 10.1.

There are techniques for which the simple mention of their name gives rise to an enormous number of results. This is the case, for example, of HgP or GAD, which in addition to being techniques that have been established for many years in the analysis of membrane filters, have important applications in all types of porous materials, not only membranes, as we have already mentioned in the corresponding chapters.

Other terms, less linked to porosimetry, but more applicable in the study of materials, such as NMR or SR, give even wider search results. Therefore, for most of these terms, a new search has been presented in the table in which the name of the technique is accompanied by the word "membrane" in order to restrict the number of results. These results, shown in the second column of the table, and also in Fig. 10.4, which we will discuss below, are indicated with an asterisk (*). Finally, to further refine results (since many of the references obtained in the previous searches have little to do with the subject of this book), a third column has been included by adding the search terms "membrane" and "pore size". The terms from the latter search constitute the third column of the table and are indicated by the symbol (†).

Tab. 10.1: Number of publications found in Scopus for each keyword combination used.

		Porometry *	160	Porometry † 108
		Porosimetry *	747	Porosimetry † 362
LLDP	34			BP † 156
CFP	136			
GLDP	16			
EP	21			
Displacement porosimetry	171			
PmP	103			PmP † 61
ThP	286			ThP † 24
HgP	9,337	HgP *	431	HgP † 220
		MIP *	174	MIP † 85
GAD	11,355	GAD *	364	GAD † 81
				N₂ adsorption † 390
		NMR *	21,786	NMR † 281
		Synchrotron radiation *	1211	Synchrotron radiation† 15
		ELLP *	854	ELLP † 43

It should be noted that the table does not include the results corresponding to the terms "microscopy", "SEM" or "AFM", which, due to their widespread use in materials science, give rise to a huge number of references that would distort the purpose of the table, a simple comparison of the relative interest that the porosimetric techniques mentioned here have aroused in the world of research. For some techniques such as GLDP, HgP or GAD, searches with alternative names (capillary flow porometry, "CFP"; mercury intrusion porometry, "MIP" and "nitrogen adsorption", respectively) commonly used by some authors have been included.

Finally, the words "porosimetry" and "porometry" (accompanied by "membrane" or "membrane pore size") have been included as search terms to check how the term porosimetry is more common in the scientific literature than its synonym porometry.

Results obtained in the Scopus search and then included in Tab. 10.1 are now presented in a more visual manner in Fig. 10.4. Ball sizes (areas really) in this figure are proportional to the number of literature references found for each term, and different colours are used to distinguish between single technique names or those accompanied by "membrane" or the most complete search "membrane pore size".

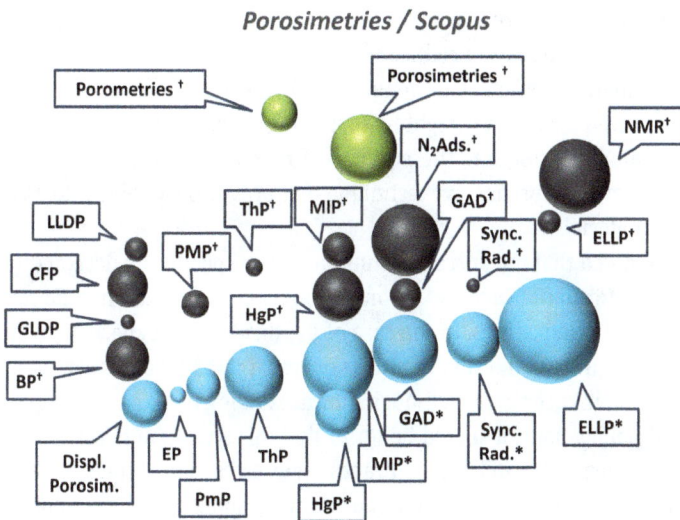

Fig. 10.4: Publications containing on their title, subject or keywords the names of the main Porosimetries, according to Scopus, from Elsevier.

From the analysis of these results, both Tab. 10.1 and Fig. 10.4, several comments can be deduced:
– First, more frequent are works that correspond to techniques that have been known for several decades and that have also found numerous applications in all types of porous materials. This is the case, as we have already mentioned, of HgP or GAD, porosimetric techniques widely used for ceramic materials, powders

and, in general, all types of porous materials. And both offer valuable information about porosity or specific pore area of such materials.

– Techniques with long experience, such as PmP and ThP, which have accumulated a significant number of references, we can see that when we narrow this search to the use of these techniques in the determination of membrane pore sizes, the number of references decreases notably. This decrease is especially large in the case of ThP, a technique that, after a strong development in its early years, has suffered a progressive decrease in publications related to membrane porosimetry, probably due to the difficult interpretation of the results and the development of other faster and more specific techniques. In the case of PmP, the number of restricted publications is still reasonably high, but some confusion in terminology also contributes to this. For example, many authors consider permporometry, the determination of fluxes through pre-wetted membranes, in experiments that would more appropriately be included as fluid displacement techniques (bubble point and derived techniques).

– The large number of publications containing terms such as ELLP or SR is somehow surprising, but, as we have already indicated, these techniques have multiple applications other than those related to the characterization of porous materials. This is noticeable when the search is restricted to "membrane pore size" so that the number of results is significantly reduced.

– NMR, too, presents many references in general, which are still many when the search range is narrowed down. This might be surprising since NMR is not the technique of choice when characterising the porosity of membranes. But there is no doubt that NMR is a very comprehensive technique that provides valuable information on all types of materials.

– Finally, it should be noted that the current techniques more specifically designed to obtain porosimetric information, such as EP and bubble point-derived ones, do not have a large number of publications. In the case of EP, this is to be expected, given the youth of the technique and the small number of researchers who publish on it. As for liquid displacement techniques, although they are adding (year by year) up to a gentle number of publications, it should be noted that this type of publications, in many cases, focus on the novelty of the membranes themselves (due to their method of manufacture, modification or configuration) or on their applications, with porosimetry being a complementary technique to which little relevance is given.

– This is noticeable in the number of references found for all techniques. Considering the huge number of publications that are currently generated (several million papers per year, both in subscription and open access journals), it seems clear that a few hundred papers over several years does not indicate an excessive production. As mentioned earlier, this is more due to the collateral role of porosimetry in general, which, although pore size data are included in almost all published works on membranes, in many cases, they are simply nominal values. And in many others, the mean pore radius is included, without further discussion or analysis of the rest of the information in the work.

10.3 Final remarks

To conclude this final chapter, we would like to give a personal, and undoubtedly partial, opinion on the different techniques studied. We believe that there is no doubt that, for MF membranes, the GLDP technique offers highly reliable results following properly standardized procedures and using commercial equipment of undoubted accuracy.

As for the UF range, the number of options is wide, but can be reasonably narrowed down. For ceramic or powdery samples, the choice could be either HgP or GAD, the latter being preferable if we also need to obtain information on the surface area of the sample.

In other types of samples, the techniques that are possibly proving to be more reliable are EP and LLDP. Both have their advantages and disadvantages, among which the non-existence (or almost, for LLDP) of commercial equipment prepared to apply them stands out. This requires design, assembly and set-up work that can significantly lengthen the time required to obtain information. But once properly set up, both techniques offer complete information on the PSD of our samples.

The other available porosimetric techniques (especially PmP and ThP) do not present a sufficient degree of confidence to be considered a competitive alternative to the two mentioned. And spectroscopic techniques such as PALS, SR or ELLP do not seem to be sufficiently specific to merit their application in membrane porosimetry, although they may be more useful for obtaining other types of information.

An interesting aspect on which research could have an impact is the extension of porosimetric techniques to the range of NF membranes, even though we are dangerously close to the limits of validity of the equations on which these techniques are based. Obviously, none of the above techniques are useful in the analysis of RO membranes, where it does not really make much sense to talk about pores.

We must also point out an avenue where further research can significantly improve the applicability of porosimetric techniques. As mentioned above, in general, membrane samples must be introduced into the corresponding measuring cells either in the form of flat samples or in pieces (as in HgP or GAD) so it is not possible to analyse devices containing such membranes as they are marketed (cartridges and spiral modules or hollow fibre cartridges). Some attempts have been made in LLDP [10] to be able to directly analyse commercial cartridges of reduced areas but being able to analyse larger areas without destroying the module or cartridge, and in conditions closer to their operational use, would be a very desirable goal.

In relation to this point, it seems clear that porosimetric techniques based on the determination of fluxes through the membrane are more suitable for the adaptation of different membrane modules and configurations for in situ or, at least, non-destructive analysis. Thus, liquid displacement porosimetry (both GLDP and LLDP) allows the adaptation of different modules for the analysis of flat membranes, small cartridges containing a reasonable number of hollow fibres or small tubular membrane modules to their measuring equipment. Certainly, the total area analysed cannot be very large (0.5 m^2 may be a suitable limit) since the pre-wetting of the samples would require working with a large amount of wetting liquid and we also have the limitation of the range in the flow metres, which are easier to find for medium-low flow rates with sufficient precision.

The rest of the porosimetric techniques requires the destruction of the cartridge or module to extract the flat sample, the tubular membrane or the fibres present inside it. Even so, any of the porosimetric techniques studied in this text can perfectly be used in the prior characterization of the membranes to be used in a certain process as well as in the so-called autopsy of the membranes after their use.

These autopsies (consisting of the characterization, among other parameters, of the PSD of the membrane after a long period of use in a given process) are fundamental when analysing the importance of fouling in this process. By comparing the PSD of the clean membrane and the already fouled sample, we can determine the extent of fouling as well as its reversible or irreversible character and even study the effectiveness of the various cleaning protocols applied in daily use [9].

Finally, we cannot rule out the appearance and development in the coming years of new porosimetric characterization techniques that improve the performance of the current ones, even at the risk of rendering the contents of this book obsolete.

References

[1] Tung K.-.L., Chang K.-.S., Wu T.-.T., Lin N.-.J., Lee K.R., Lai J.-.Y. Recent advances in the characterization of membrane morphology. Curr Opin Chem Eng, 4 (2014) 121–127. http://dx.doi.org/10.1016/j.coche.2014.03.002.
[2] Zamani F., Jayaraman P., Akhondi E., Krantz W.B., Fane A.G., Chew J.W. Extending the uppermost pore diameter measurable via Evapoporometry. J Membrane Sci, 524 (2017) 637–643. https://doi.org/10.1016/j.memsci.2016.11.082.
[3] AlMarzooqi F.A., Bilad M.R., Mansoor B., Arafat H.A. A comparative study of image analysis and porometry techniques for characterization of porous membranes. J Mater Sci, 51 (2016) 2017–2032. https://doi.org/10.1007/s10853-015-9512-0.
[4] Mitchell J., Beau J., Webber W., Strange J.H. Nuclear magnetic resonance cryoporometry. Phys Rep, 461 (2008) 1–36. https://doi.org/10.1016/j.physrep.2008.02.001.
[5] Petrov O.V., Furó I. NMR cryoporometry: Principles, applications and potential. Prog Nucl Magn Reason Spectrosc, 54 (2009) 97–122. https://doi.org/10.1016/j.pnmrs.2008.06.001.
[6] Enninful H.R.N.B., Enke D., Valiullin R. Advanced NMR Cryoporometry. Chemie Ingenieur Technik, 95 (2023) 1713–1729. https://doi.org/10.1002/cite.202300060.
[7] Herrero C., Prádanos P., Calvo J.I., Tejerina F., Hernández A. Flux Decline in Protein Microfiltration: Influence of Operative Parameters. J Colloid Interface Sci, 187 (1997) 344–351. https://doi.org/10.1006/jcis.1996.4662.
[8] Jacob J., Prádanos P., Calvo J.I., Hernández A., Jonsson G. Fouling kinetics and associated dynamics of structural modifications. Colloids Surf A: Physicochem Eng Aspects, 138 (1998) 173–183. https://doi.org/10.1016/S0927-7757(97)00082-4.

[9]   Almécija M.C., Guadix A., Calvo J.I., Guadix E.M. Changes in structure and performance during diafiltration of binary protein solutions due to repeated cycles of fouling/alkaline cleaning. Food Bioprod Process, 105 (2017) 117–128. http://dx.doi.org/10.1016/j.fbp.2017.07.003.

[10] Peinador R.I., Darbouret D., Paragot C., Calvo J.I. Automated Liquid–Liquid Displacement Porometry (LLDP) for the Non-Destructive Characterization of Ultrapure Water Purification Filtration Devices. Membranes, 13 (2023) 660. https://doi.org/10.3390/membranes13070660.

[11] Pyo M., Kim D., Kim H., Jeong S., Lee E.-J. A study of the measurement conditions for determining liquid entry pressure values for hydrophobic membranes. Available at SSRN: http://dx.doi.org/10.2139/ssrn.4576006

[12] Dytnerskii Y.I., Dmitriev A.A., Mchedlishvili B.V., Potokin I.L. Study of the Porous Structure and Selective Properties of Membranes Obtained by Plasma Polymerization in a Glow–Discharge. Colloid J USSR, 44(6) (1982) 1024–1028.

List of figures

https://doi.org/10.1515/9783110792195-011

Index

https://doi.org/10.1515/9783110792195-012

www.ingramcontent.com/pod-product-compliance
Lightning Source LLC
Chambersburg PA
CBHW061344210326
41598CB00035B/5878

* 9 7 8 3 1 1 0 7 9 2 1 8 8 *